River Mixing

River Mixing

J. C. Rutherford
NIWA Ecosystems
National Institute of Water and Atmospheric Research
Hamilton, New Zealand

JOHN WILEY & SONS
Chichester · New York · Brisbane · Toronto · Singapore

Telephone National Chichester (0243) 779777
International +44 243 779777

Other Wiley Editorial Offices

John Wiley and Sons, Inc., 605 Third Avenue,
New York, NY 10158-0012, USA

Jacaranda Wiley Ltd, 33 Park Road, Milton
Queensland 4064, Australia

John Wiley and Sons (Canada) Ltd, 22 Worcester Road,
Rexdale, Ontario M9W 1L1, Canada

John Wiley and Sons (SEA) Pte Ltd, 37 Jalan Permimpin #05-04,
Block B, Union Industrial Building, Singapore 2057

Library of Congress Cataloging-in-Publication Data

Rutherford, J. C.
River mixing / by J. C. Rutherford.
p. cm.
Includes bibliographical references and index.
ISBN 0-471-94282-0
1. Streamflow. 2. Diffusion. 3. Mixing. I. Title.
GB1207.R88 1994
551.48′3—dc20 93-33033
CIP

British Library Cataloguing in Publication Data

A catalogue record for this book is available from the British Library

ISBN 0-471-94282-0

Typeset in 10/12pt Times from author's disks by
Mathematical Composition Setters Ltd, Salisbury, Wiltshire
Printed and bound in Great Britain by Biddles Ltd, Guildford, Surrey

To Susan, William and Sally
for their encouragement and help

Contents

Introduction

In several branches of aquatic science the need arises to predict the concentration of solutes in a river. For example: the waste discharger needs to quantify effluent mixing to design and operate an outfall in a manner which complies with water pollution criteria; the water resources scientist needs to understand river mixing to check that an outfall complies with its permit and to design sensible chemical and biological monitoring programmes; the fisheries scientist needs to understand the mixing of effluents and tributary inflows to assess their effects on resident and migratory fish; the aquatic botanist needs to understand water exchange between ecologically sensitive bankside regions and the main flow; and the modeller needs to understand and quantify the rate of mass transfer between the bed and the overlying flow to predict tracer concentrations in the pore water and overlying river water. Clearly this list could be expanded. My background is that of an engineer working with mathematical models of water quality and this book is written for the engineer, scientist, ecologist or student who is not a specialist in fluid mechanics but who wishes to understand a little better the fluid mechanics of mixing and who needs simple predictive techniques for analysing important river mixing problems. In writing this book I have set out to do five things.

Firstly, the theoretical basis for the commonly used mixing models is outlined. I am not a specialist in fluid mechanics and at times have struggled with the complexities of this subject. I believe, however, that the practitioner needs to acquire a good understanding of the mixing models being used. The mechanics of mixing are complex and defy a complete mathematical description. Consequently, few river mixing problems can be solved exactly using purely theoretical methods. There are, however, a number of simplifications and semi-empirical techniques which lead to tolerably accurate models for a number of important river mixing problems. These semi-empirical models need to be calibrated using field or laboratory data to check their validity and to evaluate model coefficients. It is important for the practitioner to have an understanding of the assumptions each model embodies and for this a working knowledge of mixing theory is essential. There are a surprising number of published examples where inappropriate models have been used to analyse river tracer data and dispersion coefficients have been reported which are very misleading. By illustrating which mixing processes are important in a particular situation and how they are modelled it is hoped that the reader will get a better feel for the subject, will not select inappropriate models or mixing coefficients, and will be better able to assess the reliability of predictions.

Secondly, various well-known but simplified mixing models are described. Where possible the strengths and weaknesses of the simplified mixing models are assessed by comparing their performance with that of more sophisticated models. When addressing practical mixing problems my experience has been that it is always advisable to start by making an initial analysis using simple mixing models together with a range of coefficients selected, if necessary, from previously published work. Such an analysis can be done fairly quickly and cheaply. The resulting tracer concentration estimates must be ascribed a high uncertainty, but they help to clarify what is important and what can be ignored in a particular problem. In some cases the initial analysis may be all that is required to solve a water quality problem (e.g. predicted tracer concentrations may be unacceptably high under even the most favourable mixing conditions). In other cases it may indicate that additional mixing studies are required (e.g. concentrations may be acceptable provided the mixing conditions assumed in the modelling can be confirmed). In the latter situation additional field, mathematical, or physical model studies (or ideally a combination of all three) should be conducted. The initial analysis will help plan such studies.

Thirdly, a range of worked examples is given to illustrate the practicalities of making mixing predictions. The emphasis is placed on using simple models such as can conveniently be solved using nomographs or a small computer. Many of these examples are based on real mixing problems which I have tackled; others are hypothetical and are identified as such. I have attempted to quantify any shortcomings in the simplified methods presented here so that the practitioner can put upper and lower bound limits on mixing estimates. For several problems I have summarised solutions to the mixing equations in the form of nomographs. These serve two purposes. Firstly, they show the general form of the solution and so help the reader understand how the tracer plume behaves. Secondly, they can be used to help make quantitative calculations. Some readers may regard nomographs as old technology for the latter role and once familiar with the particular mixing equation can write small computer programs for themselves to make tracer predictions.

Fourthly, values of mixing coefficients are collated from published and some unpublished studies. Where possible I have gone back to the original references (to try and avoid transcription errors) and in some cases the original authors have been kind enough to supply supplementary information. As well as being tabulated, coefficients are examined graphically in relation to bulk flow parameters. The physical reasoning behind the resulting relationships is not always clear, but if nothing else the graphs summarise the published information in a way which helps the reader select values when making preliminary calculations and which highlights the variability in published information.

Fifthly, techniques for estimating model coefficients from field tracer data are outlined. Often the largest source of uncertainty in mixing predictions lies not in the approximate nature of the mixing models employed but in the shortage of reliable field data with which to determine model coefficients (e.g. depth, velocity and diffusivity). Although the starting point of investigations should be some preliminary mixing calculations, these often highlight the need for further work including field tracer tests.

Attention is focused in this book on the mixing of neutrally buoyant tracers. This can be partially justified by the fact that vertical mixing is rapid in most rivers so that buoyancy effects are largely confined to the near-field mixing region close to the outfall. There is, however, a brief discussion in Chapters 2 and 3 of the effects of stratification and buoyancy on vertical and transverse mixing rates although this is a difficult problem for which only approximate theories exist. The first five chapters consider tracer mixing in the water column. Chapter 6 considers the interesting but difficult problem of mixing between the river bed and the overlying water which is important in, for example, river deoxygenation and heavy metal transport.

The estimation of tracer concentrations requires a framework within which the mixing processes can be quantified together with input data which includes the location and mass inflow rate of tracer, the river geometry, velocities and rates of mixing. There are three types of framework for studying river mixing: field studies, mathematical models, and physical models. There is no best method and the greatest confidence can be placed in results duplicated using more than one approach (Fischer *et al* 1979). The emphasis in this book is placed on mathematical models and field techniques used in combination. Many of the mathematical models described, however, have been developed and tested in laboratory physical models.

There are several excellent publications covering particular aspects of river mixing. For example, Holly (1985), El-Hadi *et al* (1984) and Chatwin and Allen (1985) provide reviews of mixing theory and outline several important mixing models. They do not, however, describe in detail how to solve particular mixing problems, nor do they offer guidance on how to estimate model coefficients. Fischer (1973) reviews river mixing theory and summarises some published coefficients but does not present solution methods or worked examples. Fischer *et al* (1979) comes closest to meeting the needs of the practitioner. This book includes a review of mixing theory, worked examples for some important river mixing problems and a collation of published coefficients. Because this book also covers jets and plumes, estuary and reservoir mixing there are, however, some important gaps. Sayre (1973) and Sayre

(1979) give a very thorough exposition together with useful worked examples and a collation of published model coefficients but only for a narrow range of river mixing problems. I have attempted to draw together the important elements from these and other important works.

Acknowledgements

As a young engineer, I was encouraged by Professor Sandy Harris to prepare a slim volume entitled *Handbook on Mixing in Rivers* (Rutherford 1980) to assist practitioners to make preliminary calculations of waste mixing. The present book has grown out of the original handbook, aided by helpful comments and suggestions from its users together with a broadening of my experience in tackling river mixing problems. Sandy Harris cautioned that if a practitioner's handbook contained more than two equations it would not meet its objectives. I have chosen not to follow that sound advice and the risk is entirely mine, but I believe that the practitioner needs to acquire a working understanding of the mixing equations.

Professor Ian Wood, University of Canterbury, has been an inspiration and source of advice and encouragement to a whole generation of engineers interested in mixing. I am particularly grateful to him for helpful comments on aspects of the work described in this book and for encouraging me to continue its preparation. Dr Roger Nokes, Department of Engineering Science, University of Auckland, was kind enough to review an early draft of this book and to make a large number of very helpful suggestions. He generously re-ran his computer programs and produced several of the figures which appear in Chapter 2.

My colleague, Graham McBride, and I have collaborated on several river modelling studies over the last 15 years. I gratefully acknowledge the ideas and advice that he has given over that time, and the time he has spent reading and commenting on the manuscript of this book. Several other colleagues have either contributed data, been involved with collaborative studies, or have commented on the manuscript and I would particularly like to thank Mr Bryan Williams, Mr Ben Liley, Dr Bryce Cooper, Dr Rob Bell and Dr Bob Wilcock. Mr John Cole, Water Research Centre, Medmenham, was kind enough to make available unpublished tracer data from several rivers in the United Kingdom. Finally I would like to thank Frank Bailey, Michelle Dunn and Merylene Osborne who have worked so hard on the manuscript.

Chapter 1

Mechanisms and Models of River Mixing

1.1 INTRODUCTION

When a tracer is introduced into a river we notice that two things happen to it. Firstly, it is carried away from the point of discharge by the current, a process termed advection. Secondly, it spreads out, a process which is termed either diffusion or dispersion depending on which mechanisms are operating. This chapter describes the physics and mathematics of advection, molecular diffusion, turbulent diffusion and dispersion. The main objectives are to help the reader understand these processes and to derive the important mixing equations. The derivation of several equations is given in detail to emphasise the roles of time and space averaging. This should help the reader understand the physical significance of each of the terms and coefficients in the mixing equations and reduce the risk of selecting inappropriate models or coefficients when tackling practical problems.

Chapter 1 is principally concerned with the theoretical basis of the mixing equations and the practicalities of solving mixing problems are discussed in Chapters 2, 3 and 4. Some readers may prefer to become familiar with the practical aspects of river mixing before returning to this more theoretical chapter. Others may want to explore the physics and mathematics of river mixing before considering the practical applications. A summary of the key points in Chapter 1 is given at the outset to help the reader see where the chapter is heading and to serve as a subsequent checklist.

1.2 SUMMARY

1. The two fundamental processes involved in tracer mixing are advection and molecular diffusion. Molecular diffusion can be modelled satisfactorily using a gradient diffusion equation (Fick's law). Molecular

diffusion coefficients are properties of the solvent and solute and are almost constant. Typically they are of the order of $10^{-9}\ \mathrm{m^2\,s^{-1}}$.

2. In turbulent flow a tracer mixes more rapidly than can be accounted for just by molecular diffusion. The reason is that small-scale turbulent eddies continually erode the edges of the tracer cloud, increase local concentration gradients and hence allow molecular diffusion to proceed more rapidly. The combined effect of molecular diffusion and turbulent velocity fluctuations is termed turbulent diffusion. In stationary homogenous turbulence, turbulent diffusion can be modelled satisfactorily using Fick's law provided sufficient time has elapsed since the time of injection. Thus asymptotically the tracer flux resulting from turbulent diffusion is proportional to the spatial gradient of the ensemble mean tracer concentration. In the x direction

$$J_x = -e_{\mathrm{m}} \frac{\partial \langle c \rangle}{\partial x}$$

where J_x = mass flux in the x direction; $\langle c \rangle$ = ensemble mean concentration and e_{m} = turbulent diffusion coefficient. The ensemble mean concentration is the average concentration which would be expected over a large number of identical tracer experiments. Turbulent diffusion coefficients are properties of the flow and are not constant. They vary from channel to channel and with location and direction within a given channel. Typically turbulent diffusion coefficients are of the order of $10^{-3}\ \mathrm{m^2\,s^{-1}}$.

3. In the most general problem, advection and turbulent diffusion occur in each of the three coordinate directions. The three-dimensional advection/diffusion equation (in advection form) is

$$\frac{\partial c}{\partial t} + u_x \frac{\partial c}{\partial x} + u_y \frac{\partial c}{\partial y} + u_z \frac{\partial c}{\partial z} = \frac{\partial}{\partial x}\left(e_x \frac{\partial c}{\partial x}\right) + \frac{\partial}{\partial y}\left(e_y \frac{\partial c}{\partial y}\right) + \frac{\partial}{\partial z}\left(e_z \frac{\partial c}{\partial z}\right)$$

where x, y and z = longitudinal, vertical and transverse distances; c = ensemble mean tracer concentration; u_x, u_y and u_z = velocities; and e_x, e_y and e_z = turbulent diffusion coefficients.

4. The advection/diffusion equation describes the behaviour of the ensemble mean concentration. Because of the random nature of turbulence, concentrations fluctuate about the ensemble mean. At present it is not possible to predict extreme concentrations theoretically but empirical methods can be used to establish parameters such as the peak to mean concentration ratio.

5. In the near-field mixing region close to a non-steady point source, advection and mixing are important in all three coordinate directions and the full three-dimensional advection/diffusion equation is required to predict tracer concentrations. In some near-field mixing problems, however, the advection/diffusion equation can be simplified. For example, if the source

is steady then longitudinal concentration gradients become negligibly small. If the x axis is aligned with the flow then velocities u_y and u_z can be neglected. If, in addition, the tracer originates from a steady vertical line source then vertical concentration gradients are small everywhere and attention can be focused on transverse mixing. On the other hand if the tracer originates from a steady transverse line source then transverse concentration gradients are small everywhere and attention can be focused on vertical mixing. Simplification of the mixing equations must be done with great care to retain the important terms.

7. In most rivers, tracer originating from a point source mixes vertically long before it mixes transversely. In the mid-field mixing region vertical concentration gradients are small and may not be important ecologically. The full three-dimensional advection/dispersion equation can be averaged over the depth to yield the two-dimensional equation

$$\frac{\partial s}{\partial t} + v_x \frac{\partial s}{\partial x} + v_z \frac{\partial s}{\partial z} = \frac{1}{h}\frac{\partial}{\partial x}\left(hk_x \frac{\partial s}{\partial x}\right) + \frac{1}{h}\frac{\partial}{\partial z}\left(hk_z \frac{\partial s}{\partial z}\right)$$

where s = depth-averaged tracer concentration; v_x and v_z = depth-averaged velocities; h = local depth; and k_x and k_z = longitudinal and transverse dispersion coefficients. The dispersion coefficients quantify the effects on tracer mixing of vertical variations in velocity.

8. Tracer originating from a non-steady point source eventually mixes across the channel. In the far-field vertical and transverse concentration gradients may not be important ecologically and so the three-dimensional advection/diffusion equation can be averaged over both the depth and the width to yield the one-dimensional equation

$$\frac{\partial S}{\partial t} + V_x \frac{\partial S}{\partial x} = \frac{1}{A}\frac{\partial}{\partial x}\left(AK_x \frac{\partial S}{\partial x}\right)$$

where S = cross-sectional averaged concentration; V_x = cross-sectional averaged velocity; A = cross-sectional area; and K_x = longitudinal dispersion coefficient.

9. Figure 1.9 shows three types of tracer source and sketches the concentration distributions observed at different distances below each one. Table 1.2 gives approximate formulae for estimating the length of the near-field and mid-field mixing regions and also gives the dimensionality of the mixing equations in each region.

10. Averaging over the depth and/or width reduces the dimensionality of the advection/diffusion equation and simplifies its solution. When averaging is performed it is important to account for the effects of non-uniformities of velocity in the coordinate directions over which averaging is being carried out. Non-uniformities of velocity (velocity shear) contribute to the spreading of tracer and when the advection/diffusion equation is

averaged, their effects must be quantified by additional dispersion terms. Dispersion is not a fundamental physical process but arises as an artifact of averaging. The correct modelling of dispersion plays an important part in the development and use of mixing models. Under certain conditions Fick's law can be used to model dispersion. The most important requirement is that sufficient time must have elapsed since tracer injection for an equilibrium to have become established between velocity shear and turbulent diffusion.

11. If the channel is straight and uniform then it is best to write the mixing equations in rectangular Cartesian coordinates. In most natural rivers, however, the channel meanders and varies in shape and it is natural to rewrite the mixing equations in curvilinear coordinates. The curvilinear equations can be simplified by replacing transverse distance by cumulative flow, thereby giving the streamtube model.

1.3 CONTINUUM HYPOTHESIS

Before discussing mixing processes we need to have a clear conceptual picture of liquids and dissolved tracers. At very small length scales (typically a few angstroms) a solution consists of a mixture of discrete solvent and solute molecules. At these very small length scales variables such as velocity and tracer concentration have violently non-uniform distributions which are difficult to handle mathematically. Attempts to describe the mechanics of liquids at the molecular level using statistical methods have not been very successful (Batchelor 1967). In river mixing problems, however, the smallest length scales of interest are much larger than the distance between molecules. It is convenient to adopt the hypothesis that the fluid is a continuum in which velocities and tracer concentrations are some sort of average of the values at the molecular level and are well-behaved functions of time and location. This is known as the continuum hypothesis. One consequence of the continuum hypothesis is that when analysing turbulent flows we cannot model explicitly variations of velocity and concentration at molecular length scales. This does not present major problems because it has been found that it is the velocity variations with large length scales (i.e. the large eddies) which effect most of the momentum and mass transport in turbulent flows (Townsend 1956, Tennekes and Lumley 1972).

Within the continuum framework it is often useful to imagine a small parcel of fluid whose dimensions are comparable with the smallest length scale of interest and which has characteristic properties such as velocity and tracer concentration. A collection of such parcels is termed a tracer cloud.

1.4 ADVECTION

Advection is the bodily movement of a parcel of fluid resulting from an imposed current. Advection transports any tracer which may be dissolved or suspended in the fluid and is clearly important in rivers because it carries tracer downstream away from a fixed source. Some workers use the term convection to describe this process (e.g. Holley 1969) but this terminology is potentially confusing and is not used here because in the atmospheric and geothermal publications convection implies a flow induced by buoyancy. In rivers water moves largely as a result of gravity (channel slope) and buoyancy has little effect, except sometimes very close to a source of cooling water or dense tracer.

The amount of tracer transported per unit time per unit area perpendicular to the current is termed the advective flux and is the product of the velocity and the tracer concentration. Considering a small parcel of fluid and transport in the x direction

$$I_x = u_x c \tag{1.1}$$

where $I_x(x, y, z, t)$ = advective flux in the x direction; $u_x(x, y, z, t)$ = velocity in the x direction; and $c(x, y, z, t)$ = tracer concentration. Under the continuum hypothesis u_x, c and hence I_x are assumed to have well-defined values for each parcel of fluid. Note that I_x varies with both location and time.

It is important to appreciate that pure advection does not cause any change of concentration within a tracer cloud. To illustrate this point, consider fluid parcels in a flow where the velocity is uniform (does not vary with x, y or z). If we ignore molecular diffusion then advection causes the fluid parcels to move bodily downstream at a constant rate without undergoing any distortion or spreading. This is, of course, a hypothetical example because uniform flow rarely occurs in nature and the tracer cloud is affected by molecular and turbulent diffusion. A Lagrangian coordinate system is one which travels downstream at the mean velocity (as opposed to a Eulerian system which remains fixed in space). We often use a Lagrangian coordinate system whose origin lies at the centre of a tracer cloud. The advantage of a Lagrangian coordinate system is that concentration changes only as a result of diffusion and dispersion and not as a result of advection. This holds for both steady and unsteady tracer sources. By comparison, in an Eulerian coordinate system advection causes the concentration at a fixed point downstream from an unsteady source to vary with time, which complicates the analysis.

1.5 MOLECULAR DIFFUSION

1.5.1 Molecular Diffusion in a Stagnant Fluid

If neutrally buoyant tracer is introduced very carefully into stagnant water at a point far distant from any boundaries then it will be seen to spread out slowly at the same rate in all three coordinate directions. This spreading results from random molecular motion within the fluid (the so-called Brownian motion) and is termed molecular diffusion. Attempts to calculate the rates of molecular diffusion theoretically have been unsuccessful for solids and fluids and have met with only limited success in gases. In 1855, however, the German physiologist Fick published a paper in which he drew an analogy between the diffusion of salt in water with the diffusion of heat along a metal rod. He made the hypothesis that the net rate of transfer of tracer from a region of high concentration to a region of low concentration should proceed at a rate proportional to the concentration gradient between the two regions. This has become known as Fick's first law. It is a particular case of the more general gradient diffusion law which governs the flow of electricity along a conducting wire (Ohm's law) and the flow of heat along a metal rod (Fourier's law). In one dimension Fick's first law can be written

$$J_x = -e_m \frac{\partial c}{\partial x} \tag{1.2}$$

where J_x = molecular diffusive flux in the x direction (i.e. the transfer rate of tracer in the x direction per unit area by molecular diffusion); c = tracer concentration; $\partial c/\partial x$ = tracer concentration gradient in the x direction; and e_m = molecular diffusion coefficient. The negative sign in equation (1.2) arises because tracer diffuses from a region of high concentration to a region of low concentration.

The gradient diffusion law is based on a hypothesis rather than on a complete theoretical analysis but laboratory studies show that the relationship between flux and the concentration gradient is remarkably linear. The constant of proportionality between flux and concentration gradient, the molecular diffusion coefficient (also termed the molecular diffusivity), must of course be determined empirically. Typical values of e_m for solutes in water are in the range $0.5–2.0 \times 10^{-9}\,m^2\,s^{-1}$. The molecular diffusion coefficient is a property of the fluid and for a given solvent, solute, concentration and temperature, e_m is constant. In contrast turbulent diffusion coefficients (which are discussed shortly) are much larger than molecular diffusion coefficients and are properties of the flow rather than the fluid.

1.5.2 Molecular Diffusion in Laminar Flow

We now consider molecular diffusion in steady laminar flow. In laminar flow viscous forces are strong relative to inertial forces and fluid elements follow smooth paths (streamlines) which are reproducible and do not cross each other. The balance between viscous and inertial forces can be quantified using the Reynolds number which is defined

$$Re = \frac{uL}{\nu} \tag{1.3}$$

where u = a characteristic velocity; L = a characteristic length scale; and $\nu = \mu/\rho$ = the kinematic viscosity of the fluid. At low Reynolds numbers the flow is laminar, at high numbers it is fully turbulent and at intermediate numbers it is transitional. The appropriate length and velocity scales and the critical Reynolds numbers depend on the nature of the mixing problem. In pipe flow u = mean velocity, L = pipe diameter, flows are laminar provided the Reynolds number is less than about 2000 and fully turbulent above about 10 000. In river channels u = mean velocity, L = mean depth, flows are laminar at Reynolds numbers below 500 and turbulent above about 2000. For boundary layers u = free stream velocity, L = boundary layer thickness and the boundary layer remains laminar at Reynolds numbers below about 600.

To model tracer concentrations, we need to combine the effects of advection and molecular diffusion. This is done by assuming that the two processes are independent and additive. A Lagrangian coordinate frame of reference is adopted which moves with the fluid at the mean velocity. Within this Lagrangian coordinate system molecular diffusion is assumed to occur as if the fluid were stationary and hence the diffusive flux is given by the product of the concentration gradient (measured in the Lagrangian framework travelling at the mean velocity) and the molecular diffusivity. Having made this assumption it is possible to derive the equation governing tracer concentration based on the conservation of mass.

1.5.3 Derivation of the Molecular Diffusion Equation

We shall adopt a rectangular Cartesian coordinate system with axes x, y and z. Imagine a very small rectangular parcel of fluid which moves at the mean velocity (Figure 1.1). Conservation of mass requires that the rate of change of mass within this parcel is balanced by the rates of transfer of mass across the boundaries. Note that advection does not transport any material across the parcel boundaries because the parcel is travelling at the mean velocity. Thus

$$\frac{M_{t+\Delta t} - M_t}{\Delta t} = (J_x - J_{x+\Delta x})\Delta y \Delta z + (J_y - J_{y+\Delta y})\Delta x \Delta z + (J_z - J_{z+\Delta z})\Delta x \Delta y \tag{1.4}$$

where M_t and $M_{t+\Delta t}$ = tracer mass in the moving parcel at times t and $t + \Delta t$ respectively; J_x, J_y and J_z = diffusive fluxes (averaged over the time interval

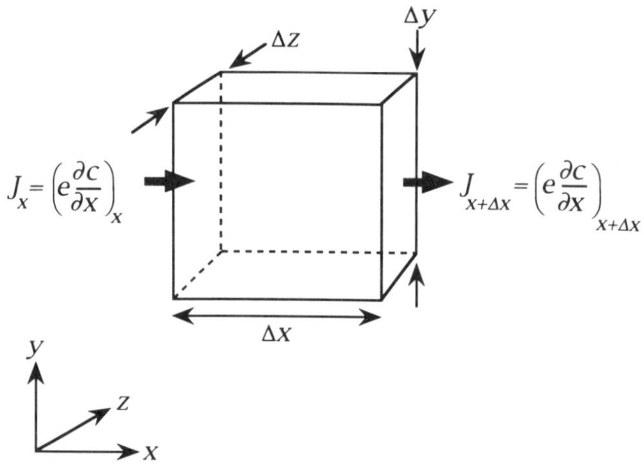

Figure 1.1. Diffusive fluxes into and out of a small fluid element

Δt) entering the parcel through the moving boundaries located at x, y, z; $J_{x+\Delta x}$, $J_{y+\Delta y}$ and $J_{z+\Delta z}$ = diffusive fluxes (averaged over the time interval Δt) leaving the parcel through the moving boundaries located at $x+\Delta x$, $y+\Delta y$, $z+\Delta z$; Δx, Δy and Δz = dimensions of the fluid parcel; and Δt = time interval. Using a Taylor's series expansion and ignoring second order and higher terms (because Δt is very small)

$$M_{t+\Delta t} = M_t + \left(\frac{\mathrm{D}M}{\mathrm{D}t}\right)_t \Delta t \tag{1.5}$$

Note that in equation (1.5) we use the material operator $\mathrm{D}/\mathrm{D}t$ to denote the rate of change with time in a Lagrangian coordinate system travelling at the mean velocity. The mass of tracer within the parcel at time t is

$$M_t = c_t \Delta x \Delta y \Delta z \tag{1.6}$$

where c_t = average tracer concentration within the parcel at time t. Consequently equation (1.5) can be rewritten

$$\frac{M_{t+\Delta t}}{\Delta t} - M_t = \left(\frac{\mathrm{D}c}{\mathrm{D}t}\right)_t \Delta x \Delta y \Delta z \tag{1.7}$$

Now we need to evaluate the diffusive flux into and out of the moving parcel. Because diffusion is defined in a Lagrangian coordinate system then strictly we need to know the concentration gradients across each of the parcel boundaries as they move downstream. These gradients are difficult to evaluate. As the time increment Δt is very small, however, we can assume that the parcel is at rest. The concentration gradients across the boundaries can then be determined using Eulerian operators. Thus

$$J_{x+\Delta x} = J_x + \left(\frac{\partial J_x}{\partial x}\right)_x \Delta x \tag{1.8a}$$

$$J_{y+\Delta y} = J_y + \left(\frac{\partial J_y}{\partial y}\right)_y \Delta y \tag{1.8b}$$

$$J_{z+\Delta z} = J_z + \left(\frac{\partial J_z}{\partial z}\right)_z \Delta z \tag{1.8c}$$

In the limit as Δt, Δx, Δy and Δz go to zero these approximations are valid and present no difficulties when using analytical solutions to the resulting advection/diffusion equation. It is worth noting, however, that problems with numerical dispersion and oscillatory solutions can arise when using discrete numerical methods to solve the advection/diffusion equation. These problems stem from the fact that diffusion is defined in a moving Lagrangian coordinate system whereas advection is defined in a fixed Eulerian coordinate system.

Using equation (1.2) to evaluate the diffusive fluxes in equation (1.8) and rearranging

$$(J_x - J_{x+\Delta x})\Delta y \Delta z = \frac{\partial}{\partial x}\left(e_{\mathrm{m}} \frac{\partial c}{\partial x}\right)\Delta x \Delta y \Delta z \tag{1.9a}$$

$$(J_y - J_{y+\Delta y})\Delta x \Delta z = \frac{\partial}{\partial y}\left(e_{\mathrm{m}} \frac{\partial c}{\partial y}\right)\Delta x \Delta y \Delta z \tag{1.9b}$$

$$(J_z - J_{z+\Delta z})\Delta x \Delta y = \frac{\partial}{\partial z}\left(e_{\mathrm{m}} \frac{\partial c}{\partial z}\right)\Delta x \Delta y \Delta z \tag{1.9c}$$

Combining equations (1.4), (1.7) and (1.9), noting that e_{m} is independent of x, y and z and simplifying gives

$$\frac{\mathrm{D}c}{\mathrm{D}t} = e_{\mathrm{m}}\left(\frac{\partial^2 c}{\partial x^2} + \frac{\partial^2 c}{\partial y^2} + \frac{\partial^2 c}{\partial z^2}\right) \tag{1.10}$$

The material derivative is (Batchelor 1967)

$$\frac{\mathrm{D}c}{\mathrm{D}t} = \frac{\partial c}{\partial t} + u_x \frac{\partial c}{\partial x} + u_y \frac{\partial c}{\partial y} + u_z \frac{\partial c}{\partial z} \tag{1.11}$$

Equation (1.10) can now be rewritten to yield the well-known three-dimensional Fickian diffusion equation in rectangular Cartesian coordinates

$$\frac{\partial c}{\partial t} + u_x \frac{\partial c}{\partial x} + u_y \frac{\partial c}{\partial y} + u_z \frac{\partial c}{\partial z} = e_{\mathrm{m}}\left(\frac{\partial^2 c}{\partial x^2} + \frac{\partial^2 c}{\partial y^2} + \frac{\partial^2 c}{\partial z^2}\right) \tag{1.12}$$

Provided the velocities and the molecular diffusion coefficient are known and suitable initial and boundary conditions for concentrations can be specified, then equation (1.12) can be solved to predict the changes of tracer concentration with time and location.

1.5.4 Properties of the Molecular Diffusion Equation

A very important solution to equation (1.12) describes the spreading of a conservative tracer slug of mass M released at $x=y=z=0$ and $t=0$ in an unbounded channel. If u_x, u_y, u_z and e_m are constant then

$$c(x, y, z, t) = \frac{M}{(4\pi e_m t)^{3/2}} \exp\left(-\frac{(x-u_x t)^2 + (y-u_y t)^2 + (z-u_z t)^2}{4 e_m t}\right) \tag{1.13}$$

At a given time the tracer concentration predicted using equation (1.13) is distributed along the x, y and z axes in a Gaussian bell-shaped curve (Figure 1.2). As will be discussed later, in field and laboratory studies tracer distributions are often measured which are approximately Gaussian and equations such as (1.12) and (1.13) can be used to model such data.

Another important feature of equations (1.12) and (1.13) relates to the spatial variance of the tracer distribution. The spatial variance in the x direction is defined

$$\sigma_x^2(t) = \frac{1}{M} \iiint (x-\bar{x})^2 c(x, y, z, t)\, \mathrm{d}x\, \mathrm{d}y\, \mathrm{d}z \tag{1.14a}$$

$$\bar{x}(t) = \frac{1}{M} \iiint x c(x, y, z, t)\, \mathrm{d}x\, \mathrm{d}y\, \mathrm{d}z \tag{1.14b}$$

$$M = \iiint c(x, y, z, t)\, \mathrm{d}x\, \mathrm{d}y\, \mathrm{d}z \tag{1.14c}$$

where $\bar{x} = x$ coordinate of the centroid of the tracer distribution and $M =$ tracer mass. The variances in the y and z directions are defined in an analogous fashion. It is a straightforward exercise to show that for equation (1.13)

$$\sigma_x^2 = \sigma_y^2 = \sigma_z^2 = 2 e_m t \tag{1.15}$$

It is a feature of solutions to equation (1.12) that the variance increases linearly with time. In field studies it is often observed that the variance of observed tracer distribution increases with time at a linear rate, which suggests that such data can be modelled by equations such as (1.12).

Tracer distributions do not need to be Gaussian for the variance to increase linearly with time. Consider an experiment where equation (1.12) applies but the initial tracer distribution along the x axis is markedly skewed because of the way the tracer was introduced into the channel. During the initial phases of the experiment the tracer distribution will be skewed but the variance of the

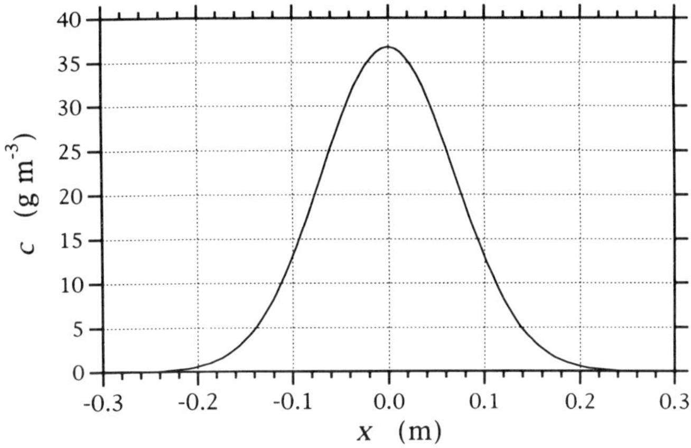

Figure 1.2. Tracer distribution along the x axis at $y = z = 0$ predicted using equation (1.13) with $M = 1$ g, $t = 1000$ hours, $e_m = 2 \times 10^{-9}$ m^2 s^{-1} and $u_x = u_y = u_w = 0$

tracer distribution will always increase linearly with time. The skewness of the tracer distribution will decrease gradually with time and at asymptotically long times the distribution will become Gaussian.

The linear increase of variance is a necessary but not a sufficient condition for the validity of equation (1.12). In other words there may be other more complex diffusion equations whose solutions show linear growth of variance. One such model is discussed briefly in Chapter 4 (see equation 4.22). There have been several theoretical investigations of non-Fickian diffusion models but few, if any, of these other equations are used in practical mixing problems.

A third important feature of equation (1.12) is that it is linear so that solutions for a number of slug injections can be superimposed to construct solutions to more complex problems. Superposition is discussed in more detail in Chapter 3 (Section 3.7).

Finally, it is worth noting that there are analytical solutions to equation (1.12) available for a large number of heat conduction problems (Carslaw and Jaegar 1959, Crank 1975) which can be adapted readily for analysing diffusion problems. A summary of these is given in the Appendix.

1.6 TURBULENCE

1.6.1 Features of Turbulent Flow

Most river and open channel flows are characterised by high Reynolds numbers [see equation (1.3)] and are turbulent except in a very thin, viscous sub-layer near the bed (there is a more detailed discussion of the viscous sub-layer in Chapter 6). One important feature of turbulent flow is that tracer spreads far more rapidly than in laminar flow. In turbulent flow viscous forces are weak relative to inertial forces. The velocity at a point exhibits random short-term fluctuations even if the long-term average velocity is constant. This is in marked contrast with laminar flow in which the velocity at a point varies smoothly and predictably over time. Particles released into a turbulent flow follow highly irregular paths and particles released from the same point at different times may follow significantly different paths. By comparison particles released from the same point in a steady laminar flow will follow the same path.

Turbulent flows are not, however, entirely chaotic and velocities at adjacent locations are correlated. The correlations are very strong if the points are close together but decrease as the separation between points increases until at points a long way apart the correlations become negligibly small. Spectral analysis is used to quantify the correlation between velocities at two (or more) points as a function of the distance separating the points. A typical spectrum may have pronounced peaks at given separations, say $\lambda_1, \lambda_2, \lambda_3 \ldots \lambda_n$, and this gives rise to the concept that within the flow there are eddies whose characteristic length scales are $\lambda_1, \lambda_2, \lambda_3 \ldots \lambda_n$. Everyone is familiar with the large-scale eddies which disturb the water surface in flowing water. Eddies also become obvious when a tracer such as dye is introduced into the flow from a steady point source. The largest separation at which the velocities remain correlated provides an estimate of the largest turbulent length scale within the flow and hence the size of the largest eddies. In rivers the length scales of the largest vertical and horizontal eddies cannot be larger than, and are often found to be comparable with, the channel depth and the channel width, respectively.

Turbulent flows are not self-sustaining and in the absence of an external energy input, turbulence decays and the flow reverts to being laminar. In rivers the energy input is provided by gravity (channel slope), although very close to an outfall it may be provided by a pressure gradient. In rivers turbulence is generated by velocity shear (Tennekes and Lumley 1972) and hence originates in regions where there are strong velocity gradients, notably near the bed, banks and obstacles within the flow. Energy enters the flow at fairly large length scales but there are strong non-linearities in the Navier–Stokes equations governing turbulent flow which cause energy to be transferred to smaller eddies. This gives rise to a wide range of eddy sizes within turbulent flow. Eventually the energy reaches the smallest eddies and is dissipated as heat. The smallest length scales present in turbulent river flows have been shown to be of the order 10^{-4} to 10^{-3} m (Fischer *et al* 1979, Chatwin and Allen 1985). As is intuitively obvious it tends to be the largest eddies which effect most of the momentum and mass transfer in turbulent flows (Townsend 1956, Tennekes and Lumley 1972).

Turbulence is termed stationary if characteristics such as the mean velocity,

velocity variance and the correlations between velocities remain constant with time (in turbulent flow stationarity is the equivalent of steadiness in laminar flow). Turbulence is homogeneous if the velocity fluctuations and correlations do not vary with location and is isotropic if they are the same in all coordinate directions. A body of theory has been developed for diffusion in stationary, homogeneous turbulence (Batchelor 1953, Townsend 1956). Turbulence in natural channels is seldom homogeneous or stationary and consequently theoretical results cannot always be applied directly. Despite this difficulty the theoretical results provide insights into the form which the turbulent mixing equations should take and they also help to guide experimental work.

1.6.2 Role of Averaging in Turbulent Flows

It is necessary to make a slight digression at this point to discuss the role of averaging in studies of turbulent flows. Consider again equation (1.12). This equation is exact and in theory it could be used to model tracer concentrations in turbulent flow. We would need to know how the velocity varies with time and space (down to very small length and time scales) and to have suitable initial and boundary conditions. Given this information and a very large computer we could try and predict the movement of each separate fluid parcel. From time to time we could determine the average tracer concentration over any length scale we were interested in and hence could determine the average concentration of the tracer cloud. An important feature of turbulent flow, however, is that velocities vary randomly with time and it is simply not practicable to try and make deterministic predictions of velocity fluctuations or to make sufficient velocity measurements to address complex mixing problems in this way. We can, however, make the problem more tractable by examining the behaviour of certain space and time averages of velocity, tracer concentration and advective flux.

There are three different sorts of average which are used when studying turbulence: ensemble, time and space averages. Consider first the ensemble average. Imagine conducting an experiment in which a known mass of tracer is released and the tracer concentration is measured at a fixed point x_o, y_o, z_o and time t_o after injection. Now repeat the same experiment several times, each time measuring the concentration at the same point and at the same time after release. Because the flow is turbulent each concentration measurement will be different, giving rise to a series of concentrations $c_i(x_o, y_o, z_o, t_o)$ where $i = 1, 2, \ldots N$ and N is the total number of experiments. Then averaging over all the experiments yields an estimate of the ensemble average concentration at x_o, y_o, z_o and t_o.

$$\langle c(x_o, y_o, z_o, t_o) \rangle = \frac{1}{N} \sum_{i=1}^{i=N} c_i(x_o, y_o, z_o, t_o) \qquad (1.16)$$

The angled brackets $\langle\ \rangle$ are customarily used to denote an ensemble average. Thus an ensemble average is taken over a large number of independent experiments or realisations and can be calculated for concentration, velocity or various velocity and concentration products.

In some situations the statistical properties of the concentration are independent of time over some time window T. Tracer concentrations below a steady source are an example. In such situations the ensemble average concentration at a fixed point can be approximated by the time average

$$\langle c(x_o, y_o, z_o, t_o)\rangle \approx \frac{1}{T}\int_{\tau=0}^{T} c(x_o, y_o, z_o, \tau)\,d\tau \tag{1.17}$$

where $0 < \tau < T$. Alternatively, in a region where the tracer is well-mixed but rapidly time-varying, it may be valid to approximate the ensemble average by the spatial average of the concentration over the volume V.

$$\langle c(x_o, y_o, z_o, t_o)\rangle \approx \frac{1}{V}\iiint c(x, y, z, t_o)\,dx\,dy\,dz \tag{1.18}$$

where x_o, y_o, z_o lies within the volume V. The assumption that the ensemble, time and space averages are identical is known as the ergodic hypothesis.

Theoreticians insist that the ensemble average should always be used when analysing turbulent mixing (Lumley and Panofsky 1964). The main reason is that separate realisations are independent so that ensemble averages are unbiased. By comparison measurements made at small intervals of space and time may not be independent and hence the resulting average may not be a reliable estimate of the true mean value. When addressing practical problems, however, it is not always feasible to conduct the large number of trials which would seem to be required to carry out ensemble averaging and consequently time and space averaging are often used despite their theoretical shortcomings. It is clearly desirable to make the averaging period (or volume) as large as possible to reduce the possibility of bias.

In situations where the period of averaging T is large compared with the time-scale of the turbulent fluctuations in concentration then equation (1.17) can be expected to provide a reliable estimate of the ensemble average. If, on the other hand, T is comparable with or smaller than the time-scale of turbulent fluctuations, then the ergodic hypothesis is invalid and equation (1.17) will yield poor estimates of the ensemble average concentration. This can present a serious problem in estuarine flows where it is desirable to make the averaging period T as large as possible, but tidal action restricts the time window over which the underlying characteristics of the flow are constant. In estuaries the maximum averaging period which can safely be used is $T = 60$–90 seconds (Chatwin and Allen 1985). Fortunately this problem arises less frequently when dealing with a steady river flow and steady tracer input. The same difficulty arises with spatial averaging. To obtain an unbiased estimate of the space-averaged concentration it is necessary to collect a number of samples at sufficiently large separation to ensure that they are independent. This implies sampling over a large area, but in many practical problems there may not be a large area where the concentration is constant except for random turbulent fluctuations. It will always be possible to determine a space or time average from experimental data but it is important to remember that these estimates may pertain only to the particular set of conditions sampled and may not give a robust estimate of typical conditions.

1.6.3 Extreme Concentrations

When assessing the ecological impact of a toxic pollutant the average concentration is clearly important but some knowledge of concentration extremes is also desirable. As pointed out by Fischer *et al* (1979), if a few high concentrations kill an organism then even a large number of low concentrations will not bring it back to life. Ideally we would like to be able to predict the probability density function of the concentration at various points in the flow because we could then say something about both mean and extreme concentrations. If, in addition, we know the time spectrum of the concentration fluctuations, then we also know something about how long high and low concentrations persist. Unfortunately the existing models for turbulent flows predict only the average concentration and do not furnish estimates of the probability density functions or the time spectra of concentrations, although current research is addressing this topic. The practitioner must rely on empirical studies to give an indication of the concentration fluctuations likely to be encountered. Much of the

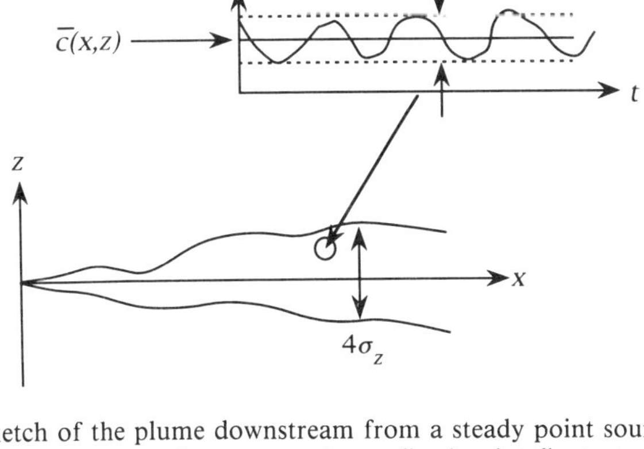

Figure 1.3. Sketch of the plume downstream from a steady point source in turbulent flow showing how concentration measured at a fixed point fluctuates with time

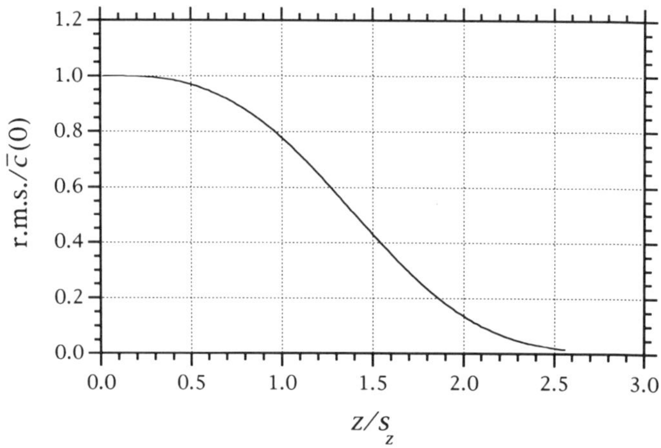

Figure 1.4. R.m.s. deviation of tracer concentration fluctuations over time as a function of distance from the centreline typical of experimental measurements downstream from a steady point source. The r.m.s. deviation is non-dimensionalised using the mean centreline concentration and the transverse distance is non-dimensionalised using the transverse spatial variance of the plume

experimental work on concentration extremes has been done in the atmosphere, but similar mechanisms operate in air and water.

Consider the problem of concentrations in the x–z plane downstream from a steady point source discharging tracer into an unbounded channel with flow in the x direction (see Figure 1.3). A continuously recording probe placed at a fixed location within the plume measures concentrations c_i where $i = 1, 2, 3 \ldots N$ at short intervals of time. The mean and root mean square (r.m.s.) deviation[1] over time are calculated at a number of points across the plume at a fixed distance downstream from the source. Figure 1.4 shows the typical distribution which is found in experimental studies. The r.m.s. deviation is normalised using the mean centreline concentration and the transverse distance is normalised using plume width (here taken as the transverse standard deviation[2] of the plume). At the centreline of the plume the r.m.s.

[1] The r.m.s. deviation is defined

$$\text{r.m.s.} = \sqrt{\frac{1}{N}\sum_{i=1}^{N}(c_i - \bar{c})^2} \qquad \text{where} \qquad \bar{c} = \frac{1}{N}\sum_{i=1}^{N} c_i$$

[2] The transverse standard deviation is defined

$$\sigma_z^2 = \frac{\int_{-\infty}^{\infty}(z - \bar{z})^2 c \, dz}{\int_{-\infty}^{\infty} c \, dz} \qquad \text{where} \qquad \bar{z} = \frac{\int_{-\infty}^{\infty} z \, dz}{\int_{-\infty}^{\infty} c \, dz}$$

deviation over time attains its maximum value which is comparable with the mean centreline concentration. Either side of the centreline the absolute value of the r.m.s. deviation decreases and it vanishes outside the plume. Csanady (1967) has gone some way towards predicting the distribution of the r.m.s. deviation within the plume.

Knowing the r.m.s. deviation we are able to say something about the maximum concentration. The absolute maximum concentration is not easy to measure, predict or interpret and a more useful measure is the concentration which is only exceeded a certain percentage of the time (say 10, 5 or 1%). The probability density function of the concentration is found to be skewed because concentrations must be non-negative and there are a few very high concentrations. This type of data can be described by either a log-normal or a Poisson distribution (Csanady 1973). For the Poisson distribution the concentrations exceeded 10 and 1% of the time are, respectively, 2.3 and 4.6 times the mean concentration (Freund 1974). Thus near the centreline of the plume we would expect the peak to mean ratio to lie in the range two to five. This accords well with experimental results in atmospheric and lake studies in which the peak to mean ratio is typically three (Csanady 1967). It might be unwise to assume that tracer plumes in rivers behave in exactly the same way because the flows are rather different, but clearly the equivalent of Figure 1.4 could be derived from experimental results in rivers.

Figure 1.4 indicates that towards the edge of the plume the ratio of the r.m.s. deviation to the centreline concentration decreases. The local concentration also decreases and it is important to understand how the peak to mean ratio behaves locally. Experimental work in the atmospheric boundary layer shows that the local peak to mean ratio increases from about three at the centreline (as discussed in the previous paragraph) to about 20–50 at the edge of the plume, with occasional values as high as 100. These ratios are high at the edge of the plume because for much of the time there is no tracer present, but occasionally an eddy containing an appreciable amount of tracer is present at the sampling point for a short period of time. It is found in atmospheric studies that the peak to mean ratio at the edge of the plume is highest close to the source and decreases with distance downstream so that at very large distances from the source the peak to mean ratio reaches a lower limit of the order of two. Again it must be stressed that these figures may not be directly applicable to river mixing problems. It is not immediately obvious why the peak to mean ratio should decrease with distance from the source. Presumably close to a point source the plume contains a small number of fluid parcels containing high tracer concentrations among other parcels substantially free from tracer. Further away from the source diffusion has had time to mix the tracer between adjacent fluid parcels so that concentrations are less variable across the plume.

In the remainder of this book we will discuss models and coefficients for predicting the ensemble mean tracer concentration. It is important to remember, however, that the tracer concentration will fluctuate about this mean value. An initial assessment can be made of the importance of these fluctuations using the ideas and results discussed in the preceding paragraphs but to make more accurate predictions of concentration fluctuations it will be necessary to conduct site-specific experimental studies.

Example 1.1

Short-term temperature fluctuations measured in the Waikato River, New Zealand, downstream from the Huntly power station. Cooling water with an excess temperature averaging +8°C is discharged from a bankside outfall which mixes the effluent with 50–80% of the river flow within about 300 m of the outfall. Vertical mixing is complete within 200–300 m. Temperatures were measured every second for 30 minutes at four sections 250–550 m downstream from the outfall. Results are summarised in Table 1.1.

Temperature fluctuations were found to have a periodicity of the order of three minutes at each site. The distribution of temperature was approximately normal and the peak to mean ratio was found to lie in the range 1.05–1.12. This is markedly different from the experimental results for a point source discussed in the text where the tracer distribution is skewed and the minimum peak to mean ratio is typically three. The difference arises because the Huntly cooling water has a maximum excess temperature of +8°C at the point of discharge (which places an upper limit on the peak to mean ratio) and there is rapid vertical mixing together with significant entrainment of ambient river water close to the outfall.

For measurements 40 m from the left bank (where temperatures are close to the maximum) the peak to mean ratio decreased with distance

Table 1.1. Temperature fluctuations with time measured in the Waikato River below the Huntly power station. During Experiments 1 and 2 initial dilution was with 80 and 45% of river flow, respectively. Transverse distances are measured from the left bank. The peak temperature occurs close to the left bank. The r.m.s. deviation is calculated over 30 minutes. All temperatures are °C above ambient

Longitudinal distance (m)	Transverse distance (m)	Transect peak temperature (°C)	Local mean temperature (°C)	r.m.s. time deviation (°C)	Peak to mean ratio
Expt. 1					
250	40	2.3	2.0	0.063	1.071
350	40	2.2	2.0	0.050	1.058
550	40	2.2	2.0	0.053	1.053
Expt. 2					
450	40	4.5	3.0	0.063	1.048
450	70	4.5	2.0	0.100	1.115

downstream from the outfall Table 1.1. This is in accord with the experimental results discussed in the text. For measurements made near the edge of the plume (70 m from the left bank) the absolute value of the r.m.s. deviation was nearly twice that measured at the other sites and the local peak to mean ratio was also significantly higher. Along the edge of the plume there is a high temperature gradient and turbulent eddies in this region cause larger temperature flucuations than elsewhere in the plume where the temperature gradients are smaller.

1.6.4 Lagrangian Coordinate Systems

It is clear from the derivation of equation (1.12) that it is best to describe diffusion in a Lagrangian coordinate system which moves at the mean velocity. In steady laminar flow there is little difficulty defining the mean velocity to use. In stationary turbulent flow, however, there are two possible Lagrangian coordinate systems to choose from: one which moves at the mean velocity of an individual tracer cloud and one which moves at the ensemble mean velocity.

Imagine conducting a number of identical experiments in each of which an instantaneous tracer release is made. Figure 1.5 shows two tracer clouds each measured at time t after release. Because of the random nature of turbulence, each tracer cloud follows a slightly different path and spreads in a slightly different manner. The broken axes in Figure 1.5 are placed through the

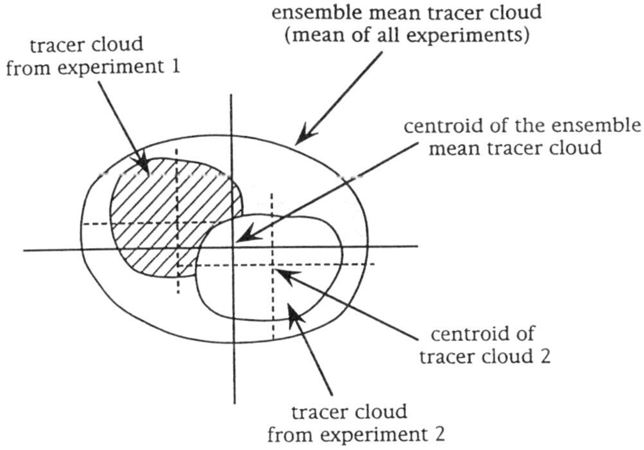

Figure 1.5. Sketch showing tracer clouds from two different experiments measured at the same time after injection. The broken axes mark the centroid of each tracer cloud. The ensemble mean tracer cloud represents the average of a large number of identical experiments. The solid coordinate axes move at the ensemble mean velocity

centroids of the individual tracer clouds. These broken axes move at the mean velocity of the individual tracer clouds. If we conducted a large number of identical experiments and averaged the results we would describe the ensemble mean tracer cloud. The solid axes in Figure 1.5 are drawn through the centroid of the ensemble mean tracer cloud. It is clear that the dimensions of the ensemble average tracer cloud are larger than the dimensions of the individual tracer clouds. The corollary is that the ensemble average concentration is smaller than the average within an individual tracer cloud. The biggest differences occur in the early stages of diffusion when the tracer cloud is small relative to the size of the turbulent eddies and once the size of the tracer cloud exceeds the largest turbulent eddies then differences between the averages decrease. In Figure 1.5 the solid axes move at the ensemble mean velocity (i.e. the average velocity that would be determined from a large number of realisations of the flow). It is usual to quantify the rate of diffusion from the ensemble-averaged dimensions of the tracer cloud for the reasons discussed in Section 1.6.2.

1.6.5 Measures of Spread

When studying diffusion we are ultimately aiming to predict tracer concentrations. A convenient way to approach this problem is to examine the distances separating tracer particles. If tracer mass M is spread over a distance L in each coordinate direction, then the principle of mass conservation tells us that the average concentration is M/L^3. An important measure of the size of a tracer cloud is its variance, which is the mean square displacement about the centre of mass. Equation 1.14 defines the variance of an individual tracer cloud about its centroid. In turbulence modelling we are concerned not with individual tracer clouds but with ensemble averages. The variance of an ensemble of dye clouds about the centre of mass of the ensemble is

$$\Sigma_x^2 = \frac{1}{M} \iiint (x - \langle \bar{x} \rangle)^2 \langle c(x, y, z) \rangle \, dx \, dy \, dz \tag{1.19a}$$

where $\langle \bar{x} \rangle$ = centre of mass of the ensemble given by

$$\langle \bar{x} \rangle = \frac{1}{M} \iiint x \langle c(x, y, z) \rangle \, dx \, dy \, dz \tag{1.19b}$$

From Figure 1.5 it is clear that the ensemble average tracer cloud covers a larger area than either of the individual tracer clouds. Thus the ensemble average variance Σ_x^2 is larger than the variance of an individual cloud σ_x^2. Indeed it can be shown (Fischer *et al* 1979) that

$$\Sigma_x^2 = \underset{[1]}{\langle \sigma_x^2 \rangle} + \underset{[2]}{\langle (\bar{x} - \langle \bar{x} \rangle)^2 \rangle} \tag{1.20}$$

This equation states that the variance of the ensemble mean tracer cloud is the sum of the mean variance of each tracer cloud about its centre of mass (term [1]) plus the mean square displacement of the centroid of each cloud from the centroid of the ensemble (term [2]). Taylor (1921) investigated diffusion in stationary homogeneous turbulence and developed a theory for predicting the rate of change of Σ_x^2. Most mixing models are based on Taylor's analysis and we make extensive use of his results in this book. [There is also a body of theory dealing with diffusion of individual tracer clouds relative to their individual centres of mass. This is termed relative diffusion and is described by Csanady (1973), Fisher *et al* (1979) and others. We shall not discuss or use relative diffusion theory in this book.] It is important to note that the ensemble mean concentrations may be lower than the mean concentration observed within an individual tracer cloud. The methods described in Section 1.6.3 can be used to account for these differences.

1.7 SOLUTE TRANSPORT IN TURBULENT FLOW

In Section 1.5.3 we outlined the derivation of the three-dimensional advection–diffusion equation in laminar flow

$$\frac{\partial c}{\partial t} + u_x \frac{\partial c}{\partial x} + u_y \frac{\partial c}{\partial y} + u_z \frac{\partial c}{\partial z} = e_{\mathrm{m}} \left(\frac{\partial^2 c}{\partial x^2} + \frac{\partial^2 c}{\partial y^2} + \frac{\partial^2 c}{\partial z^2} \right) \tag{1.12}$$

In this section we discuss the application of equation (1.12) to solute transport in stationary turbulent flow.

1.7.1 Reynolds Averaging

The famous physicist Osborne Reynolds assumed that in turbulent flow the observed velocities and concentrations can be written

$$u_x = \langle u_x \rangle + u_x' \tag{1.21a}$$

$$u_y = \langle u_y \rangle + u_y' \tag{1.21b}$$

$$u_z = \langle u_z \rangle + u_z' \tag{1.21c}$$

$$c = \langle c \rangle + c' \tag{1.21d}$$

where u_x, u_y, u_z and $c =$ observed velocities and concentration; $\langle\ \rangle$ denotes an ensemble average; and the prime denotes a deviation from the ensemble average. By definition the average values of c', u_x', u_y' and u_z' are zero. If equations (1.21) are substituted into equation (1.12) we obtain

$$\frac{\partial}{\partial t}(\langle c\rangle + c') + (\langle u_x\rangle + u_x')\frac{\partial}{\partial x}(\langle c\rangle + c') + (\langle u_y\rangle + u_y')\frac{\partial}{\partial y}(\langle c\rangle + c')$$

$$+ (\langle u_z\rangle + u_z')\frac{\partial}{\partial z}(\langle c\rangle + c')$$

$$= e_\text{m}\left[\frac{\partial^2}{\partial x^2}(\langle c\rangle + c') + \frac{\partial^2}{\partial y^2}(\langle c\rangle + c') + \frac{\partial^2}{\partial z^2}(\langle c\rangle + c')\right] \quad (1.22)$$

We then take the ensemble average of equation (1.22) and note that the ensemble average of terms such as

$$\frac{\partial c'}{\partial t},\ u_x'\frac{\partial\langle c\rangle}{\partial x},\ \langle u_x\rangle\frac{\partial c'}{\partial x},\ \frac{\partial^2 c'}{\partial x^2} \quad (1.23)$$

all vanish. The left-hand side of equation (1.22) contains the term

$$u_x'\frac{\partial c'}{\partial x} + u_y'\frac{\partial c'}{\partial y} + u_z'\frac{\partial c'}{\partial z} \quad (1.24)$$

This can be replaced by

$$\frac{\partial}{\partial x}(u_x'c') + \frac{\partial}{\partial y}(u_y'c') + \frac{\partial}{\partial z}(u_z'c') \quad (1.25)$$

because the flow is stationary incompressible continuity requires that

$$\frac{\partial u_x'}{\partial x} + \frac{\partial u_y'}{\partial y} + \frac{\partial u_z'}{\partial z} = 0 \quad (1.26)$$

Making these simplifications we are left with

$$\overset{[1]}{\frac{\partial\langle c\rangle}{\partial t}} + \overset{[2]}{\left(\langle u_x\rangle\frac{\partial\langle c\rangle}{\partial x} + \langle u_y\rangle\frac{\partial\langle c\rangle}{\partial y} + \langle u_z\rangle\frac{\partial\langle c\rangle}{\partial z}\right)}$$

$$= e_\text{m}\overset{[3]}{\left(\frac{\partial^2\langle c\rangle}{\partial x^2} + \frac{\partial^2\langle c\rangle}{\partial y^2} + \frac{\partial^2\langle c\rangle}{\partial z^2}\right)} - \overset{[4]}{\left(\frac{\partial}{\partial x}\langle u_x'c'\rangle + \frac{\partial}{\partial y}\langle u_y'c'\rangle + \frac{\partial}{\partial z}\langle u_z'c'\rangle\right)} \quad (1.27)$$

Term [1] in equation (1.27) is the time rate of change of the ensemble mean concentration. The three terms [2] quantify advection of the ensemble mean concentration by the mean velocity. The three terms [3] quantify molecular diffusion. Comparing equation (1.12) with (1.27) we see that the first seven terms are similar in form except that $\langle u_x\rangle$, $\langle u_y\rangle$, $\langle u_z\rangle$ and $\langle c\rangle$ appear in place of u_x, u_y, u_z and c. The major difference is the appearance on the right-hand side of equations (1.27) of the three new turbulent transport terms [4].

$$-\frac{\partial}{\partial x}\langle u_x'c'\rangle-\frac{\partial}{\partial y}\langle u_y'c'\rangle-\frac{\partial}{\partial z}\langle u_z'c'\rangle \tag{1.28}$$

Although the ensemble means of the velocity and concentration deviations are zero, the ensemble means of their products are in general non-zero. In other words the fluctuations of velocity and concentration are correlated in some manner (which we have yet to define) and this correlation gives rise to additional transport. The additional transport is termed turbulent diffusion or eddy diffusion.

In rivers eddy diffusion causes tracer to spread far more rapidly than could happen if spreading only occurred as a result of molecular diffusion. This does not mean that molecular diffusion plays no part in turbulent mixing. Molecular diffusion is still the fundamental process which causes tracer to diffuse from a region of high concentration to a region of low concentration. Turbulence has the effect of tearing apart tracer clouds thereby increasing local concentration gradients and accelerating molecular diffusion. It is the combination of turbulent eddies and molecular diffusion which causes rapid tracer mixing in turbulent flow.

In situations where the ergodic hypothesis is valid then time averages can be used in place of ensemble averages. Equation (1.27) then becomes

$$\frac{\partial \bar{c}}{\partial t}+\left(\bar{u}_x\frac{\partial \bar{c}}{\partial x}+\bar{u}_y\frac{\partial \bar{c}}{\partial y}+\bar{u}_z\frac{\partial \bar{c}}{\partial z}\right)$$

$$=e_{\mathrm{m}}\left(\frac{\partial^2\bar{c}}{\partial x^2}+\frac{\partial^2\bar{c}}{\partial y^2}+\frac{\partial^2\bar{c}}{\partial z^2}\right)-\frac{\partial}{\partial x}\,(\overline{u_x'c'})-\frac{\partial}{\partial y}\,(\overline{u_y'c'})-\frac{\partial}{\partial z}\,(\overline{u_z'c'}) \tag{1.29}$$

where the overbar denotes a time average. It may appear strange that we have time-averaged and yet produced equation (1.29) which predicts the rate of change of concentration with time. The key to successfully applying the Reynolds decomposition is that the averaging period T used is large compared with the time-scale of the turbulent fluctuations but is small compared with the time-scale of the change in concentration which we wish to study. We have problems using the ergodic hypothesis if the time-scales of the turbulent fluctuations are comparable with the time-scales of the concentration change we are interested in, as can occur in estuarine flows, but we rarely encounter this problem in rivers.

1.7.2 The Problem of Closure

In deriving equations (1.27) and (1.29) we have not introduced any approximations. Unfortunately the averaging process has created a new problem because now turbulent transport of the mean concentration is governed by terms involving the unknown velocity and concentration fluctuations. This is the problem of closure which plagues the subject of turbulence modelling. Before we can solve equations (1.27) and (1.29) we need to relate the turbulent fluxes to some property of the average concentration. We shall see in the next section that this can indeed be done and that under certain conditions the turbulent fluxes are proportional to the spatial gradients of the ensemble mean concentration.

1.7.3 Taylor's Analysis of Turbulent Diffusion

In 1921 Taylor published a classic paper in which he made a theoretical analysis of the spreading of a cloud of tracer particles released into stationary, homogeneous turbulence (Taylor 1921). This analysis remains today the key to quantifying turbulent diffusion. Taylor adopted a Lagrangian coordinate system travelling at the mean velocity and examined the rate of longitudinal mixing. If a tracer particle is released at the origin then after a time t it will be located at a longitudinal distance x from the moving origin where

$$x(t) = \int_{\tau=0}^{t} u_x'(\tau)\,\mathrm{d}\tau \tag{1.30}$$

and u_x' = the turbulent velocity fluctuation about the mean defined in equation (1.21a). If N particles are released then because of the random nature of the turbulent velocity fluctuations they will be at different locations x_i (where $i = 1, 2, 3 \ldots N$) after time t. Now it follows from the conservation of mass that the ensemble mean variance of the resulting tracer cloud is identically equal to the ensemble mean square displacement of tracer particles. Thus

$$\Sigma_x^2 = \langle (x - \bar{x})^2 \rangle \tag{1.31a}$$

Because a Lagrangian coordinate system is being used

$$\langle \bar{x} \rangle = 0 \tag{1.31b}$$

and so

$$\Sigma_x^2 = \langle x^2 \rangle \tag{1.31c}$$

In other words the ensemble mean size of the tracer cloud is the same as the ensemble mean variance of the displacement of the individual particles from the origin of the Lagrangian coordinate system.

We can rewrite equation (1.30)

$$x^2(t) = \left[\int_{t_1=0}^{t} u_x'(t_1)\, \mathrm{d}t_1\right]\left[\int_{t_2=0}^{t} u_x'(t_2)\, \mathrm{d}t_2\right]$$

$$= \int_{t_1=0}^{t}\int_{t_2=0}^{t} u_x'(t_1)u_x'(t_2)\, \mathrm{d}t_1\, \mathrm{d}t_2 \tag{1.32}$$

In equation (1.32) the product

$$u_x'(t_1)u_x'(t_2) \tag{1.33}$$

is the autocovariance between the velocity of a particle at time t_1 and the velocity of the same particle at a later time t_2. Taking the ensemble average of equation (1.32)

$$\langle x^2\rangle = \int_{t_1=0}^{t}\int_{t_2=0}^{t} \langle u_x'(t_1)u_x'(t_2)\rangle\, \mathrm{d}t_1\, \mathrm{d}t_2 \tag{1.34}$$

The Lagrangian autocorrelation function is defined

$$R_x(t_2 - t_1) = \frac{\langle u_x'(t_1)u_x'(t_2)\rangle}{\langle u_x'^2\rangle} \tag{1.35}$$

where $\sqrt{\langle u_x'^2\rangle}$ = intensity of turbulence.

Because the turbulence is assumed to be stationary, the autocorrelation can only depend on the time difference $t_2 - t_1$. Combining equations (1.34) and (1.35)

$$\langle x^2\rangle = \langle u_x'^2\rangle \int_{t_1=0}^{t}\int_{t_2=0}^{t} R_x(t_2 - t_1)\, \mathrm{d}t_2\, \mathrm{d}t_1 \tag{1.36}$$

which can be rewritten

$$\langle x^2\rangle = 2\langle u_x'^2\rangle \int_{s=0}^{t} (t - s)R_x(s)\, \mathrm{d}s \tag{1.37}$$

The Lagrangian autocorrelation function cannot be predicted theoretically but its general form is known (Figure 1.6). By definition

$$R_x(0) = 1 \tag{1.38}$$

and at large separations the velocity fluctuations become independent so that

$$R_x(\infty) = 0 \tag{1.39}$$

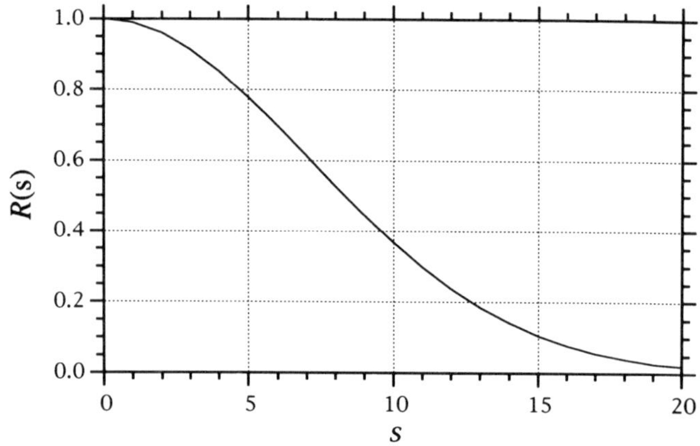

Figure 1.6. Schematic representation of the Lagrangian autocorrelation function typical of turbulent flow

Two limiting cases can now be examined. At very small values of t, $R_x = 1$ and equation (1.37) becomes

$$\langle x^2 \rangle = \langle u_x'^2 \rangle t^2 \tag{1.40}$$

Thus at small times, the variance increases at a rate proportional to t^2. At very large t, $R_x = 0$ and equation (1.37) becomes

$$\langle x^2 \rangle \to 2\langle u_x'^2 \rangle T_x t + \text{constant} \tag{1.41}$$

where

$$T_x = \int_{s=0}^{\infty} R_x(s)\, \mathrm{d}s \tag{1.42}$$

T_x is known as the integral Lagrangian time-scale and is a measure of the time taken for a particle to forget its original velocity. The relationship in (1.41) is a very important result which implies that some time after the tracer has been released into turbulent flow the variance of the tracer cloud increases linearly with time. The Lagrangian time-scale tells us how long we have to wait before this happens. Equation (1.41) also states that the rate at which the variance of the tracer cloud grows increases with the square of the turbulent intensity (i.e. spreading occurs more rapidly in highly turbulent flows).

1.7.4 Fickian Model of Turbulent Diffusion

In Section 1.5.4 it was shown that one of the features of the Fickian diffusion model is that the variance of a tracer cloud increases linearly with time. Taylor's analysis demonstrates that in stationary homogeneous turbulence the variance of a tracer cloud increases linearly with time at asymptotically large times. This suggests that in stationary homogeneous turbulence turbulent diffusion can also be modelled using Fick's law provided sufficient time has elapsed since tracer injection. By analogy with equation (1.12) we can write

$$\frac{\partial\langle c\rangle}{\partial t}+\langle u_x\rangle\frac{\partial\langle c\rangle}{\partial x}+\langle u_y\rangle\frac{\partial\langle c\rangle}{\partial y}+\langle u_z\rangle\frac{\partial\langle c\rangle}{\partial z}$$
$$=(e_{\mathrm{m}}+e_{\mathrm{t}})\left(\frac{\partial^2\langle c\rangle}{\partial x^2}+\frac{\partial^2\langle c\rangle}{\partial y^2}+\frac{\partial^2\langle c\rangle}{\partial z^2}\right) \qquad (1.43)$$

where e_{t} = turbulent diffusion coefficient (eddy diffusivity) which is assumed to be isotropic and homogeneous. The effects of turbulent dispersion and molecular diffusion are assumed to be additive although this is largely academic because invariably $e_{\mathrm{t}} \gg e_{\mathrm{m}}$. Comparing equations (1.43) and (1.29)

$$e_{\mathrm{t}}\frac{\partial^2\langle c\rangle}{\partial x^2}=-\frac{\partial}{\partial x}\langle u_x'c'\rangle \qquad (1.44a)$$

$$e_{\mathrm{t}}\frac{\partial^2\langle c\rangle}{\partial y^2}=-\frac{\partial}{\partial y}\langle u_y'c'\rangle \qquad (1.44b)$$

$$e_{\mathrm{t}}\frac{\partial^2\langle c\rangle}{\partial z^2}=-\frac{\partial}{\partial z}\langle u_z'c'\rangle \qquad (1.44c)$$

By analogy with equation (1.2) the turbulent diffusive fluxes are

$$J_x=\langle u_x'c'\rangle=-e_{\mathrm{t}}\frac{\partial\langle c\rangle}{\partial x} \qquad (1.45a)$$

$$J_y=\langle u_y'c'\rangle=-e_{\mathrm{t}}\frac{\partial\langle c\rangle}{\partial y} \qquad (1.45b)$$

$$J_z=\langle u_z'c'\rangle=-e_{\mathrm{t}}\frac{\partial\langle c\rangle}{\partial z} \qquad (1.45c)$$

Equations (1.43)–(1.45) constitute the Fickian model for turbulent diffusion.

There is some conjecture about the validity of this model. In his analysis, Taylor (1921) stopped after showing that at asymptotically large times the variance of the tracer cloud increases linearly with time and he did not go so far as to postulate equation (1.43). Fischer *et al* (1979) argue that the linear growth of variance is a necessary condition for the validity of equation (1.43) but that it is not a sufficient condition which establishes beyond doubt the validity of the Fickian model for turbulent diffusion. Despite these theoretical shortcomings the model is widely used to address practical mixing problems and there is a large body of empirical evidence to support its use in river mixing problems. There are difficulties applying the model in the near-field when buoyancy and momentum are important (Tennekes and Lumley 1972) but these stem from the fact that the turbulence is neither homogeneous nor stationary. In the mid and far-fields the Fickian model provides an economical and tolerably accurate model for turbulent diffusion. Fickian models are used extensively in practical mixing problems when detailed measurements of turbulence are not available and hence the use of more sophisticated models is not feasible.

The Fickian model of turbulent diffusion ensures closure of the turbulence equations provided we can estimate the value of the eddy diffusivity. The eddy diffusivity e_t is in some ways analogous to the molecular diffusivity e_m but there are three important differences. Firstly, as a rough guide, e_t is of the order $10^{-3}\ m^2 s^{-1}$ whereas e_m is typically $10^{-9}\ m^2 s^{-1}$. Secondly, whereas the molecular diffusivity is a property of the fluid, the eddy diffusivity is a property of the flow and varies with the velocity, turbulence and geometry of the flow. Thirdly, equations (1.43)–(1.45) only hold when $t \gg T_x$ where T_x is the integral Lagrangian time-scale defined in equation (1.42).

1.7.5 Eddy Viscosity

Equations (1.45) are far older than Taylor's analysis and it is worthwhile making a digression to discuss momentum transfer in turbulent flows. Turbulent eddies transfer both momentum and mass and experimental work shows that in many flows the rates of transfer are almost the same. This assumption is known as the Reynolds analogy and holds provided the tracer is neutrally buoyant and the concentration is low. Consider the case of laminar two-dimensional flow. The velocity is assumed uniform across and along the channel. The no-slip condition requires that the velocity at the bed is zero and this causes a vertical velocity gradient (Figure 1.7). In laminar flow random molecular movement causes a small exchange of flow between adjacent fluid layers whereas in turbulent flow a much larger exchange is effected by turbulent eddies. As the mean velocity of the adjacent layers differs, any vertical exchange of fluid causes a transfer of momentum. From Newton's second law this results in a shear stress. In plane laminar shear flow the viscous shear stress is

$$\tau = \rho \nu \frac{\partial u_x}{\partial y} \qquad (1.46)$$

where $\nu = \mu/\rho$ = kinematic viscosity of the fluid, $u_x(y)$ = velocity parallel to the bed, y = height above the bed and τ = viscous shear stress. In turbulent flow eddies transfer momentum vertically more rapidly than in laminar flow and the shear stresses are markedly higher. The shear stresses which arise from turbulent velocity fluctuations are called turbulent or Reynolds stresses. By applying the Reynolds decomposition to the equations of motion (see, for

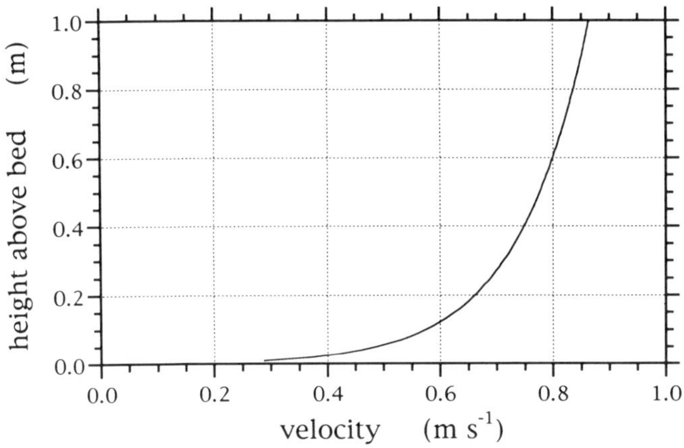

Figure 1.7. Sketch showing the logarithmic velocity profile predicted from equation (2.13) with $u^* = 0.05\ \mathrm{m\,s^{-1}}$, $y_0 = 0.001$ m and $h = 1.0$ m

example, Hinze 1972 or Schlichting 1968) the Reynolds stress is found to be

$$\tau_t = \rho\langle u_x' u_y' \rangle \tag{1.47}$$

where u_x' and u_y' = turbulent velocity fluctuations in the x and y directions and τ_t = turbulent shear stress (the Reynolds stress). Note that equation (1.47) for the Reynolds stress contains the term

$$\langle u_x' u_y' \rangle \tag{1.48}$$

and is similar to equations (1.45) for the eddy diffusivities which contain terms such as

$$\langle u_x' c' \rangle \tag{1.49}$$

The Reynolds stresses arise from correlations between turbulent velocity fluctuations whereas eddy diffusivities arise from correlations between turbulent velocity and concentration fluctuations. When the Reynolds stresses appear in the momentum equations they introduce the problem of closure discussed earlier. In 1877 Boussinesq suggested that by analogy with equation (1.46) the Reynolds stresses can be related to the gradient in the mean velocity. Thus

$$\tau_t = \rho\nu_t \frac{\partial\langle u_x \rangle}{\partial y} \tag{1.50}$$

where $\langle u_x \rangle$ = ensemble mean velocity and ν_t = the eddy viscosity. Whereas the kinematic viscosity ν is a property of the fluid, the eddy viscosity ν_t is a property of the flow and depends on the intensity of turbulence and the length scales of the dominant eddies.

Reynolds pointed out that the turbulent eddies which transport momentum also transport mass. Equation (1.50) states that the rate of transfer of momentum is proportional to the velocity gradient and by analogy the rate of transfer of mass is proportional to the concentration gradient. This analogy leads to

$$J_y = -e_t \frac{\partial \langle c \rangle}{\partial y} \tag{1.51}$$

where J_y = vertical flux of tracer mass; $\langle c \rangle$ = ensemble mean concentration and e_t = vertical turbulent diffusion coefficient. Note that e_t and ν_t have the same units (i.e. $m^2 s^{-1}$). Reynolds analogy is that for tracer mass

$$e_t = \nu_t \tag{1.52}$$

In some situations (e.g. when dealing with buoyant tracers or high sediment concentrations) Reynolds analogy breaks down and it is necessary to write

$$e_t = Sc\ \nu_t \tag{1.53}$$

where Sc = turbulent Schmidt number, which must be determined experimentally. Typically in rivers $0.3 < Sc < 1.0$.

1.7.6 Prandtl's Mixing Length Hypothesis

In 1925 Prandtl published a theory for predicting the eddy viscosity in turbulent shear flow (see, for example, Hinze 1972) which has become known as the mixing length hypothesis. This hypothesis is described in more detail in Section 2.2 but the two main results are stated here. Prandtl noted that the kinetic theory of gases gives rise to an expression for viscosity

$$\nu_g = \tfrac{1}{3} L_g u_m \tag{1.54}$$

where ν_g = kinematic viscosity of the gas, L_g = mean free path of the gas molecules (which is a measure of the distance molecules travel between collisions) and u_m = a typical molecular velocity. He drew an analogy between the kinetic theory of gases and the behaviour of eddies in turbulent flow and suggested that

$$\nu_t = u_t L_m \tag{1.55}$$

where ν_t = eddy viscosity, L_m = Prandtl's mixing length and u_t = a turbulent velocity.

Prandtl's theory has two deficiencies. Firstly, it is by no means certain that eddies interact in the same way as gas molecules and, secondly, there is a large range of eddy sizes in turbulent flow so that it is not always easy to decide on a value for L_m. Nevertheless equation (1.55) is a very useful equation which indicates that the eddy viscosity depends on a turbulent length scale and a turbulent velocity. Indeed, this equation can be derived purely on dimensional grounds. We may choose not to accept Prandtl's mixing length hypothesis in its entirety but can still retain equation (1.55) and estimate u_t and L_m experimentally. For example, equation (1.41) indicates that some time after its release, the variance of a tracer cloud increases linearly with time. Equations (1.52) and (1.55) imply the existence of a constant eddy diffusivity which can be evaluated from equation (1.41) to be

$$e_t = \frac{1}{2}\frac{d}{dt}\langle x^2 \rangle \tag{1.56}$$

Comparing equations (1.56) and (1.41)

$$e_t = \langle u_x'^2 \rangle T_x \tag{1.57}$$

where T_x is the integral Lagrangian time-scale defined by equation (1.42). We can define an integral length scale

$$L_x = \sqrt{\langle u_x'^2 \rangle}\, T_x \tag{1.58}$$

Substituting equation (1.58) into (1.57)

$$e_t = \sqrt{\langle u_x'^2 \rangle}\, L_x \tag{1.59}$$

There is a close similarity between equations (1.55) and (1.59) and they are identical if

$$e_t = \nu_t \tag{1.60a}$$

$$L_x = L_m \tag{1.60b}$$

$$u_t = \sqrt{\langle u_x'^2 \rangle} \tag{1.60c}$$

The time and length scales T_x, L_x and L_m cannot be determined theoretically but they can be estimated experimentally by examining the correlation structure of the turbulent velocity and concentration fluctuations or by determining the point at which tracer clouds begin to increase linearly.

In the second part of his analysis, Prandtl postulated values for u_t and L_m and suggested that

$$\nu_t = y^2 \frac{\partial \langle u_x \rangle}{\partial y} \tag{1.61}$$

where y = height above the bed and $\langle u_x \rangle$ = ensemble mean longitudinal velocity. Equation (1.61) has been validated by experiments for boundary layer flows but does not hold universally. For example, at the centre of a pipe the radial velocity gradient is zero and equation (1.61) implies that the eddy viscosity (and hence the eddy diffusivity) is zero. Experimental results show, however, that the eddy diffusivity at the centre of the pipe is about 80% of the cross-sectional averaged value (Rodi 1980). In rivers turbulence is mostly generated by friction at the bed and the mixing length hypothesis can be applied with some confidence. In Chapter 2 it is used to derive an equation for the vertical eddy diffusivity.

1.7.7 Diffusion in Non-homogeneous Turbulence

In Section 1.7.4 it was argued that equation (1.43) provides a reasonable description of turbulent diffusion in stationary homogeneous turbulence. The turbulent diffusion coefficient e_t was assumed to be homogeneous (i.e. not to vary spatially) and isotropic (i.e. the same in all three coordinate directions) in line with Taylor's analysis. Because the diffusion coefficient is isotropic, the diffusive flux in any direction is the product of the concentration gradient in that direction and the diffusion coefficient. One ramification is that if we rotate the coordinate axes then the diffusion equation is the same in the new coordinate system as in the old. It is left as an exercise for the reader to confirm this. The molecular diffusion coefficient is a property of the fluid and so is homogeneous and isotropic. The turbulent diffusion coefficient, however, is a property of the flow and consequently may vary spatially and with direction. For example, in river channels where the ratio of width/depth is large there are good reasons for expecting the transverse diffusivity to be larger than the vertical diffusivity. We will make an intuitive extension of equation (1.43) to the non-isotropic case and then discuss its subtleties.

In a straight channel the x axis is aligned with the main flow, the y axis is taken as vertical and the z axis is taken as transverse across the channel. The flow is assumed stationary and the turbulent diffusion coefficient non-isotropic and non-uniform. It seems logical to rewrite equation (1.43) as

$$\frac{\partial c}{\partial t}+\frac{\partial}{\partial x}(u_x c)+\frac{\partial}{\partial y}(u_y c)+\frac{\partial}{\partial z}(u_z c)=\frac{\partial}{\partial x}\left(e_x\frac{\partial c}{\partial x}\right)+\frac{\partial}{\partial y}\left(e_y\frac{\partial c}{\partial y}\right)+\frac{\partial}{\partial z}\left(e_z\frac{\partial c}{\partial z}\right) \tag{1.62}$$

Equation (1.62) is written in the so-called conservation form. Continuity requires that for incompressible flow

$$\frac{\partial u_x}{\partial x}+\frac{\partial u_y}{\partial y}+\frac{\partial u_z}{\partial z}=0 \tag{1.63}$$

Expanding the spatial derivatives on the left-hand side of equation (1.62) and using equation (1.63) yields the advection form

$$\frac{\partial c}{\partial t}+u_x\frac{\partial c}{\partial x}+u_y\frac{\partial c}{\partial y}+u_z\frac{\partial c}{\partial z}=\frac{\partial}{\partial x}\left(e_x\frac{\partial c}{\partial x}\right)+\frac{\partial}{\partial y}\left(e_y\frac{\partial c}{\partial y}\right)+\frac{\partial}{\partial z}\left(e_z\frac{\partial c}{\partial z}\right) \tag{1.64}$$

If the diffusion coefficients are constant then equation (1.64) becomes

$$\frac{\partial c}{\partial t}+u_x\frac{\partial c}{\partial x}+u_y\frac{\partial c}{\partial y}+u_z\frac{\partial c}{\partial z}=e_x\frac{\partial^2 c}{\partial x^2}+e_y\frac{\partial^2 c}{\partial y^2}+e_z\frac{\partial^2 c}{\partial z^2} \tag{1.65}$$

Equations (1.62)–(1.64) are the basis for most of the mixing calculations in this book and are of fundamental importance. The molecular diffusivity is assumed to be negligibly small compared with the turbulent diffusivity and is omitted.

One of the subtleties of equations (1.62), (1.64) and (1.65) appears if the coordinate axes are rotated to new coordinate axes $\alpha(x, y, z)$, $\beta(x, y, z)$ and $\gamma(x, y, z)$. To make this transformation the chain rule of differentials is used

$$\frac{\partial}{\partial x} = \frac{\partial}{\partial \alpha}\frac{\partial \alpha}{\partial x} + \frac{\partial}{\partial \beta}\frac{\partial \beta}{\partial x} + \frac{\partial}{\partial \gamma}\frac{\partial \gamma}{\partial x} \tag{1.66a}$$

$$\frac{\partial}{\partial y} = \frac{\partial}{\partial \alpha}\frac{\partial \alpha}{\partial y} + \frac{\partial}{\partial \beta}\frac{\partial \beta}{\partial y} + \frac{\partial}{\partial \gamma}\frac{\partial \gamma}{\partial y} \tag{1.66b}$$

$$\frac{\partial}{\partial z} = \frac{\partial}{\partial \alpha}\frac{\partial \alpha}{\partial z} + \frac{\partial}{\partial \beta}\frac{\partial \beta}{\partial z} + \frac{\partial}{\partial \gamma}\frac{\partial \gamma}{\partial z} \tag{1.66c}$$

It is left as an exercise for the reader to show that equation (1.64) becomes

$$\begin{aligned} &\frac{\partial c}{\partial t} + u_\alpha \frac{\partial c}{\partial \alpha} + u_\beta \frac{\partial c}{\partial \beta} + u_\gamma \frac{\partial c}{\partial \gamma} \\ &= e_{\alpha\alpha}\frac{\partial^2 c}{\partial \alpha^2} + e_{\beta\beta}\frac{\partial^2 c}{\partial \beta^2} + e_{\gamma\gamma}\frac{\partial^2 c}{\partial \gamma^2} + e_{\alpha\beta}\frac{\partial^2 c}{\partial \alpha \partial \beta} + e_{\alpha\gamma}\frac{\partial^2 c}{\partial \alpha \partial \gamma} + e_{\beta\gamma}\frac{\partial^2 c}{\partial \beta \partial \gamma} \end{aligned} \tag{1.67}$$

where u_α, u_β and u_γ are velocities in the new coordinate directions (which are functions of u_x, u_y, u_z, α, β and γ) while $e_{\alpha\alpha}$, $e_{\beta\beta}$, $e_{\gamma\gamma}$, $e_{\alpha\beta}$, $e_{\alpha\gamma}$ and $e_{\beta\gamma}$ are diffusion coefficients in the new coordinate directions (which are functions of e_x, e_y, e_z, α, β and γ). The important thing to note about equation (1.67) is the appearance of the three cross-derivative dispersion terms at the end. These arise because turbulent diffusion cannot in general be described by the vector (e_x, e_y, e_z) but requires at least a second-order tensor (Dagan 1969). In equations (1.62), (1.64) and (1.65) there are no cross-derivative terms because the coordinate axes happen to coincide with the principal axes of the diffusion tensor.

An obvious question is whether we can always arrange for the coordinate axes and the principal axes of the diffusion tensor to coincide. Unfortunately in natural channels this is not possible, particularly near the banks. In a straight, infinitely wide, rectangular channel the flow is two-dimensional, the isovels are everywhere horizontal and the only velocity shear occurs in the vertical direction. Consequently one principal axis of the diffusion tensor is everywhere vertical. There is no difficulty in this situation aligning the coordinate axes with the principal axes of the diffusion tensor. In an irregular channel, or a channel of finite width where friction at the banks is important, the isovels are no longer horizontal (see Figure 1.8). Prandtl's mixing length hypothesis furnishes estimates of the eddy diffusivity along axes normal to the isovels but

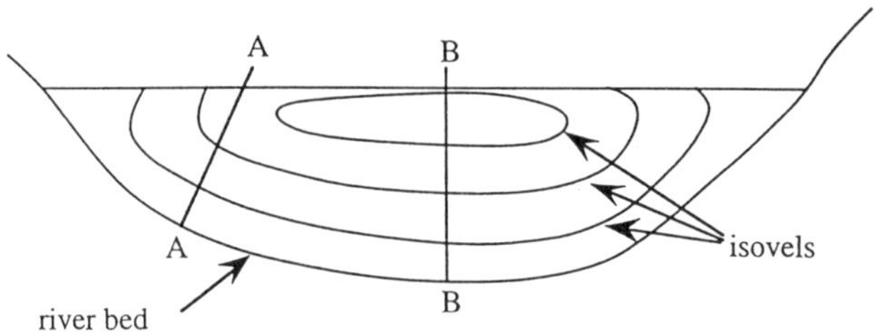

Figure 1.8. Diagram showing the isovels in a natural channel. Prandtl's mixing length hypothesis furnishes estimates of the eddy diffusivity along axes (such as A–A and B–B) which are orthogonal to the isovels

this direction is no longer fixed. Strictly speaking the cross-product terms in equation (1.67) should be retained, especially near the banks. This greatly complicates the solution of the mixing equation. In most practical problems, therefore, the cross-products are simply neglected (Fischer *et al* 1979, Chatwin and Allen 1985). This does not appear to cause serious problems when modelling tracer concentrations in rivers and estuaries. The eddy diffusivities e_x, e_y and e_z are commonly estimated by fitting solutions to measured tracer concentrations and the calibration process is able to compensate for any slight deficiencies in the model structure. It should be remembered, however, that near the banks and in regions where there are secondary currents equations (1.62), (1.64) and (1.65) provide only an approximate description of tracer mixing.

1.8 SIMPLIFYING THE MIXING EQUATIONS

1.8.1 Introduction

The three-dimensional advection diffusion equation (1.64) forms the basis for analysing mixing problems in rivers. One practical difficulty is that the full equation requires more information on water depths, velocities and diffusion coefficients than can conveniently be gathered during field experiments in natural channels. In addition the three-dimensional equation is difficult to solve in a complex natural channel. In particular mixing problems, however, some of the terms in equation (1.64) are negligibly small and the problem can be simplified. Figure 1.9 and Table 1.2 summarise a number of common river mixing problems.

Firstly, consider mixing downstream from a point source. It is convenient to define three mixing zones: the near-field, mid-field and far-field. Table 1.2 gives approximate formulae for estimating the length of each mixing zone as a guide to the reader: further details of how these formulae were derived are given in Chapter 2 (near-field) and Chapter 3 (mid-field). Table 1.2 also indicates the dimensionality of the mixing equations in each problem. In the

Table 1.2. A summary of important river mixing problems and the terms required to model them

Type of Source			Type of Solution (b)	Terms required (a)		Dimensions
Time	Space (a)			Advection	Dispersion	
instantaneous	point source	near-field	$0 < x < u_x h^2/2e_y$	x (c)	x, y, z	3
		mid-field	$u_x h^2/2e_y < x < v_x b^2/2k_z$	x	x, z	2
		far-field	$v_x b^2/2k_z < x$	x	x	1
steady	point source	near-field	$0 < x < u_x h^2/2e_y$	x (c)	y, z	3 (e)
		mid-field	$u_x h^2/2e_y < x < v_x b^2/2k_z$	x	z	2 (e)
		far-field	$v_x b^2/2k_z < x$	— (d)	— (d)	0 (d)
instantaneous	z-line source	near-field	$0 < x < u_x h^2/2e_y$	x	x, y	2
		far-field	$v_x b^2/2k_z < x$	x	x	1
steady	z-line source	near-field	$0 < x < u_x h^2/2e_y$	x	y	2 (e)
		far-field	$v_x b^2/2k_z < x$	— (d)	— (d)	0 (d)
instantaneous	y-line source	mid-field	$0 < x < v_x b^2/2k_z$	x	x, z	2
		far-field	$v_x b^2/2k_z < x$	x	x	1
steady	y-line source	mid-field	$0 < x < v_x b^2/2k_z$	x	z	2 (e)
		far-field	$v_x b^2/2k_z < x$	— (d)	— (d)	0 (d)

Notes: (a) x = longitudinal, y = vertical, z = transverse. (b) e_y = vertical diffusion coefficient, u_x = longitudinal velocity, v_x = depth-averaged velocity, k_z = transverse dispersion coefficient. (c) On a small scale y, z advection may be present in the prototype. (d) Concentration is constant (fully-mixed). (e) The dimensionality can be reduced by one if the coordinate system used travels downstream at mean velocity. Tracer is assumed to be conservative.

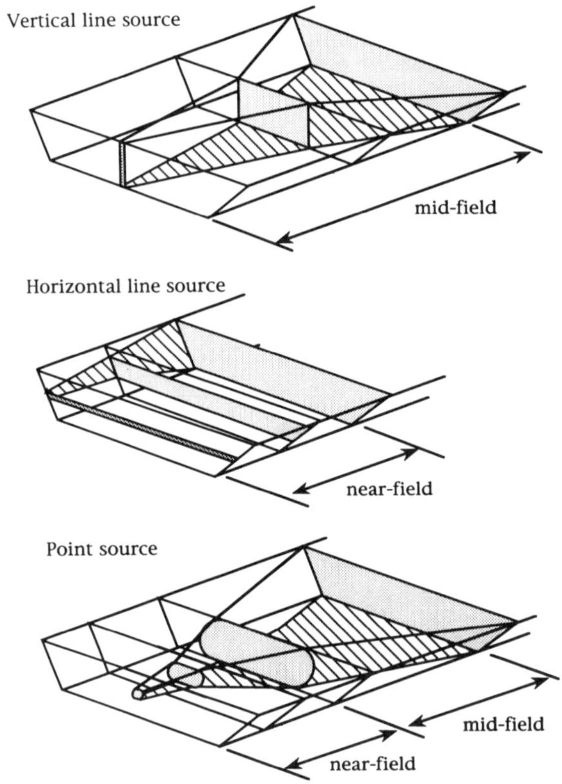

Figure 1.9. Diagram showing the general characteristics of tracer plumes downstream from three different types of tracer source. Below a point source both vertical and transverse concentration gradients are significant in the near-field, transverse concentration gradients are significant in the mid-field and tracer is well-mixed across the channel in the far-field. If the source is steady then longitudinal concentration gradients are small. Tracer is vertically well-mixed everywhere downstream from a vertical line source and transversely well-mixed everywhere downstream from a transverse line source. In practice momentum and buoyancy may affect tracer mixing in the near-field

near-field, tracer released from an instantaneous point source diffuses vertically, transversely and longitudinally as it is advected downstream and the full three-dimensional mixing equation is required. (In the near-field buoyancy and momentum may also be important depending on the nature of the source.) In the near-field it is possible to align the x axis with the flow so that $u_y = u_z = 0$. (It may not be possible, however, to align the x axis with the main flow throughout the mid-field of a natural channel because the thalweg (i.e. the main flow path) tends to meander from one side of the channel to the other.) In this situation equation (1.64) becomes

$$\frac{\partial c}{\partial t} + u_x \frac{\partial c}{\partial x} = \frac{\partial}{\partial x}\left(e_x \frac{\partial c}{\partial x}\right) + \frac{\partial}{\partial y}\left(e_y \frac{\partial c}{\partial y}\right) + \frac{\partial}{\partial z}\left(e_z \frac{\partial c}{\partial z}\right) \tag{1.68a}$$

If the tracer source is steady then the first term in equation (1.68a) vanishes and the longitudinal concentration gradients become negligibly small so that

$$u_x \frac{\partial c}{\partial x} = \frac{\partial}{\partial y}\left(e_y \frac{\partial c}{\partial y}\right) + \frac{\partial}{\partial z}\left(e_z \frac{\partial c}{\partial z}\right) \tag{1.68b}$$

If tracer originates from a steady transverse line source then transverse concentration gradients are everywhere small and

$$u_x \frac{\partial c}{\partial x} = \frac{\partial}{\partial y}\left(e_y \frac{\partial c}{\partial y}\right) \tag{1.68c}$$

Equation (1.68c) is used extensively in Chapter 2 for vertical mixing in the near-field.

The mid-field is the region in which vertical concentration gradients are small and attention can be focused on transverse and longitudinal changes in the depth-averaged concentration. The problem can be simplified by averaging equation (1.64) over the depth (as is discussed in Section 1.8.2) to give a two-dimensional model. In the mid-field it is not always possible to align the x axis with the main flow and both longitudinal and transverse advection are important. For an instantaneous source both transverse and longitudinal dispersion are important in the mid-field, but for a steady source the latter can be neglected.

In the far-field transverse concentration gradients become unimportant, attention is focused on the cross-sectional averaged concentration and equation (1.64) is averaged over the cross-section (as is discussed in Section 1.8.6) to yield a one-dimensional model. If the source is steady, the tracer is conservative and there are no tributary inflows, then the tracer concentration in the far-field is constant. If the source is non-steady (e.g. following an instantaneous discharge) then the resulting one-dimensional equation can be used to model advection and longitudinal dispersion.

1.8.2 Depth Averaging: Part One

In most rivers the aspect ratio (width/depth) is large and tracer becomes well-mixed vertically long before becoming well-mixed transversely. Consequently vertical mixing is only important close to the tracer source. In the mid-field we can neglect vertical concentration gradients and focus our attention on transverse and longitudinal changes of the depth-averaged concentration. This requires only a two-dimensional equation. Some care must be taken when reducing the dimensionality of a mixing problem. If the mixing equations are simplified incorrectly the resulting model is unlikely to predict the behaviour of tracer accurately and the only way to make the model match field data is by distorting the model coefficients. There are several published examples of mixing equations being incorrectly simplified [despite excellent expositions of the correct procedures given by (Holly 1975) and (Yotsukura 1977)] and as a result some published mixing coefficients are seriously in error. To derive equations governing the depth-averaged tracer concentration it is not permissible simply to neglect all the terms in equation (1.64) which vary with depth. The correct procedure for simplifying the mixing equations is to integrate over the depth term by term taking careful account of any depth variations of velocity and concentration. The derivation is given here in some detail and although the calculus may be a little daunting the reader is urged to persevere in order to appreciate that spatial variations of longitudinal and transverse velocity are very important in mixing tracer.

The three-dimensional advection/diffusion equation in rectangular Cartesian coordinates is

$$\frac{\partial c}{\partial t}+\frac{\partial}{\partial x}(u_x c)+\frac{\partial}{\partial y}(u_y c)+\frac{\partial}{\partial z}(u_z c)=\frac{\partial}{\partial x}\left(e_x\frac{\partial c}{\partial x}\right)+\frac{\partial}{\partial y}\left(e_y\frac{\partial c}{\partial y}\right)+\frac{\partial}{\partial z}\left(e_z\frac{\partial c}{\partial z}\right) \tag{1.62}$$

We integrate equation (1.62) from the bed $y=b(x,z)$ to the surface $y=a(x,z)$ assuming steady flow. Extensive use is made of Leibnitz's rule.

$$\int_{b(x,z)}^{a(x,z)}\frac{\partial}{\partial x}\mathrm{f}(x,y)\,\mathrm{d}z=\frac{\partial}{\partial x}\int_{b(x,z)}^{a(x,z)}\mathrm{f}(x,y)\,\mathrm{d}y-\mathrm{f}(x,a)\frac{\partial a}{\partial x}+\mathrm{f}(x,b)\frac{\partial b}{\partial x} \tag{1.69}$$

Integrating equation (1.62) term by term leads to

$$\begin{aligned}&\int_b^a\frac{\partial c}{\partial t}\,\mathrm{d}y+\frac{\partial}{\partial x}\int_b^a u_x c\,\mathrm{d}y-(u_x c)_a\frac{\partial a}{\partial x}+(u_x c)_b\frac{\partial b}{\partial x}+(u_y c)_a-(u_y c)_b\\&\quad+\frac{\partial}{\partial z}\int_b^a u_z c\,\mathrm{d}y-(u_z c)_a\frac{\partial a}{\partial z}+(u_z c)_b\frac{\partial b}{\partial z}\\&=\frac{\partial}{\partial x}\int_b^a e_x\frac{\partial c}{\partial x}\,\mathrm{d}y-\left(e_x\frac{\partial c}{\partial x}\right)_a\frac{\partial a}{\partial x}+\left(e_x\frac{\partial c}{\partial x}\right)_b\frac{\partial b}{\partial x}+\left(e_y\frac{\partial c}{\partial y}\right)_a-\left(e_y\frac{\partial c}{\partial y}\right)_b\\&\quad+\frac{\partial}{\partial z}\int_b^a e_z\frac{\partial c}{\partial z}\,\mathrm{d}y-\left(e_z\frac{\partial c}{\partial z}\right)_a\frac{\partial a}{\partial z}+\left(e_z\frac{\partial c}{\partial z}\right)_b\frac{\partial b}{\partial z}\end{aligned} \tag{1.70}$$

Rearranging equation (1.70) gives

$$\int_b^a \frac{\partial c}{\partial t}\,\mathrm{d}y + \frac{\partial}{\partial x}\int_b^a u_x c\,\mathrm{d}y + \frac{\partial}{\partial z}\int_b^a u_z c\,\mathrm{d}y$$

$$[1]$$

$$= \frac{\partial}{\partial x}\int_b^a e_x \frac{\partial c}{\partial x}\,\mathrm{d}y + \frac{\partial}{\partial z}\int_b^a e_z \frac{\partial c}{\partial z}\,\mathrm{d}y + \left[c\left(u_x \frac{\partial a}{\partial x} - u_y + u_z \frac{\partial a}{\partial z}\right)\right]_a$$

$$[2] \qquad\qquad [3]$$

$$- \left[c\left(u_x \frac{\partial b}{\partial x} - u_y + u_z \frac{\partial b}{\partial z}\right)\right]_b - \left(e_x \frac{\partial c}{\partial x}\frac{\partial a}{\partial x} - e_y \frac{\partial c}{\partial y} + e_z \frac{\partial c}{\partial z}\frac{\partial a}{\partial z}\right)_a$$

$$[4]$$

$$+ \left(e_x \frac{\partial c}{\partial x}\frac{\partial b}{\partial x} - e_y \frac{\partial c}{\partial y} + e_z \frac{\partial c}{\partial z}\frac{\partial b}{\partial z}\right)_b \tag{1.71}$$

At both the water surface and the river bed the fluxes of water and tracer normal to the boundaries are zero. We can use these boundary conditions to show that terms [1]–[4] are all identically zero. Some readers may be happy to accept this statement without question and proceed directly to equation (1.76).

The assumption of zero flow normal to the boundary is called the kinematic boundary condition and in steady flow requires (Yih 1969)

$$u_x \frac{\partial \eta}{\partial x} + u_y + u_z \frac{\partial \eta}{\partial z} = 0 \tag{1.72}$$

where $\eta(x, z)$ = coordinates of the boundary. In equation (1.71) setting $\eta(x, z) = a(x, z)$ eliminates term [1] and setting $\eta(x, z) = b(x, z)$ eliminates term [2]. The tracer flux normal to both boundaries must also be zero. The normal vectors to the water surface and the bed are, respectively:

$$\boldsymbol{n}_a = \frac{\partial a}{\partial x}\boldsymbol{i} - \boldsymbol{j} + \frac{\partial a}{\partial z}\boldsymbol{k} \tag{1.73a}$$

$$\boldsymbol{n}_b = \frac{\partial b}{\partial x}\boldsymbol{i} - \boldsymbol{j} + \frac{\partial b}{\partial z}\boldsymbol{k} \tag{1.73b}$$

where $\boldsymbol{i}, \boldsymbol{j}$ and $\boldsymbol{k}$ = unit normal vectors in the x, y and z directions respectively; and $\boldsymbol{n}_a$ and $\boldsymbol{n}_b$ = normal vectors to the water surface and the bed. The tracer flux vector at any point is

$$\boldsymbol{J} = e_x \frac{\partial c}{\partial x}\boldsymbol{i} + e_y \frac{\partial c}{\partial y}\boldsymbol{j} + e_z \frac{\partial c}{\partial z}\boldsymbol{k} \tag{1.74}$$

As the flux normal to a boundary is zero, it follows that

$$\boldsymbol{J}\cdot\boldsymbol{n}_a=\boldsymbol{J}\cdot\boldsymbol{n}_b=0 \tag{1.75}$$

Substituting equations (1.73) and (1.74) into (1.75) indicates that terms [3] and [4] of equation (1.71) are zero.

The integrals over the depth can be written

$$\frac{\partial}{\partial x}\int_a^b uc\,\mathrm{d}y=\frac{\partial}{\partial x}(h\,\overline{uc}) \tag{1.76}$$

where the overbar denotes a depth average and h = local depth. Substituting equation (1.76) in (1.71) gives

$$h\frac{\partial\bar{c}}{\partial t}+\frac{\partial}{\partial x}(h\,\overline{u_xc})+\frac{\partial}{\partial z}(h\,\overline{u_zc})=\frac{\partial}{\partial x}\left(h\,\overline{e_x\frac{\partial c}{\partial x}}\right)+\frac{\partial}{\partial z}\left(h\,\overline{e_z\frac{\partial c}{\partial z}}\right) \tag{1.77}$$

The velocities and concentrations can be expressed as the sum of a depth-average and a vertical deviation. This is another application of Reynolds' decomposition introduced earlier

$$u_x=v_x+u_x' \tag{1.78a}$$

$$u_z=v_z+u_z' \tag{1.78b}$$

$$c=s+c' \tag{1.78c}$$

where

$$v_x=\frac{1}{h}\int_b^a u_x\,\mathrm{d}y \tag{1.79a}$$

$$v_z=\frac{1}{h}\int_b^a u_z\,\mathrm{d}y \tag{1.79b}$$

$$s=\frac{1}{h}\int_b^a c\,\mathrm{d}y \tag{1.79c}$$

By definition the depth average of the deviations c', u_x' and u_z' are zero. The depth-average of a product can be expressed as follows

$$\overline{u_xc}=\overline{(v_x+u_x')(s+c')}=v_xs+\overline{u_x'c'}$$

since

$$\overline{u_x'}=\overline{c'}=0 \tag{1.80}$$

If we assume that e_x and e_z do not vary with depth then

$$\overline{e_x \frac{\partial c}{\partial x}} = e_x \frac{\partial s}{\partial x} \tag{1.81a}$$

$$\overline{e_z \frac{\partial c}{\partial z}} = e_z \frac{\partial s}{\partial z} \tag{1.81b}$$

Making these substitutions into equation (1.77) yields

$$h \frac{\partial s}{\partial t} + \frac{\partial}{\partial x}(hv_x s) + \frac{\partial}{\partial z}(hv_z s) = \frac{\partial}{\partial x}\left(-h\,\overline{u_x' c'} + he_x \frac{\partial s}{\partial x}\right) + \frac{\partial}{\partial z}\left(-h\,\overline{u_y' c'} + he_z \frac{\partial s}{\partial z}\right) \tag{1.82}$$

In equation (1.82)

$$v_x s \quad \text{and} \quad v_y s \tag{1.83a}$$

are the advective fluxes whereas

$$e_x \frac{\partial s}{\partial x} \quad \text{and} \quad e_z \frac{\partial s}{\partial z} \tag{1.83b}$$

are the turbulent diffusive fluxes in the longitudinal and transverse directions respectively. The terms

$$-\overline{u_x' c'} \quad \text{and} \quad -\overline{u_z' c'} \tag{1.83c}$$

arise as a result of depth averaging and quantify the additional transport which results from non-uniformities over the depth of velocity and concentration. This additional transport is termed dispersion. A better name is shear dispersion to emphasise that it arises from non-uniformities of velocity. The next section describes the physical processes which give rise to shear dispersion.

1.8.3 Dispersion in Turbulent Shear Flow

In uniform flow (where the velocity is everywhere the same) advection moves a tracer cloud bodily downstream without causing any distortion or spreading. Velocity is seldom uniform in a natural channel because of boundary friction. What then is the effect on the tracer cloud of velocity shear?

Figure 1.10 shows a river channel in which there is pronounced vertical velocity shear arising from bed friction. At time $t = t_0$ a pulse of tracer is introduced into the flow from a vertical line source. Tracer is carried downstream by the current and spreads out as a result of turbulent diffusion. Velocity shear causes tracer near the bed to be carried downstream more

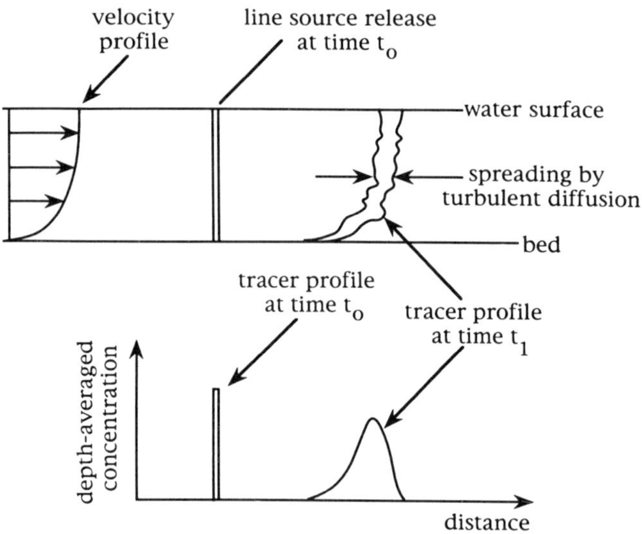

Figure 1.10. Sketch showing the combined effects of vertical velocity shear and turbulent diffusion on longitudinal spreading. A sudden release of tracer is made from a vertical line source at time t_0. At a later time t_1 the tracer cloud has been advected downstream, distorted by vertical velocity shear and smeared by turbulent diffusion

slowly than tracer near the surface and the initial straight line of tracer becomes distorted. Turbulent diffusion acts to smear and thicken the line of tracer. If we calculate the depth-averaged tracer concentration at $t = t_1$ we see that there has been considerable longitudinal spreading. Part of this spreading can be attributed to turbulent diffusion but most is attributable to vertical velocity shear. The terms in (1.83c) quantify the longitudinal and transverse tracer fluxes which result from vertical shear in the longitudinal and transverse velocities.

1.8.4 Taylor's Analysis of Turbulent Shear Flow

At this stage the dispersion terms (1.83c) are unknown and consequently equation (1.82) is intractable. We encountered a similar closure problem in Section 1.7.1. It would be possible to solve equation (1.82) if a relationship could be found between the dispersive flux and some property of the depth-averaged concentration. Taylor's asymptotic analysis of shear dispersion in a pipe furnishes such a relationship.

If turbulent diffusion was negligibly small then velocity shear would distort the tracer profile as shown in Figure 1.10 and would continue to distort it indefinitely. If the difference in velocity over the depth was Δu_x then after time

t tracer would be spread along the channel over a distance of the order

$$L = t\Delta u_x \tag{1.84}$$

Equation (1.84) suggests that the length of the cloud increases linearly with time. The length of the cloud is linearly related to the standard deviation (i.e. the square root of the variance) and so equation (1.84) implies that the variance of the tracer cloud increases as the square of time. In practice the variance increases at this rate only for a short period immediately after the tracer is released. This is the behaviour suggested by Taylor's analysis at small times [see equation (1.40)]. As the cloud evolves the rate of spreading decreases until at asymptotically long times the variance increases linearly with time [see equation (1.41)]. The reason for this change of behaviour is the interaction between velocity shear and turbulent diffusion. Vertical velocity shear distorts the initially straight line of tracer. This distortion increases the vertical concentration gradient and hence promotes vertical diffusion. Vertical diffusion counteracts the effects of vertical velocity shear and is the reason why equation (1.84) is not valid.

Taylor, whose 1921 paper is the key to quantifying turbulent diffusion, also elucidated the interaction between velocity shear and diffusion in steady laminar and then stationary turbulent pipe flow (Taylor 1953, 1954). He addressed the effects of radial velocity shear and radial diffusion on longitudinal mixing in a pipe, but subsequently his analysis was extended to the effects of both vertical (Elder 1959) and transverse velocity shear (Fischer 1966) on longitudinal mixing in open channels.

We will examine the effects of vertical velocity shear in open channel flow when tracer originates from an instantaneous transverse line source so that transverse concentration gradients are negligibly small. The channel is assumed to be uniform and the x axis is aligned with the flow so that $u_y = u_z = 0$. The contribution made to the longitudinal spreading of the tracer cloud by turbulent diffusion is small compared with the effects of vertical velocity shear and consequently the term

$$\frac{\partial}{\partial x}\left(e_x \frac{\partial c}{\partial x}\right) \tag{1.85}$$

can be neglected. Making these assumptions equation (1.64) becomes

$$\frac{\partial c}{\partial t} + u_x \frac{\partial c}{\partial x} = \frac{\partial}{\partial y}\left(e_y \frac{\partial c}{\partial y}\right) \tag{1.86}$$

The velocity and concentration are now written in terms of their depth averages and deviations following equation (1.78) to give

$$c(x, y, t) = s(x, t) + c'(x, y, t) \tag{1.87a}$$

$$u_x(x, y) = v_x(x) + u_x'(x, y) \tag{1.87b}$$

Substituting from equation (1.87) in (1.86) and simplifying gives

$$\frac{\partial}{\partial t}(s+c')+(v_x+u_x')\frac{\partial}{\partial x}(s+c')=\frac{\partial}{\partial y}\left(e_z\frac{\partial c'}{\partial y}\right) \tag{1.88}$$

Note that s does not vary with y and so does not appear in the last term. Next, equation (1.88) is transformed into a Lagrangian coordinate system which travels downstream at mean velocity.

$$\xi = x - v_x t \tag{1.89a}$$

$$\tau = t \tag{1.89b}$$

Making use of the chain rule

$$\frac{\partial}{\partial x}=\frac{\partial}{\partial \xi} \tag{1.90a}$$

$$\frac{\partial}{\partial t}=\frac{\partial}{\partial \tau}-v_x\frac{\partial}{\partial \xi} \tag{1.90b}$$

Equation (1.88) then becomes

$$\frac{\partial s}{\partial \tau}+\frac{\partial c'}{\partial \tau}+u_x'\frac{\partial s}{\partial \xi}+u_x'\frac{\partial c'}{\partial \xi}=\frac{\partial}{\partial y}\left(e_y\frac{\partial c'}{\partial y}\right) \tag{1.91}$$

A general solution to equation (1.91) cannot be found because u_x' is a function of y. Taylor argued, however, that after tracer has been in the flow for a sufficiently long time an equilibrium becomes established between vertical velocity shear and turbulent diffusion. Once this equilibrium is established, the vertical distribution of the concentration deviation becomes independent of both τ and ξ. Thus

$$c' = c'(y) \tag{1.92}$$

In addition, the mean concentration only changes slowly so that

$$\frac{\partial s}{\partial \tau}=0 \tag{1.93}$$

Fischer *et al* (1979) determined the orders of magnitude of each of the terms in equation (1.91) and showed that it is justified to make these assumptions, while Sayre (1968a) showed using numerical experiments that at asymptotically large times equation (1.92) is valid. Equation (1.91) can then be simplified to

$$u_x'\frac{\partial s}{\partial \xi}=\frac{\partial}{\partial y}\left(e_y\frac{\partial c'}{\partial y}\right) \tag{1.94}$$

At the bed and water surface the vertical diffusive flux is zero

$$\frac{\partial c'}{\partial y} = 0 \qquad \text{at} \qquad y = 0 \text{ and } h \tag{1.95}$$

Integrating equation (1.94) twice with respect to y, making use of the boundary condition in equation (1.95) and noting that s is independent of y

$$c'(y) = \frac{\partial s}{\partial \xi} \int_{y'=0}^{y} \frac{\partial y'}{e_y} \int_{y''=0}^{y'} u_x'(y'')\,\mathrm{d}y'' + c'(0) \tag{1.96}$$

The depth-averaged dispersive flux in the longitudinal direction is

$$\frac{1}{h} \int_{y=0}^{h} u_x' c'\,\mathrm{d}y = \frac{\partial s}{\partial \xi} \frac{1}{h} \int_{y=0}^{h} u_x'\,\mathrm{d}y \int_{y'=0}^{y} \frac{\mathrm{d}y'}{e_y} \int_{y''=0}^{y'} u_x'\,\mathrm{d}y'' \tag{1.97}$$

Equation (1.97) implies that the longitudinal dispersive flux is proportional to the longitudinal gradient of the depth-averaged concentration. Thus at asymptotically long times longitudinal dispersion can be described using Fick's law

$$J_\xi = -k_x \frac{\partial s}{\partial \xi} \tag{1.98}$$

where k_x is the longitudinal dispersion coefficient, which from (1.97) is

$$k_x = \frac{1}{h} \int_{y=0}^{h} u_x'\,\mathrm{d}y \int_{y'=0}^{y} \frac{\mathrm{d}y'}{e_y} \int_{y''=0}^{y'} u_x'\,\mathrm{d}y'' \tag{1.99}$$

By applying a mass balance to a water parcel which occupies the entire depth it is straightforward to show that

$$\frac{\partial s}{\partial \tau} = k_x \frac{\partial^2 s}{\partial \xi^2} \tag{1.100}$$

where k_x is assumed constant. Reverting to Cartesian coordinates

$$\frac{\partial s}{\partial t} + v_x \frac{\partial s}{\partial x} = k_x \frac{\partial^2 s}{\partial x^2} \tag{1.101}$$

It is important to appreciate that equation (1.101) is an asymptotic result which only applies once an equilibrium becomes established between velocity shear and diffusion. In stationary homogeneous turbulence such an equilibrium is likely once an individual tracer particle has sampled the entire flow field. The time-scale for a tracer particle to sample the entire flow field in this analysis is the time-scale for vertical mixing. On dimensional grounds this is

$$T_y = \frac{h^2}{e_y} \tag{1.102}$$

where h = depth and e_y = vertical turbulent diffusion coefficient.

If turbulence is non-stationary or non-homogeneous then we have difficulties applying Taylor's analysis. In this situation there may be insufficient time for an equilibrium to become established between velocity shear and turbulent diffusion before the tracer cloud is carried into a different part of the channel with a different turbulent regime.

1.8.5 Depth Averaging: Part Two

In Section 1.8.2 we found that equation (1.82) contained unknown dispersive fluxes. Taylor's analysis, as outlined in Section 1.8.4 shows that at asymptotically large times the longitudinal dispersive flux is proportional to the longitudinal gradient in depth-averaged concentration. Thus

$$-\overline{u_x'c'} = k_x \frac{\partial s}{\partial x} \tag{1.103}$$

where k_x = the longitudinal dispersion coefficient which accounts for the effects on the depth-averaged tracer concentration of depth variations in the longitudinal velocity. By analogy with equation (1.103) the transverse dispersive flux is

$$-\overline{u_z'c'} = k_z \frac{\partial s}{\partial z} \tag{1.104}$$

where k_z = the transverse dispersion coefficient which accounts for the effects on the depth-averaged tracer concentration of depth variations in the transverse velocity. Incorporating equations (1.103) and (1.104) into (1.82) yields

$$\frac{\partial s}{\partial t} + \frac{\partial}{\partial x}(hv_x s) + \frac{\partial}{\partial z}(hv_z s) = \frac{\partial}{\partial x}\left[h(e_x + k_x)\frac{\partial s}{\partial x}\right] + \frac{\partial}{\partial z}\left[h(e_z + k_z)\frac{\partial s}{\partial z}\right] \tag{1.105}$$

Invariably in river channels $k_x \gg e_x$ and $k_z \gg e_z$. Using the continuity equation (1.63), equation (1.105) can be rewritten in advection form as

$$\frac{\partial s}{\partial t} + v_x \frac{\partial s}{\partial x} + v_z \frac{\partial s}{\partial z} = \frac{1}{h}\frac{\partial}{\partial x}\left(hk_x \frac{\partial s}{\partial x}\right) + \frac{1}{h}\frac{\partial}{\partial z}\left(hk_z \frac{\partial s}{\partial z}\right) \tag{1.106}$$

Equation (1.106) is used extensively to solve river mixing problems in the mid-field and its practical application is described in Chapter 3.

1.8.6 Cross-section Averaging

A long way downstream from its point of injection tracer becomes spread across the whole channel. Beyond this point attention can be focused on the rate at which the cross-sectional averaged concentration is advected downstream and dispersed longitudinally. Longitudinal dispersion is only of interest when the input of tracer is non-steady (e.g. below a slug injection) because below a steady source the fully mixed concentration of a conservative tracer is constant. An equation governing the cross-sectional averaged concentration can be derived by integrating equation (1.106) across the channel which leads to

$$A\frac{\partial S}{\partial t}+\frac{\partial}{\partial x}(AV_xS)=\frac{\partial}{\partial x}\left(-A\,\overline{v_x's'}+Ak_x\frac{\partial S}{\partial x}\right) \tag{1.107}$$

where S = cross-sectional averaged concentration, A = cross-sectional area of the channel and V_x = cross-sectional averaged velocity. The approach is similar to that used in Sections 1.8.2 and 1.8.5 and is left as an exercise for the reader.

$$S(x)=\frac{1}{b}\int_{z=0}^{b} s(x,z)\,\mathrm{d}z \tag{1.108a}$$

$$V_x(x)=\frac{1}{b}\int_{z=0}^{b} v_x(x,z)\,\mathrm{d}z \tag{1.108b}$$

$$s(x,z)=S(x)+s'(x,z) \tag{1.108c}$$

and

$$v_x(x,z)=V_x(x)+v_x'(x,z) \tag{1.108d}$$

where b = channel width. It is assumed that k_x is constant across the channel. The overbar denotes a width-average. The term

$$\overline{-v_x's'} \tag{1.109}$$

is the longitudinal dispersive tracer flux resulting from variations in velocity across the channel. Taylor's analysis of turbulent shear flow (Section 1.8.4) suggests that at asymptotically large times this flux is proportional to the longitudinal gradient in cross-sectional averaged concentration. Thus

$$\overline{-v_x's'}=K_x\frac{\partial S}{\partial x} \tag{1.110}$$

where K_x = a constant. Making this substitution in equation (1.107) and noting that invariably $K_x \gg k_x$ yields

$$\frac{\partial S}{\partial t} + V_x \frac{\partial S}{\partial x} = \frac{1}{A} \frac{\partial}{\partial x} \left(A K_x \frac{\partial S}{\partial x} \right) \tag{1.111}$$

where K_x = the longitudinal dispersion coefficient which accounts for the effects on the cross-sectional averaged tracer concentration of variations of velocity across the channel cross-section. The practical application of equation (1.111) is discussed in Chapter 4.

1.9 CHANNEL CURVATURE

1.9.1 Cartesian Models

Where the channel is straight and uniform it is sensible to write the depth-averaged mixing equation in rectangular Cartesian coordinates (i.e. x–z co-ordinates). This is clearly the best option in man-made channels such as canals and laboratory flumes and is sometimes satisfactory in the mid-field of natural rivers. In such situations the depth-averaged tracer concentration can be predicted using

$$\frac{\partial s}{\partial t} + v_x \frac{\partial s}{\partial x} + v_z \frac{\partial s}{\partial z} = \frac{1}{h} \frac{\partial}{\partial x} \left(h k_x \frac{\partial s}{\partial x} \right) + \frac{1}{h} \frac{\partial}{\partial z} \left(h k_z \frac{\partial s}{\partial z} \right) \tag{1.106}$$

In many rivers, however, the channel meanders, the banks are not straight and the cross-section is highly non-uniform. There are several difficulties using equation (1.106) in such channels. Firstly, there are good reasons for believing that in rivers the longitudinal and transverse dispersion coefficients are non-homogeneous ($k_x \neq k_y$). In a meandering channel it is not possible to arrange for the cartesian x axis always to be aligned with the main flow direction nor the cartesian z axis transversely across the main flow. Consequently, k_x and k_z are not always the longitudinal and transverse dispersion coefficients. Secondly, in a meandering channel the x and z axes are no longer always aligned with the principal axes of the diffusion tensor. If the longitudinal and transverse diffusivities are different then (as discussed in Section 1.7.7) it is necessary to include cross-derivative dispersion terms in the mixing equation (see equation 1.67). Both of these problems disappear if $k_x = k_y$. This assumption can safely be made if the tracer source is steady because in this situation longitudinal dispersion is unimportant. Thirdly, at the banks the usual boundary condition is that the tracer flux normal to the bank is zero. In a meandering channel the normal flux is a complicated function involving derivatives in both x and z directions which makes solving equation (1.106) difficult.

One method to overcome these problems is to align the x axis with the main flow path. The x axis is then no longer straight but follows the meander pattern of the channel. The z axis is taken everywhere orthogonal to the x axis. Distance x is measured in the local downstream direction and distance z is measured in the local transverse direction. This leads to a curvilinear coordinate system (Figure 1.11).

As a first approximation the fact that the downstream and transverse directions vary from place to place along the channel can be ignored and equation (1.106) can be retained. This approach works well in gently meandering rivers (Rutherford and Williams 1992). Equation (1.106) is, however, no longer exact because curvature distorts the coordinate system.

1.9.2 Curvilinear Models

A more rigorous approach is to transform the equation governing tracer concentration from the rectangular Cartesian coordinate system to a curvilinear system. Consider the problem of predicting the depth-averaged tracer concentration in a meandering channel. Figure 1.11 shows such a channel in plan view together with curvilinear coordinate axes denoted α (longitudinal) and β (transverse). The α and β lines are curved, at right angles to each other (orthogonal) and the α lines are parallel with the local depth-averaged velocity vectors. These two assumptions mean that several terms disappear from the mixing equation during the coordinate transformation leading to a simpler model. Because of curvature and/or variations in channel width, longitudinal distances for a given increment in α vary depending on the value of β. Thus in Figure 1.11 the distance L_{AB} is not in general equal to the distance L_{CD}. Similarly transverse distances for a given increment in β vary depending on the value of α. Thus the distance L_{AC} is not in general equal to the distance L_{BD}. Metric coefficients are defined which relate distances measured along the

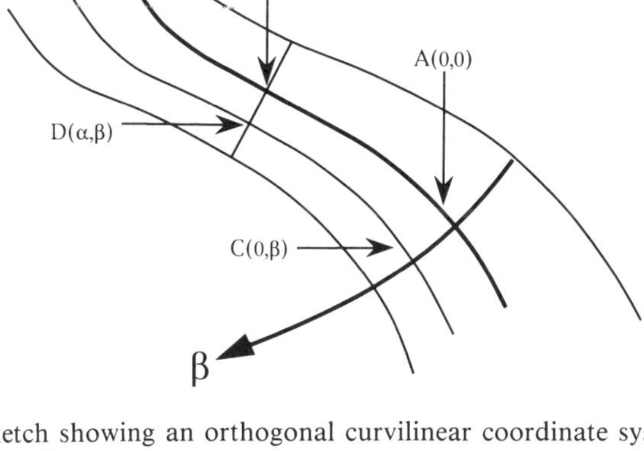

Figure 1.11. Sketch showing an orthogonal curvilinear coordinate system in a meandering river channel

coordinate axes to distances measured along other α and β lines. The metric coefficients are functions of α and β. In Figure 1.11 the origin of the curvilinear coordinate system is located at A. The coordinates of points A, B, C and D are respectively $(0, 0)$, $(\alpha, 0)$, $(0, \beta)$, (α, β). The average value of the longitudinal metric coefficient m_α along the line C-D is

$$\overline{m_\alpha} = \frac{L_{\mathrm{CD}}}{L_{\mathrm{AB}}} \tag{1.112}$$

or, more generally

$$L_{\mathrm{CD}} = \int_{\alpha'=0}^{\alpha} m_\alpha(\alpha', \beta)\, \mathrm{d}\alpha' \tag{1.113}$$

Similarly the average value of the transverse metric coefficient m_β along the line B-D is

$$\overline{m_\beta} = \frac{L_{\mathrm{BD}}}{L_{\mathrm{AC}}} \tag{1.114}$$

or, more generally

$$L_{\mathrm{BD}} = \int_{\beta'=0}^{\beta} m_\beta(\alpha, \beta')\, \mathrm{d}\beta' \tag{1.115}$$

The rectangular Cartesian coordinate system is a special case in which $m_\alpha = m_\beta = 1$. In curved channels the metric coefficients are typically of the order of 0.8–1.2 (Holly 1985).

The derivation of the advection/dispersion equation in curvilinear coordinates is given in some detail by Chang (1971) and Yotsukura (1977) and so is not repeated here. The depth-averaged equations of continuity and tracer mass balance for unsteady flow are simply stated

$$m_\alpha m_\beta \frac{\partial h}{\partial t} + \frac{\partial}{\partial x}(m_\beta h v_\alpha) + \frac{\partial}{\partial z}(m_\alpha h v_\beta) = 0 \tag{1.116}$$

$$\frac{\partial s}{\partial t} + \frac{v_\alpha}{m_\alpha}\frac{\partial s}{\partial \alpha} + \frac{v_\beta}{m_\beta}\frac{\partial s}{\partial \beta} = \frac{1}{h m_\alpha m_\beta}\left[\frac{\partial}{\partial \alpha}\left(\frac{m_\beta}{m_\alpha} h k_\alpha \frac{\partial s}{\partial \alpha}\right) + \frac{\partial}{\partial \beta}\left(\frac{m_\alpha}{m_\beta} h k_\beta \frac{\partial s}{\partial \beta}\right)\right] \tag{1.117}$$

where α and β = longitudinal and transverse curvilinear distances; h = local water depth; v_α and v_β = depth-averaged longitudinal and transverse velocities; s = depth-averaged concentration; k_α and k_β = longitudinal and transverse dispersion coefficients (which quantify the effects of turbulent diffusion and depth variations in longitudinal and transverse velocity); and m_α and m_β = metric coefficients.

1.9.3 Streamtube Model

It is possible to simplify equation (1.117) in the situation where the river flow is steady. Firstly, the time varying term disappears. We can also define a new variable, the cumulative discharge

$$q(\alpha, \beta) = \int_{\beta' = -\infty}^{\beta} m_\beta h v_\alpha \, \mathrm{d}\beta' \tag{1.118}$$

where $q(\alpha, \beta)$ = cumulative discharge; $h(\alpha, \beta)$ = local depth; $v_\alpha(\alpha, \beta)$ = local depth-averaged longitudinal velocity; and $m_\beta(\alpha, \beta)$ = transverse metric coefficient. If the left and right banks are located at β_l and β_r then $q(\alpha, \beta_l) = 0$ and $q(\alpha, \beta_r) = Q$ (for all values of α) where Q = total river flow.

Now equation (1.117) can be integrated with respect to β starting from the left bank. The approach is similar to that outlined in Section 1.8.2. Use is made of Leibnitz's rule (equation 1.69), the chain rule (equation 1.66) and the boundary condition $v_\alpha = v_\beta = 0$ at the left and right banks. The interested reader is referred to Yotsukura (1977) for full details of the derivation. For an unsteady tracer source in steady river flow the conservation of mass equation becomes

$$\frac{\partial s}{\partial t} + \frac{v_\alpha}{m_\alpha}\frac{\partial s}{\partial \alpha} = \frac{1}{h m_\alpha m_\beta}\frac{\partial}{\partial \alpha}\left(\frac{m_\beta}{m_\alpha} h k_\alpha \frac{\partial s}{\partial \alpha}\right) + \frac{v_\alpha}{m_\alpha}\frac{\partial}{\partial q}\left(m_\alpha h^2 v_\alpha k_\beta \frac{\partial s}{\partial q}\right) \tag{1.119}$$

If the tracer source is steady the unsteady term (term 1) vanishes and the longitudinal dispersion term (term 3) becomes negligibly small (Yotsukura and Sayre 1976) giving

$$\frac{\partial s}{\partial \alpha} = \frac{\partial}{\partial q}\left(m_\alpha h^2 v_\alpha k_\beta \frac{\partial s}{\partial q}\right) \tag{1.120}$$

Equations (1.119) and (1.120) contain the product

$$D(\alpha, q) = m_\alpha h^2 v_\alpha k_\beta \tag{1.121}$$

D is called the factor of diffusion and has the units $\mathrm{m^5\,s^{-2}}$. Thus D/Q^2 has the units $\mathrm{m^{-1}}$ and is an inverse length scale. The factor of diffusion varies across the channel because h and v_α are both low near the banks. The great advantage of equations (1.119) and (1.120) is that they do not contain the transverse velocity v_β explicitly although they still take account of cross-channel water movements. The practical application of the stream tube model is described in Chapter 3.

NOTATION

a, b	location of the water surface and river bed
A	cross-sectional area
b	channel width
c	tracer concentration
$\bar{c}$	mean concentration
c'	concentration deviation from the ensemble or depth average
$\langle c \rangle$	ensemble mean tracer concentration
c_i	realisation i of the tracer concentration
c_t	average concentration in a parcel at time t
D	factor of diffusion
e_{m}	molecular diffusion coefficient
e_t	turbulent diffusion coefficient (or eddy diffusivity)
e_x, e_y, e_z	eddy diffusivities in the x, y, z directions
e_{ij}	elements of the eddy diffusivity tensor
h	local depth
$\boldsymbol{i}, \boldsymbol{j}, \boldsymbol{k}$	unit normal vectors in the x, y, z directions
I_x	advective flux in the x direction
$\boldsymbol{J}$	tracer flux vector
J_x, J_y, J_z	diffusive fluxes in the x, y, z directions
J_ξ	diffusive flux in a moving coordinate system
k_α, k_β	longitudinal and transverse dispersion coefficients in depth-averaged curvilinear mixing equations
k_x, k_z	longitudinal and transverse dispersion coefficients in depth-averaged mixing equations
K_x	longitudinal dispersion coefficient in cross-sectional averaged mixing equations
L	a characteristic length scale
L_{AB}	distance along a streamline between points A and B
L_{g}	mean free path of a gas molecule
L_{L}	integral length scale
L_{m}	Prandtl's turbulent mixing length
L_x	integral length scale in the x direction
m	mass inflow rate of tracer
m_α, m_β	metric coefficients for a curvilinear coordinate system
M	mass of tracer injected
M_t	tracer mass in a fluid parcel at time t
$\boldsymbol{n}_a, \boldsymbol{n}_b$	vectors normal to the water surface and bed
N	number of realisations
Re	Reynolds number
R_x	Lagrangian autocorrelation function

q	cumulative flow
Q	total river flow
s	depth-averaged concentration
s'	deviation from the depth-averaged concentration
S	cross-sectional averaged concentration
Sc	turbulent Schmidt number $= e_t/\nu_t$
t	time
Δt	time increment
t_0	fixed time
T	time period for averaging
T_x	integral Lagrangian time scale in the x direction
T_y	time scale for vertical mixing
u	a characteristic velocity
u_m	typical molecular velocity of a gas
u_t	turbulent velocity
u_x, u_y, u_z	velocities in the x, y, z directions
$\langle u_x \rangle, \langle u_y \rangle, \langle u_z \rangle$	ensemble mean velocities in the x, y, z directions
Δu	velocity difference
u'_x, u'_y, u'_z	velocity deviation from the depth or ensemble average
v_x, v_z	depth-averaged velocities in the x and z directions
v_α, v_β	depth-averaged velocities in the α, β directions
V	volume
V_x	cross-sectional averaged velocity
x, y, z	longitudinal, vertical and transverse distances
$\bar{x}, \bar{y}, \bar{z}$	centroid of an individual tracer cloud
$\langle x \rangle, \langle y \rangle, \langle z \rangle$	centroid of an ensemble of tracer clouds
$\Delta x, \Delta y, \Delta z$	incremental distances in the x, y, z directions
x_0, y_0, z_0	fixed location
α, β, γ	rotated Cartesian coordinate axes
α, β	curvilinear coordinate directions
$\lambda_1, \lambda_2 \ldots$	eddy length scales
η	location of the water surface or the bed
μ	dynamic viscosity
ν	kinematic viscosity
ν_t	eddy viscosity
ν_g	eddy viscosity of a gas
ξ	longitudinal distance in a moving Lagrangian coordinate system
ρ	density
$\sigma_x^2, \sigma_y^2, \sigma_z^2$	tracer cloud variances in the x, y, z direction about the centre of mass of the individual tracer cloud
$\Sigma_x^2, \Sigma_y^2, \Sigma_z^2$	variances about the ensemble mean centre of mass

τ	viscous shear stress
τ_t	turbulent shear stress (Reynolds stress)
τ	time

Chapter 2
Vertical Mixing

2.1 INTRODUCTION

In rivers, the turbulence generated by bed friction is strong and the depth is generally small so that a tracer becomes vertically well-mixed within a short distance of the source. This point is illustarted in Figure 1.9 and Table 1.2. This chapter concentrates on the near-field and quantifies the rate of vertical mixing. It begins with a discussion of Prandtl's mixing length hypothesis (which was introduced briefly in Chapter 1) and shows how it provides expressions for the vertical variation of the ensemble mean velocity and the vertical diffusivity. An approximate model is then introduced for vertical mixing downstream from a steady transverse line source which assumes that the velocity and diffusivity are constant. This model admits analytical solutions and can be used to derive fairly simple solutions to a large number of vertical mixing problems. One of the advantages of the constant-coefficient model is that solutions to instantaneous point source problems can be superposed to address problems of vertical mixing downstream from sources with complex geometry and/or time-varying discharge. These solutions are useful for understanding the general behaviour of the tracer plume as well as for making approximate calculations of tracer concentration. A more realistic, but more complex, model is then introduced in which the velocity varies logarithmically and the diffusivity varies parabolically over the depth. There are slight differences between the variable- and constant-coefficient models which are discussed. For most of this chapter tracers are assumed to be neutrally buoyant, but there is a brief discussion of the effects of buoyancy on vertical mixing.

2.2 SUMMARY

This summary is placed at the beginning of the chapter to help the reader appreciate where the ensuing discussions are leading and to act as a checklist subsequently.

1. Vertical mixing occurs as a result of turbulence generated at the bed by friction. Prandtl's mixing length hypothesis provides theoretical formulae for the variation with depth of the ensemble mean velocity (logarithmic) and the vertical turbulent diffusion coefficient (parabolic). These formulae have been corroborated by laboratory studies. Vertical mixing is one of the few mixing problems where theoretical estimates of the mixing coefficient are reliable.
2. The time-scale for vertical mixing is very much smaller that the time-scale for transverse mixing so that in most rivers tracer originating from a point source becomes well-mixed over the depth long before it becomes well-mixed across the channel. This means that in many situations the problems of vertical and transverse mixing can be considered separately. A two-dimensional model can be used to analyse vertical mixing downstream from a transverse line source and to determine both concentration profiles and the distance below the source where tracer becomes well-mixed vertically (the vertical mixing distance). This estimate of the vertical mixing distance holds for steady and non-steady point and transverse line sources. To calculate tracer concentrations downstream from a point source a three-dimensional model is required.
3. An approximate model (which assumes that the velocity tracer-source and diffusivity are constant) admits analytical solutions which can be applied to a number of important river mixing problems. Because the model is linear the principle of superposition can be used to predict concentrations downstream from time-varying sources and sources with complex geometry. In addition, the effects of the river bed, water surface and banks can be modelled using the method of images. The constant coefficient model is useful for showing the general behaviour of concentration contours and for making preliminary calculations of tracer concentration. This model predicts that the optimum source location (which minimises the vertical mixing distance and also minimises concentrations at the bed and the water surface) lies at mid-depth. For a source at this depth a useful rule of thumb is that tracer becomes vertically well-mixed within a distance of about 50 times the water depth.
4. The more realistic, but more complex, variable-coefficient model of Nokes (1986) and McNulty (1983) allows the velocity to vary logarithmically and the diffusivity to vary parabolically with depth. The variable-coefficient model predicts slightly shorter vertical mixing distances (typically by 10–20%) than the constant-coefficient model for sources near the bed and water surface, but for sources near mid-depth the mixing distances are very similar. The variable-coefficient model predicts that the optimum source height lies above mid-depth and increases with increasing bed roughness.
5. For many river mixing problems buoyancy effects are small and are confined to the near-field. If the receiving flow is stratified (e.g. in an estuary)

and the vertical velocity and density structure are known, then a semi-empirical formula can be used to quantify the reduction in the vertical diffusion coefficient. If the tracer is buoyant then the laboratory results of Schiller and Sayre (1973, 1975) can be used to estimate the reduction in the vertical diffusion coefficient and the resulting increase in the vertical mixing distance.

2.3 MECHANISMS CAUSING VERTICAL MIXING

In rivers the principal mechanism causing the vertical mixing of a neutrally buoyant tracer is turbulence generated by velocity shear at the bed. Vertical mixing can be increased locally by secondary currents (notably at sharp bends) and by obstacles in the flow (such as bridge piers), although neither of these mechanisms has been quantified adequately. Experimental studies indicate that in plane shear flows (such as those generated at the river bed) Prandtl's mixing length hypothesis provides an economical and tolerably accurate model for estimating the eddy viscosity.

2.3.1 Prandtl's Mixing Length Hypothesis

In Section 1.7.5 it was shown that in plane laminar shear flow the viscous shear stress is

$$\tau = \rho \nu \frac{\partial u_x}{\partial y} \tag{2.1}$$

where ν = kinematic viscosity of water; ρ = density of water; $u_x(y)$ = longitudinal velocity; y = height above the bed; and τ = viscous shear stress. If the flow is turbulent there is an additional shear stress, the so-called Reynolds stress

$$\tau_t = \rho \langle u_x' u_y' \rangle \tag{2.2}$$

where $\langle\ \rangle$ denotes an ensemble average; τ_t = turbulent shear stress (the Reynolds stress); and u_x' and u_y' = turbulent velocity fluctuations. Generally u_x' and u_y' are unknown and progress is possible only if the Reynolds stress can be related to some property of the mean velocity. By analogy with equation (2.1) the Reynolds stress is written

$$\tau_t = \rho \nu_t \frac{\partial \langle u_x \rangle}{\partial y} \tag{2.3}$$

where ν_t = the so-called eddy viscosity and $\langle u_x \rangle$ = ensemble mean velocity. The problem now is to estimate the eddy viscosity. Prandtl argued that by analogy with the kinetic theory of gases

$$\nu_t = u_t L_m \tag{2.4}$$

where L_m = Prandtl's mixing length (a turbulent length scale) and u_t = a turbulent velocity. He then postulated that in plane turbulent shear flow

$$u_t = L_m \left| \frac{\partial \langle u_x \rangle}{\partial y} \right| \tag{2.5}$$

Substituting equation (2.5) into (2.4) gives

$$\nu_t = L_m^2 \left| \frac{\partial \langle u_x \rangle}{\partial y} \right| \tag{2.6}$$

2.3.2 The Logarithmic Velocity Profile

In general the mixing length, the turbulent velocity and the shear stress all vary with depth. Very close to the bed, however, it is reasonable to assume that the shear stress is constant and that

$$\tau = \tau_o \tag{2.7}$$

where τ_o = the bed shear stress. Near the bed the turbulent length scale is restricted by the presence of the boundary and a reasonable assumption is that L_m is proportional to the distance from the bed

$$L_m = \varkappa y \tag{2.8}$$

where $\varkappa$ = a constant of proportionality (known as von Karman's constant). Substituting equations (2.6), (2.7) and (2.8) into (2.3) gives

$$\tau_o = \rho \varkappa^2 y^2 \left(\frac{\partial \langle u_x \rangle}{\partial y} \right)^2 \tag{2.9}$$

Rearranging and integrating gives the so-called universal logarithmic velocity distribution for turbulent open channel flow

$$\langle u_x(y) \rangle = \frac{1}{\varkappa} \sqrt{\frac{\tau_o}{\rho}} \log_e \left(\frac{y}{y_o} \right) + \langle u_x(y_o) \rangle \tag{2.10}$$

where y_o = arbitrary depth at which the velocity is $\langle u_x(y_o) \rangle$. Equation (2.10) includes the angled brackets to emphasise the point that in turbulent flow the velocity fluctuates and that we are dealing with the ensemble mean velocity. The shear velocity is defined

$$u^* = \sqrt{\frac{\tau_o}{\rho}} \tag{2.11}$$

Assuming that

$$\langle u_x(y_o)\rangle = 0 \tag{2.12}$$

omitting the angle brackets for convenience and making use of equation (2.11) we can rewrite equation (2.10) as

$$\frac{u_x(y)}{u^*} = \frac{1}{\varkappa}\log_e\left(\frac{y}{y_o}\right) \tag{2.13}$$

There is a conceptual difficulty with this derivation because it assumes that the shear stress is constant over the entire depth. At the water surface, however, the shear stress must be zero and a simple gravity–stress balance requires that the shear stress varies linearly over the depth [see equation (2.16) below]. Despite this theoretical shortcoming equation (2.13) has been shown empirically to provide a very good description of the vertical variation of velocity in turbulent open channel flow.

2.3.3 The Vertical Eddy Diffusivity

We are now in a position to evaluate the vertical eddy diffusivity (turbulent diffusion coefficient). Reynolds analogy states that

$$e_y \approx \nu_t \tag{2.14}$$

where e_y = vertical eddy diffusivity and ν_t = eddy viscosity. Experimental studies show that equation (2.14) holds within 5–10% for dilute solutions of neutrally buoyant tracers in shear flows. More generally

$$e_y = Sc\ \nu_t \tag{2.15}$$

where Sc = the dimensionless turbulent Schmidt number, $0 < Sc < 1$. The eddy diffusivity for heat differs slightly from the eddy viscosity, implying a value of Sc less than unity. It is rarely possible in practical river mixing problems to estimate ν_t or e_y with sufficient accuracy to worry about slight departures from equation (2.14).

In plane turbulent shear flow the shear stress varies linearly with depth and vanishes at the surface so that

$$\tau_t = \tau_o\left(1 - \frac{y}{h}\right) \tag{2.16}$$

Substituting equations (2.14) and (2.16) in (2.3) and rearranging

$$e_y = \left[\tau_o\left(1 - \frac{y}{h}\right)\right] \Big/ \left(\rho\,\frac{\partial\langle u_x\rangle}{\partial y}\right) \tag{2.17}$$

Assuming that the velocity varies logarithmically with depth and substituting equation (2.13) into (2.17) gives

$$e_y = \varkappa u^* y\left(1 - \frac{y}{h}\right) \tag{2.18}$$

Thus the vertical diffusivity varies parabolically with the depth, vanishing at both the bed and the water surface. Jobson and Sayre (1970a) found that predicted concentration profiles were not very sensitive to the vertical distribution of e_y and for many practical problems it is sufficiently accurate to use the depth-averaged value

$$\overline{e_y} = \frac{\varkappa}{6} hu^* \approx 0.067hu^* \tag{2.19}$$

as $\varkappa = 0.4$. It can be noted in passing that Prandtl's analysis assumes the flow is two-dimensional and that turbulent eddies lie in the $x-y$ plane. Consequently eddies which cause vertical diffusion also cause longitudinal diffusion and it is reasonable to assume that

$$e_x = e_y \tag{2.20}$$

Although Prandtl's analysis provides an estimate of the longitudinal eddy diffusivity e_x it sheds little light on the transverse eddy diffusivity e_z. This point is taken up again in Chapter 3.

2.4 MEASURED VALUES OF THE VERTICAL EDDY DIFFUSIVITY

Equations (2.18) and (2.19) are derived from a theoretical analysis of plane turbulent shear flow. It is possible with some care to produce plane shear flow in uniform laboratory channels and several laboratory studies have provided experimental results which match equation (2.18) closely (Jobson and Sayre 1970a, Schiller and Sayre 1975, French 1979 and Nokes 1986). Vertical mixing in plane shear flow is one of the few river mixing problems for which the diffusion coefficient can be predicted satisfactorily on theoretical grounds.

There are surprisingly few published vertical diffusivities in rivers. This partly reflects the fact that vertical mixing is rapid and that most practical problems focus on transverse and longitudinal mixing. It also reflects the fact that it is difficult to separate the effects of turbulent diffusion, momentum and buoyancy in the near-field. The corollary is that equation (2.18) only provides reliable estimates of the vertical diffusivity below a source of passive tracer (i.e. a tracer that does not affect the flow either because of its initial momentum or its buoyancy). Many velocity profiles measured in natural channels are approximately logarithmic and in such situations

Table 2.1. Summary of published vertical diffusion coefficients

Width b (m)	Depth h (m)	Velocity u_x (m s^{-1})	Shear velocity u^x (m s^{-1})	Diffusivity		Reference.
				e_y (m^2 s^{-1} × 10^4)	e_y/hu^*	
Laboratory channels						
Unstratified flow						
2.44	0.400	—	0.049–0.136	—	0.063	Jobson and Sayre (1970b)
0.76	0.071	0.046	0.040	1.8	0.042	Schiller and Sayre (1973)
0.56	0.039–0.076	—	0.041–0.058	—	0.067*	McNulty (1983)
0.56	0.050–0.065	0.024–0.028	0.014–0.016	—	0.067*	Nokes (1986)
Waikato river						
80	2.0–2.75	0.90	0.066	86–155 (mean 120)	0.055–0.099 (mean 0.076)	Rutherford (Unpublished)
Stratified flow						
0.76	0.071	0.031	0.0037	0.01	0.004	Schiller and Sayre (1973)
0.40	0.037	0.040	0.0690	1.00	0.004	French (1979)
Estuaries: Stratified flow						
	7–18	0.11–0.18		1.0–9.0	(Sacramento)	Lung and O'Connor (1982)
	7	0.10		2.0–3.0	(James)	
	1–12	0.02		0.5–1.5	(Patuxent)	
	3–30	0.11–0.24		0.3–7.9	(Hudson)	

* McNulty (1983) and Nokes (1986) found that their experimental results closely matched predictions made assuming $e_y = 0.067u^*h$.

equation (2.18) can be expected to provide reasonable estimates of the vertical diffusivity. [I have had occasion to measure the distance below the source at which tracer becomes well-mixed vertically in a number of rivers and have found that the experimental results closely match the predicted values (see Example 2.1)]. In some parts of the channel (for example, near the banks, at the edge of the main channel and at sharp bends) observed velocity profiles depart significantly from the logarithmic form of equation (2.13). This indicates the existence of secondary currents (Henderson 1966). Schiller and Sayre (1973) observed weak secondary currents near the edges of their straight laboratory flume but found that they had a negligibly small effect on vertical mixing. This may not be true universally and secondary currents at sharp bends can be expected to accelerate vertical mixing. Obstacles which disturb the flow (such as bridge piers) increase the intensity of turbulence, promote secondary circulations and would be expected to cause locally high values of diffusivity. Nokes (1986) found in a laboratory channel that flow disturbance around the tracer injection nozzle resulted in a detectable increase in vertical mixing. It is not possible to account theoretically for the effect of secondary currents or flow disturbances on the vertical diffusivity, nor are there sufficient published data from natural rivers to quantify their effect. If reliable estimates of vertical diffusivity are required in regions of the channel where plane shear flow does not occur, then recourse will be necessary to field tests or physical model studies.

There have been several studies of vertical mixing in estuaries and a selection of results is given in Table 2.1. It is clear from these measurements that in stratified flow vertical mixing is strongly inhibited, a point which is discussed in more detail in Section 2.9.

2.5 TIME-SCALES OF VERTICAL AND TRANSVERSE MIXING

In most rivers the width is many times larger than the depth. If the vertical and transverse diffusion coefficients are equal, then it is obvious that neutrally buoyant tracer originating from a point source will mix over the depth more quickly than over the width. In practice transverse diffusivites are higher than vertical diffusivities and we need to examine the time-scales of vertical and transverse mixing to determine their relative importance. From dimensional arguments the time T taken for tracer to mix over a distance L is

$$T \propto \frac{L^2}{e} \tag{2.21}$$

where e = diffusion coefficient. By applying equation (2.21) to both vertical and transverse mixing, the ratio of mixing times is

$$\frac{T_z}{T_y} = \left(\frac{L_z}{L_y}\right)^2 \frac{e_y}{e_z} \tag{2.22}$$

where the y and z axes are vertical and transverse; L_y and L_z are the length scales; and e_y and e_z are the diffusivities in the y and z directions, respectively. For a point source located on the bed at the mid-point of a rectangular channel, $L_y = h$ and $L_z = b/2$, where h = depth and b = channel width. For a typical river channel $b/h = 50$ and $e_z = 2e_y$. Substituting these values into equation (2.22) gives

$$\frac{T_z}{T_y} = 310 \tag{2.23}$$

In such a channel vertical mixing occurs two orders of magnitude more rapidly than transverse mixing. The same general conclusion holds for neutrally buoyant tracer in most river channels, as has been shown by El-Hadi *et al* (1984).

2.6 VERTICAL MIXING FROM A STEADY TRANSVERSE LINE SOURCE

The simplest vertical mixing problem is to predict concentrations of a neutrally buoyant tracer downstream from a steady transverse line source (such as a diffuser stretching across the entire channel) in plane shear flow. Near mid-channel in a uniform channel the flow approximates plane shear flow and this model has important practical applications. The three-dimensional advection/diffusion equation derived in Chapter 1 is

$$\underset{[1]}{\frac{\partial c}{\partial t}} + \underset{[2]}{u_x \frac{\partial c}{\partial x}} + \underset{[3]}{u_y \frac{\partial c}{\partial y}} + \underset{[4]}{u_z \frac{\partial c}{\partial z}} = \underset{[5]}{\frac{\partial}{\partial x}\left(e_x \frac{\partial c}{\partial x}\right)} + \underset{[6]}{\frac{\partial}{\partial y}\left(e_y \frac{\partial c}{\partial y}\right)} + \underset{[7]}{\frac{\partial}{\partial z}\left(e_z \frac{\partial c}{\partial z}\right)} \tag{1.64}$$

In a straight rectangular channel we can align the x axis with the main flow and we will ignore any complications which arise near the banks with transverse velocity shear and secondary currents. Concentrations are steady (which eliminates term [1]), vertical and transverse velocities are zero because of our assumption of plane flow (which eliminates terms [3] and [4]), transverse concentration gradients are negligibly small because of the nature of the tracer source (which eliminates term [7]) and longitudinal diffusion is negligible because the source is steady (which eliminates term [5]). This leaves

$$u_x \frac{\partial c}{\partial x} = \frac{\partial}{\partial y}\left(e_y \frac{\partial c}{\partial y}\right) \tag{2.24}$$

In uniform plane flow u_x and e_y do not vary transversely or with distance downstream (x or z), but both vary with depth (y).

2.6.1 Approximate Constant-coefficient Model

An approximate analysis of vertical mixing below a steady transverse line source in a uniform channel can be made by assuming that both the velocity and vertical diffusivity are constant. Equation (2.24) then becomes

$$u_x \frac{\partial c}{\partial x} = e_y \frac{\partial^2 c}{\partial y^2} \tag{2.25}$$

where u_x and e_y are both constant. For a steady transverse line source at $x = 0$ and $y = 0$, equation (2.25) has the following analytical solution in a vertically unbounded flow

$$c_{\mathrm{u}}(x, y) = \frac{m}{b\sqrt{4\pi e_y x u_x}} \exp\left(-\frac{u_x y^2}{4 e_y x}\right) \tag{2.26}$$

where c_{u} = concentration in an unbounded flow; b = channel width and m = tracer mass inflow rate. The boundary conditions for this problem are that the tracer flux is zero at the bed and the water surface. Hence

$$e_y \frac{\partial c}{\partial y} = 0 \qquad \text{at } y = 0 \text{ and } y = h \tag{2.27}$$

As e_y is constant, equation (2.27) requires that the concentration gradient vanishes at the boundaries. This can be accommodated by assuming that the river bed and the water surface are perfect reflectors and using the method of images. Tracer which impinges on the boundary is reflected and can be assumed to have originated from an imaginary source outside the boundary as shown in Figure 2.1. Imaginary sources need to be located at $y = -y_{\mathrm{o}}$, $2nh - y_{\mathrm{o}}$, $y_{\mathrm{o}} - 2nh$, $y_{\mathrm{o}} + 2nh$ and $-y_{\mathrm{o}} - 2nh$ for $n = 1, 2, \ldots N$ in addition to the original source at $y = +y_{\mathrm{o}}$. Equation (2.25) is linear and the contributions from each of the virtual sources can be added together to give the overall concentration

$$\begin{aligned} c(x, y) = c_{\mathrm{u}}(x, y + y_{\mathrm{o}}) + c_{\mathrm{u}}(x, y - y_{\mathrm{o}}) + \sum_{n=1}^{N} [c_{\mathrm{u}}(x, 2nh - y - y_{\mathrm{o}}) \\ + c_{\mathrm{u}}(x, 2nh + y + y_{\mathrm{o}}) + c_{\mathrm{u}}(x, 2nh - y + y_{\mathrm{o}}) \\ + c_{\mathrm{u}}(x, 2nh + y - y_{\mathrm{o}})] \end{aligned} \tag{2.28}$$

where the source is located at $y = y_{\mathrm{o}}$, the river bed is at $y = 0$, the water surface is at $y = h$ and $c_{\mathrm{u}}(x, y)$ is the unbounded solution given by equation (2.27). In practice it is only necessary to evaluate the summation as far as $N = 6$ throughout most of the near-field, but larger values of N are required once the tracer becomes well-mixed vertically. A shorthand notation can be used to

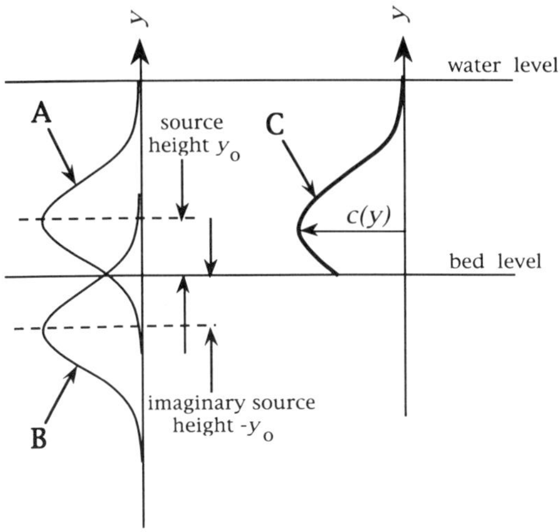

Figure 2.1. Sketch illustrating the method of images. Profile A was predicted using equation (2.26) for a source at y_o in an unbounded channel. Profile B was predicted using the same manner for an imaginary source at $-y_o$. Profile C is the sum of A and B. As the tracer becomes more fully mixed, additional imaginary sources are required (see text for details)

rewrite equation (2.28)

$$c(x, y) = \underbrace{c_u(x, y \pm y_o)}_{[1]} + \sum_{n=1}^{N} \underbrace{c_u(x, 2nh \pm y \pm y_o)}_{[2]} \tag{2.29}$$

where expression [1] contains two terms and expression [2] contains four terms for each value of n. This shorthand notation is used again later.

Figure 2.2 shows lines of equal concentration downstream from transverse line sources of conservative tracer in plane shear flow predicted using equation (2.29). Variables are expressed in non-dimensional form so that many combinations of parameters appear on the same graph.

$$c^* = \frac{c}{C_m} \tag{2.30a}$$

$$y^* = \frac{y}{h} \tag{2.30b}$$

$$x^* = \frac{xe_y}{u_x h^2} \tag{2.30c}$$

$$C_m = \frac{m}{Q} \tag{2.30d}$$

(a)

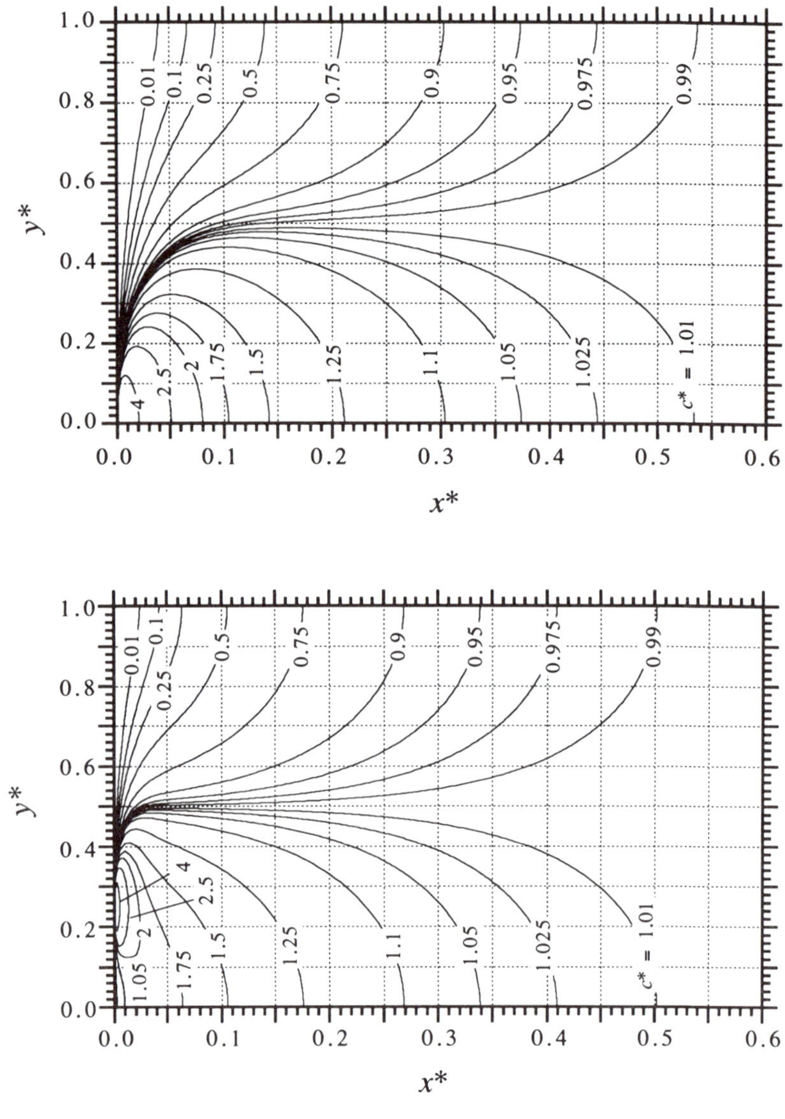

Figure 2.2. (a) Concentration contours downstream from a transverse line source predicted using the constant-coefficient model for sources at $y^* = 0.00$ and 0.25. Concentrations are symmetrical about mid-depth so these are also the contours for sources at $y^* = 1.00$ and 0.75. (b) Concentration contours downstream from a transverse line source predicted using the constant-coefficient model for sources at $y^* = 0.33$ and 0.50. Concentrations are symmetrical about mid-depth so these are also the contours for a source at $y^* = 0.67$. Note the change of scale in x^*

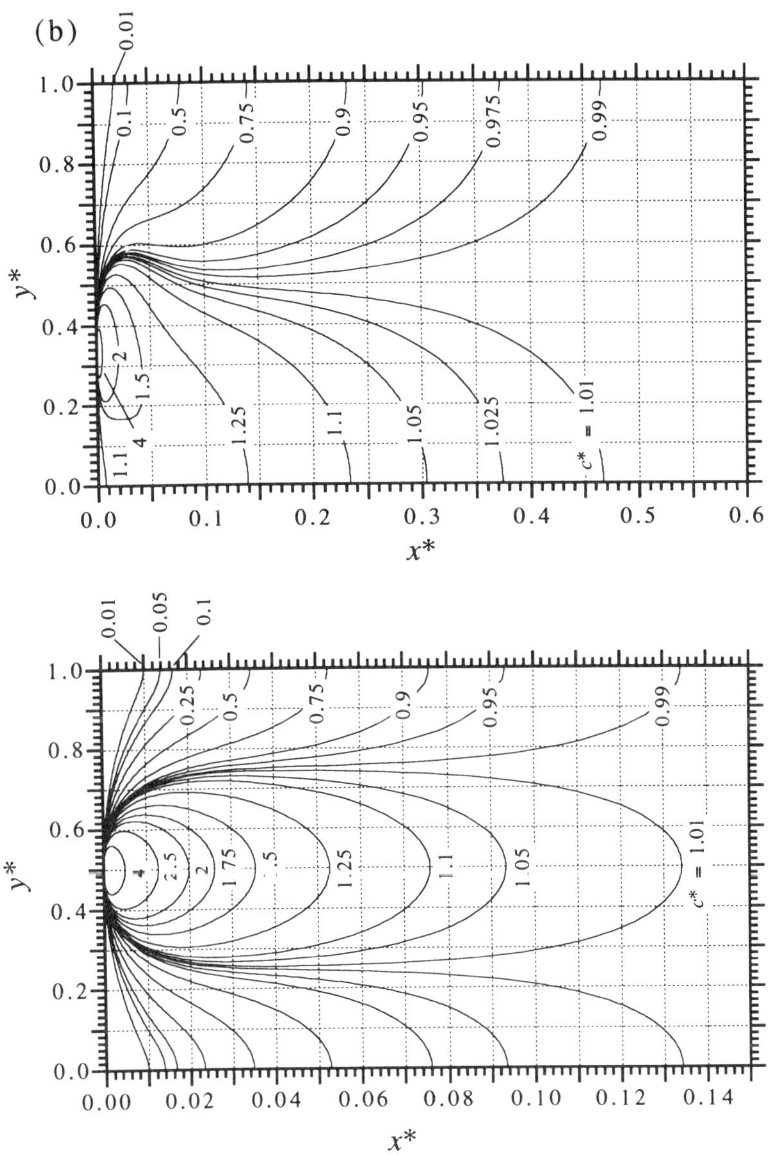

Figure 2.2. *(continued)*

where c^*, y^* and x^* = non-dimensional concentration, vertical distance and longitudinal distance, respectively; c = local concentration; h = depth; u_x = velocity (assumed constant); e_y = vertical diffusivity (assumed constant); C_m = fully mixed concentration; Q = total river flow; and m = tracer mass inflow rate. The bed and water surfaces are located at $y^* = 0$ and $y^* = 1$, respectively, but the solution is symmetrical in y^* as the velocity and diffusion coefficient have been assumed constant. A long way below the source the tracer becomes fully mixed over the depth and $c^* = 1$. The regions to the left of the $c^* = 0.01$ contour do not contain any tracer, whereas in the region to the right of the $c^* = 1.01$ and $c^* = 0.99$ contours, tracer is fully mixed.

2.6.1.1 Behaviour of the concentration maximum

Close to the source the maximum concentration is found at the same depth as the source. It can be seen in Figure 2.2 that for a source part-way between the bed and mid-depth the concentration maximum moves towards the bed with increasing distance downstream. By symmetry, the concentration maximum migrates towards the surface for sources above mid-depth. The concentration maximum migrates towards the nearest boundary because when tracer impinges on either the bed or the surface the non-flux boundary condition causes it to accumulate. When the source is at mid-depth, both boundaries are equidistant and the constant-coefficient model predicts that the maximum remains at mid-depth.

2.6.1.2 Distance for complete vertical mixing

The degree of mixing can be described by the ratio

$$P_m = \frac{c_{min}}{c_{max}} \tag{2.31}$$

where c_{max} and c_{min} are the maximum and minimum concentrations over the depth at a given distance below the source. At large distances from the source tracer is well-mixed $c_{max} = c_{min} = C_m$ and $P_m = 1$. Figure 2.3 shows the vertical mixing distance (i.e. the longitudinal distance required to attain a certain value of P_m) for several source locations. For a point source located at mid-depth, $P_m = 0.98$ (i.e. vertical mixing is 98% complete) within a distance

$$L_y = 0.134\,\frac{u_x h^2}{e_y} \tag{2.32}$$

The constant-coefficient model predicts that the vertical mixing distance is shorter for a source located at mid-depth than for a source at any other depth. (We shall see later that for the more realistic assumption of logarithmic velocity and parabolic diffusivity the optimum source location is not at

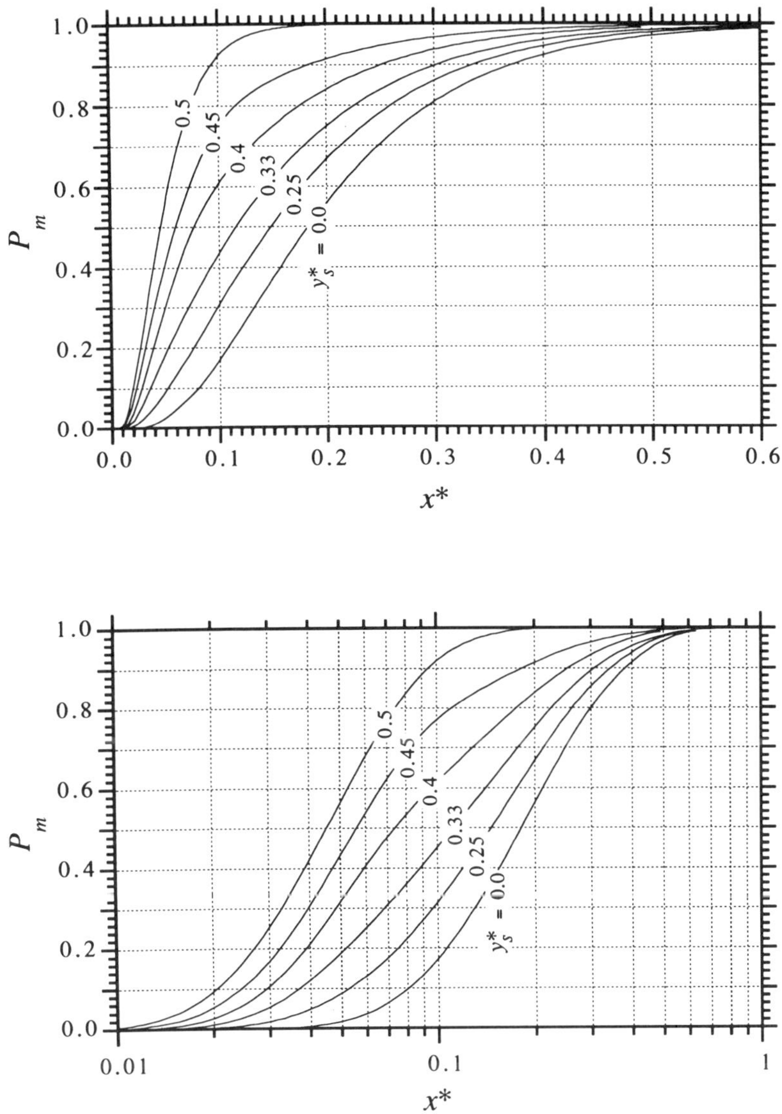

Figure 2.3. Degree of mixing as a function of non-dimensional distance downstream from a transverse line source predicted using the constant-coefficient model. The source height is y_s^*

mid-depth but is slightly closer to the surface.) This is intuitively obvious because for a source at mid-depth tracer can spread equally quickly towards both the top and bottom boundaries. For a point source at the bed or at the surface the vertical mixing distance is four times larger

$$L_y = 0.536 \frac{u_x h^2}{e_y} \tag{2.33}$$

which is the maximum possible value for the mixing distance.

It is sometimes more convenient in field studies to define L_y as the distance below the source at which concentrations differ by 5, 10 or 20% over the depth because it is difficult to measure tracer concentrations with sufficient accuracy to determine when the difference is only 2%. For several source locations values of L_y for different values of P_m can be determined from Figure 2.3 and for other source locations approximate values can be derived by interpolation.

For a point source at mid-depth, substituting equation (2.19) into (2.32) and rearranging gives

$$\frac{L_y}{h} = 2 \frac{u_x}{u^*} \tag{2.34}$$

In rivers u_x/u^* rarely exceeds 25 and so equation (2.34) indicates that

$$\frac{L_y}{h} \leqslant 50 \tag{2.35}$$

Thus as a rule of thumb, neutrally buoyant tracer becomes fully mixed vertically within 50 depths of the source.

Example 2.1

The Waikato River just downstream from Hamilton, New Zealand, has the hydraulic parameters summarised in Table 2.2. In the 1970s a multiport outfall was built to discharge treated sewage. The outfall characteristics are summarised in Table 2.3. An estimate can be made using the constant-coefficient model of how quickly effluent reaches the surface at minimum river flow.

Table 2.2. Channel parameters for the Waikato River near Hamilton

Minimum flow (Q_{min})	140 $m^3 s^{-1}$
Width (b)	95 m
Mean depth (h)	2.60 m
Minimum depth (h_{min})	1.40 m
Maximum depth (h_{max})	4.10 m
Mean velocity (u_x)	0.57 $m s^{-1}$
Channel slope (s)	1.20×10^{-4}

Table 2.3. Characteristics of the outfall at Hamilton

Outfall length (L_o)	85 m
Number of ports (n)	20
Port diameter (d)	0.229 m
Total effluent flow (q_e)	0.527 $m^3 s^{-1}$

Firstly, assume that the channel is rectangular with the mean depth shown in Table 2.2. The shear velocity is then $u^* = 0.055\ m\,s^{-1}$ and from equation (2.19) the depth-averaged vertical diffusivity is $e_y = 0.067hu^* = 0.0096\ m^2 s^{-1}$. For a source at the bed, equation (2.33) gives $L_y = 0.536u_x h^2/e_y = 215$ m. Secondly, we can note that about 40% of the channel is only 1.40 m deep. In this region $u^* = 0.041\ m\,s^{-1}$, $e_y = 0.0038\ m^2 s^{-1}$ and hence $L_y = 160$ m.

The outfall consists of risers which position the ports slightly above the mobile sand bed. The ports are angled upwards and jet momentum carries the tracer a short way up from the bed. Rather than assuming a release point on the bed a better assumption might be to assume a source at three-quarters of the depth. We require the distance x^* at which $P_m = 0.98 = c_{min}/c_{max}$. This is given by the value of x^* at which the $c^* = 0.99$ contour reaches the water surface, which is the same as the value of x^* at which the $c^* = 1.01$ contour reaches the bed. From Figure 2.2a and a source at $y^* = 0.25$ this occurs at $x^* = 0.5$. Thus $L_y = 0.5u_x h^2/e_y = 150–200$ m for the two depths considered above.

A qualitative dye test was conducted during commissioning to test the degree of transverse mixing. Dye was observed to surface some '... tens of metres below the outfall...' (Wilkie 1981) At first sight this appears to be at variance with the estimates of mixing length given above of $L_y = 150–215$ m. It is likely, however, that dye was visible at the surface before it became fully mixed vertically. No dye measurements were made but a plausible assumption is that dye was visible at concentrations between $c^* = 0.25$ and 0.50 (25 and 50% of the fully mixed concentration). Figure 2.2a indicates that the $c^* = 0.25$ contour reaches the surface at $x \approx 0.06u_x h^2/e_y \approx 20–25$ m. The $c^* = 0.50$ contours reaches the surface at $x \approx 0.104u_x h^2/e_y \approx 30–40$ m. This accords with the qualitative observations made during the dye test. Thus at low flows some tracer from this outfall reaches the surface within about 20–40 m of the outfall, but complete mixing is not attained for about 150–200 m.

Example 2.2

The Hamilton sewage effluent has a BOD_5 concentration of about 260 $g\,m^{-3}$ and an estimate was made of the extent of the 2 $g\,m^{-3}$ concentration contour at low river flow. From equation (2.30d) the fully mixed concentration is $C_m = 0.98\ \mathrm{g\,m^{-3}}$ and from equation (2.30a) $c = 2\ \mathrm{g\,m^{-3}}$ corresponds to $c^* \approx 2$. Figure 2.2a indicates that for a source at $y^* = 0.25$ the $c^* = 2$ contour extends downstream for a distance of $x^* = xe_y/u_x h^2 = 0.025$. Taking $u_x = 0.57\ \mathrm{m\,s^{-1}}$, $h = 2.60$ m and $e_y = 0.0096\ \mathrm{m^2\,s^{-1}}$ then $x \approx 10$ m. In the shallow water $h = 1.40$ m and $e_y = 0.0038\ \mathrm{m^2\,s^{-1}}$ giving $x \approx 7$ m.

Figure 2.2a indicates that the $c^* = 2$ contour does not impinge on the bed and has a maximum width of less than $0.3h$ (0.4–0.8 m). Thus the $2\ \mathrm{g\,m^{-3}}$ contour is unlikely to extend 10 m beyond the outfall nor to extend over more than 1 m of the depth.

2.6.2 Variable-coefficient Model

The previous section described a simplified model which assumes that the velocity and vertical diffusivity are both constant. In uniform natural channels, however, the velocity and the diffusivity both vary with depth. For example, in plane shear flow the velocity varies logarithmically while the vertical diffusivity varies parabolically. An important question, therefore, is how accurate are the predictions made using the constant-coefficient model?

McNulty (1983) and Nokes (1986) have solved the vertical mixing equation for a steady source allowing the velocity and diffusivity to vary over depth. McNulty used the method of moments whereas Nokes developed an eigenfunction expansion solution. They considered the case of steady two-dimensional plane shear flow. For a steady source, tracer concentration satisfies the equation

$$u_x \frac{\partial c}{\partial x} = \frac{\partial}{\partial y}\left(e_y \frac{\partial c}{\partial y}\right) \tag{2.24}$$

with the boundary conditions

$$e_y \frac{\partial c}{\partial y} = 0 \qquad \text{at } y = 0 \text{ and } y = h \tag{2.27}$$

The ensemble mean velocity (the angle brackets used to denote ensemble means are dropped for simplicity) varies logarithmically with depth

$$\frac{u_x(y)}{u^*} = \frac{1}{\varkappa} \log_e\left(\frac{y}{y_o}\right) \tag{2.13}$$

and the vertical diffusivity varies parabolically with depth

$$e_y = \varkappa u^* y\left(1 - \frac{y}{h}\right) \tag{2.18}$$

Nokes (1986) defined the non-dimensional variables

$$x' = \frac{\varkappa x}{h}\sqrt{\frac{f}{8}} \tag{2.36}$$

$$y' = \frac{y}{h} \tag{2.37}$$

where

$$\frac{u_x}{u^*} = \sqrt{\frac{8}{f}} \tag{2.38}$$

where f = Darcy friction factor. Rewriting equation (2.24) in terms of these variables

$$\chi(y')\frac{\partial c}{\partial x'} = \frac{\partial}{\partial y'}\left(\psi(y')\frac{\partial c}{\partial y'}\right) \tag{2.39}$$

where

$$\chi(y') = 1 + \frac{1}{\varkappa}\sqrt{\frac{f}{8}}\,[1 + \log_e(y')] \tag{2.40}$$

and

$$\psi(y') = y'(1 - y') \tag{2.41}$$

The same non-dimensional depth is used in both constant and variable coefficient models, $y^* = y' = y/h$. Combining equations (2.19), (2.30c), (2.36) and (2.38) it can be seen that $x' = 6x^*$.

The variable-coefficient model contains two coefficients, the von Karman constant $\varkappa$ and the Darcy friction factor f. The former does not vary widely in natural channels and to a good approximation may be assumed constant at $\varkappa = 0.40$. The friction factor on the other hand varies from about $f = 0.01$ in smooth laboratory channels to about $f - 0.30$ in shallow, cobble bed streams.

2.6.3 Comparison Between Constant and Variable-coefficient Models

Figure 2.4 compares concentration contours predicted using the variable-coefficient and constant-coefficient models for sources at $y^* = 0.65$ and 0.35. In the variable-coefficient model $\varkappa = 0.40$ and $f = 0.3$ (a rough channel) and both models have the same depth-averaged velocity and diffusivity. It can be seen that the two models predict concentration contours which follow the same general pattern, but there are three important differences.

Firstly, the concentration maxima behave rather differently. In Figure 2.4a, for a source at $y^* = 0.35$, the concentration maximum moves from the height of the source towards the river bed more rapidly in the variable-coefficient

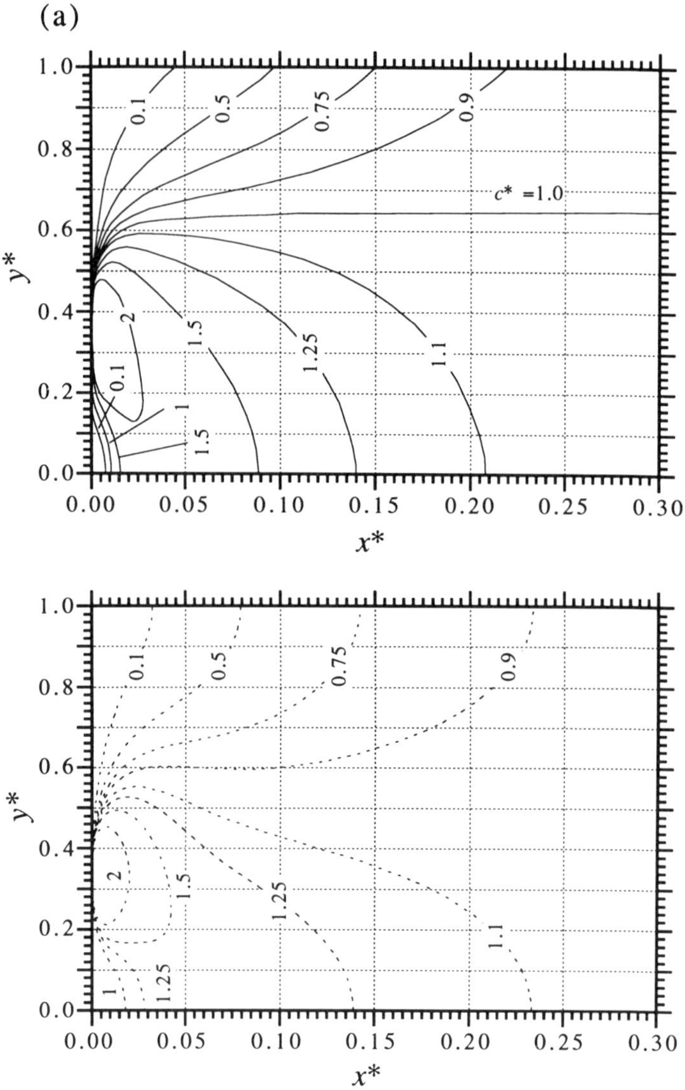

Figure 2.4. (a) Comparison of concentration contours predicted using the variable-coefficient (solid line) and constant-coefficient (broken line) models for a source at $y^* = 0.35$. The mean velocity and diffusivity are the same for both models. (b) Comparison of concentration contours predicted using the variable-coefficient (solid line) and constant-coefficient (broken line) models for a source at $y^* = 0.65$. The mean velocity and diffusivity are the same for both models. Note that for the rough channel ($f = 0.3$) considered here the optimum source height is $y^* = 0.64$

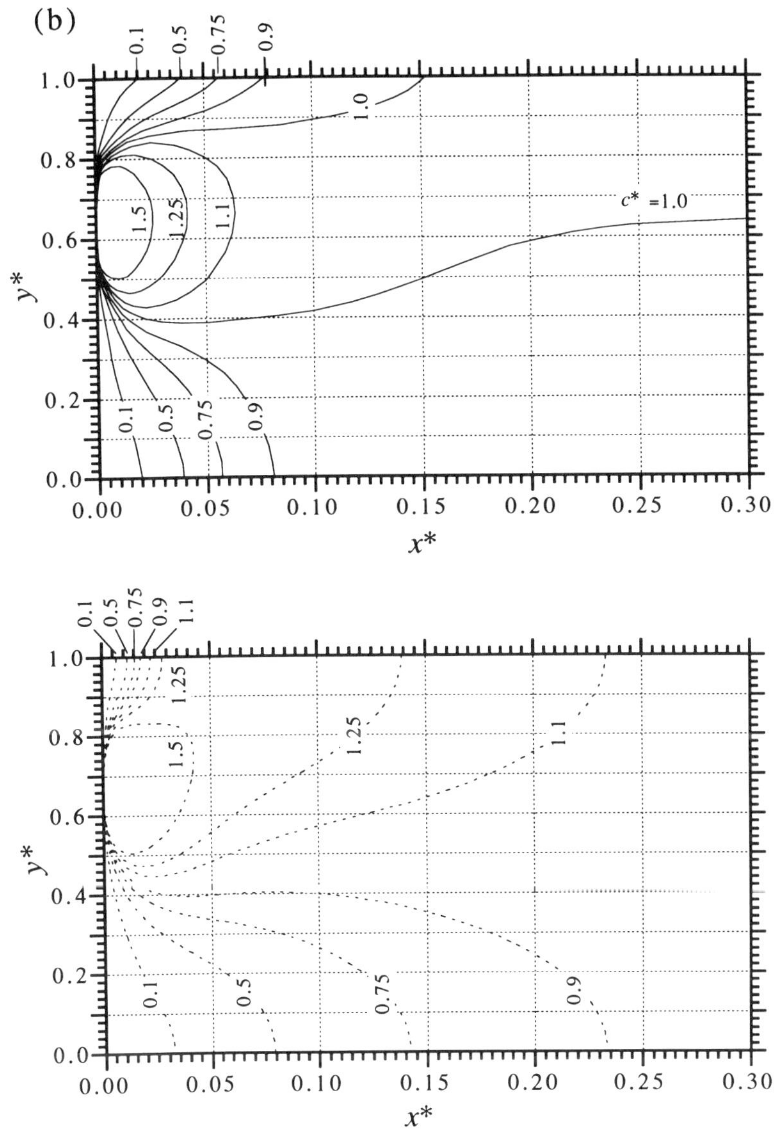

Figure 2.4. (*continued*)

model. This can be explained qualitatively as follows. Consider what happens to the concentration maximum in an unbounded channel where the diffusivity is non-uniform and the velocity is constant. If the diffusivity is higher below the concentration maximum then diffusion will be faster and hence concentrations will decrease more rapidly below the concentration maximum than above it. This causes the concentration maximum to migrate upwards towards the region of low diffusivity. In the case of the parabolic diffusivity distribution, the concentration maximum moves towards either the bed or the water surface, whichever is closer. When the tracer reaches either the water surface or the bed, the no-flux boundary condition causes it to accumulate at the boundary. This accelerates the rate at which the concentration maximum migrates towards the boundary. In the constant-coefficient model the concentration maximum also migrates towards the nearest boundary, but it does so only as a result of the no-flux boundary condition and hence only when tracer begins to impinge on the boundary. In the more realistic variable-coefficient model the concentration maximum migrates more rapidly because of both the non-uniform diffusivity and the effect of the boundaries.

It is more difficult to quantify the effects of the non-uniform velocity. Consider the situation where the diffusivity is constant, there are no boundaries, the velocity varies logarithmically and there is a concentration maximum at a certain depth. Tracer travels downstream more slowly below the concentration maximum than above it. This means there is more time for diffusion to take place below the concentration maximum. As a result, concentrations might be expected to drop more rapidly near the bed, which would cause the concentration maximum to migrate upwards. On the other hand, below the concentration maximum the concentration gradient decreases with distance more rapidly and this reduces the rate of diffusion near the bed. It is not intuitively obvious which of these two processes dominates, but Nokes (1986) has been able to demonstrate that it is the latter. Consequently, when the diffusivity is constant and the velocity varies logarithmically, the concentration maximum migrates towards the velocity minimum (i.e. towards the bed). In a natural channel the two mechanisms described in the previous paragraphs interact. There is a tendency for the concentration maximum to migrate towards the region of low velocity which lies near the bed. There is also a tendency for it to migrate towards regions of low diffusivity which lie at both the water surface and the bed. This latter process is accelerated by the no-flux boundary condition at both the surface and the bed. Consequently there is a greater tendency for the concentration maximum to migrate towards the bed. For a source at mid-depth, the variable-coefficient model predicts that the concentration maximum migrates towards the bed, whereas the constant-coefficient model predicts that the concentration maximum remains at mid-depth. Figure 2.4b compares the two models for a source at $y^* = 0.65$. In both models the concentration maximum migrates towards the water surface, but in this situation it migrates more quickly for the constant-coefficient model. This is because in the variable-coefficient model the concentration maximum is drawn towards the bed where the velocity is low and also towards the water surface where the diffusivity is low. For a source at $y^* = 0.65$ in a rough channel ($f = 0.30$) the water surface exerts a slightly stronger influence. This source height lies very close to the optimum height for a rough channel ($y^* = 0.64$) for which the concentration maximum does not migrate either up or down.

The second difference between the models relates to the optimum source depth. There is a source depth for which the concentration maximum does not migrate towards either the bed or the surface, but remains at the height of the source as the tracer cloud is carried downstream. By placing the source at this depth the concentrations at both boundaries are minimised and, as is intuitively obvious, the mixing distance is minimised. It is clear why this source depth is referred to as the optimum. For the constant-coefficient model the optimum source location is at mid-depth (see Figure 2.2) and does not vary with the model coefficients (u_x and e_y). For the variable-coefficient model, however, the optimum source location lies slightly closer to the surface and it varies with the model coefficients ($\varkappa$ and f). Figure 2.5 shows that the optimum source depth varies from $y^* = 0.53$ in a smooth channel ($f = 0.01$) to $y^* = 0.64$ in a rough channel ($f = 0.30$).

The third difference relates to the vertical mixing distance. Figure 2.6 compares the distance downstream from a point source at which there is 80% mixing (i.e. $P_m = c_{min}/c_{max} = 0.80$) for several source heights. In the variable-

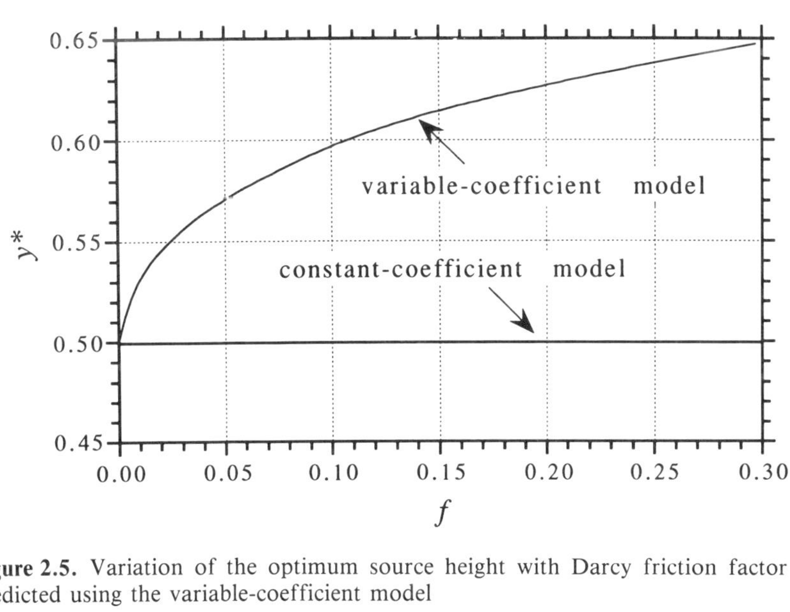

Figure 2.5. Variation of the optimum source height with Darcy friction factor (f) predicted using the variable-coefficient model

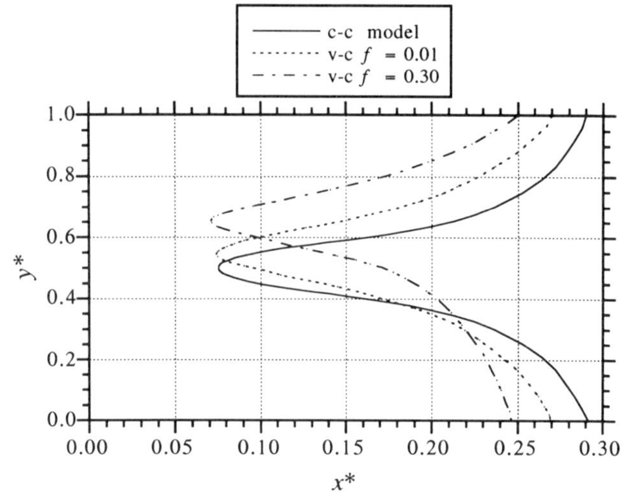

Figure 2.6. Comparison of the vertical mixing distance at which $c_{min}/c_{max} = 0.80$ predicted using the constant-coefficient (c-c) and the variable-coefficient (v-c) models for rough ($f = 0.3$) and smooth ($f = 0.01$) channels

coefficient model it is assumed that $\varkappa = 0.40$ and results are presented for very smooth ($f = 0.01$) and very rough ($f = 0.30$) channels.

For sources close to the optimum depth there is very little difference in mixing distance between models or channels. Thus equation (2.32) provides a reliable estimate of the minimum distance required to achieve complete vertical mixing even though it is derived using the approximate constant-coefficient model. Note, however, that the source height to achieve this minimum mixing distance varies (see Figures 2.5 and 2.6) and is best predicted using the variable-coefficient model. For sources close to the bed or water surface the variable-coefficient model predicts mixing distances which are shorter than the constant-coefficient model. Thus equation (2.33) provides an estimate of the maximum vertical mixing distance which is of the order of 15% too large. Taking a pragmatic view, the simple constant-coefficient model provides an accurate estimate of the minimum vertical mixing distance and a conservative estimate (i.e. a 15% overestimate) of the maximum mixing distance.

In smooth channels the variable-coefficient model almost invariably predicts a shorter mixing distance than the constant-coefficient model. The reasons for this are not intuitively obvious. In the variable-coefficient model the diffusivity is low near the bed and this might be expected to increase the mixing distance. The velocity near the bed is also low, however, which slows the downstream transport of tracer and allows more time for vertical mixing. Near the surface the diffusivity is low and here the velocity is high. This combination might also be expected to increase the mixing length. The central part

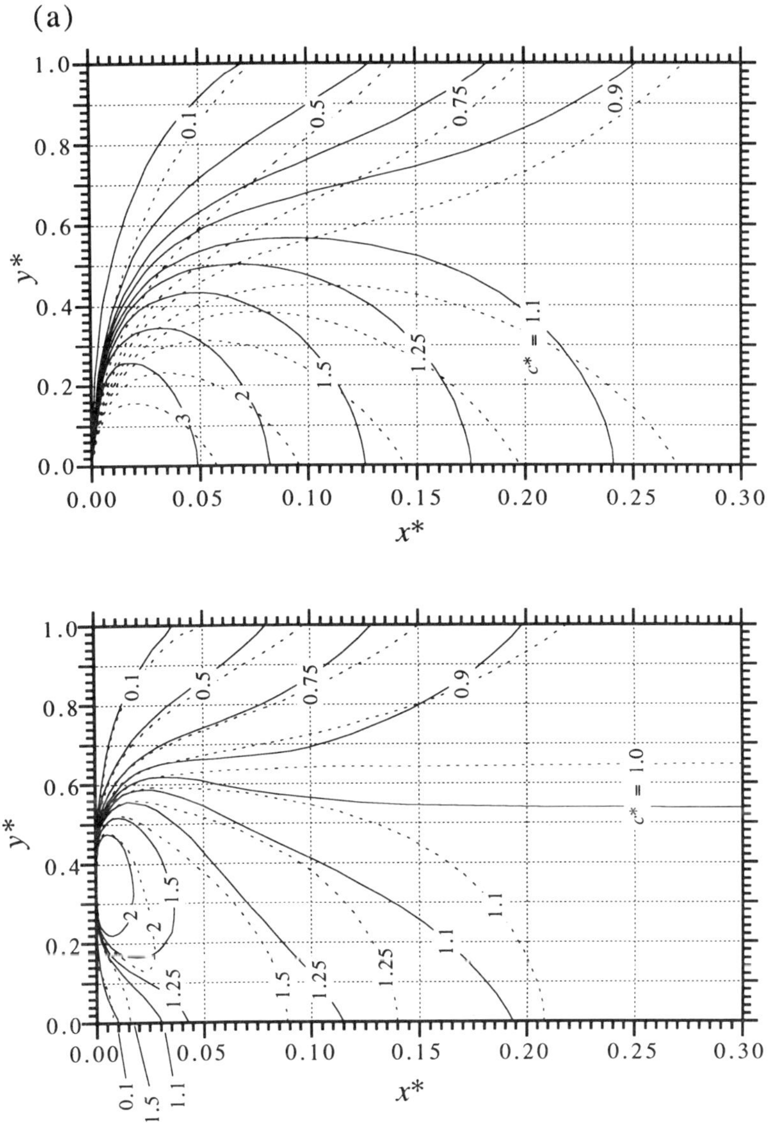

Figure 2.7. (a) Concentration contours downstream from a steady transverse line source predicted using the variable-coefficient model in a smooth (solid line) and rough (broken line) channel for sources at $y^* = 0.01$ and 0.35. (b) Concentration contours downstream from a steady transverse line source predicted using the variable-coefficient model in a smooth (solid line) and rough (broken line) channel for sources at $y^* = 0.50$ and 0.65. (c) Concentration contours downstream from a steady transverse line source predicted using the variable-coefficient model in a smooth (solid line) and rough (broken line) channel for a source at $y^* = 0.99$

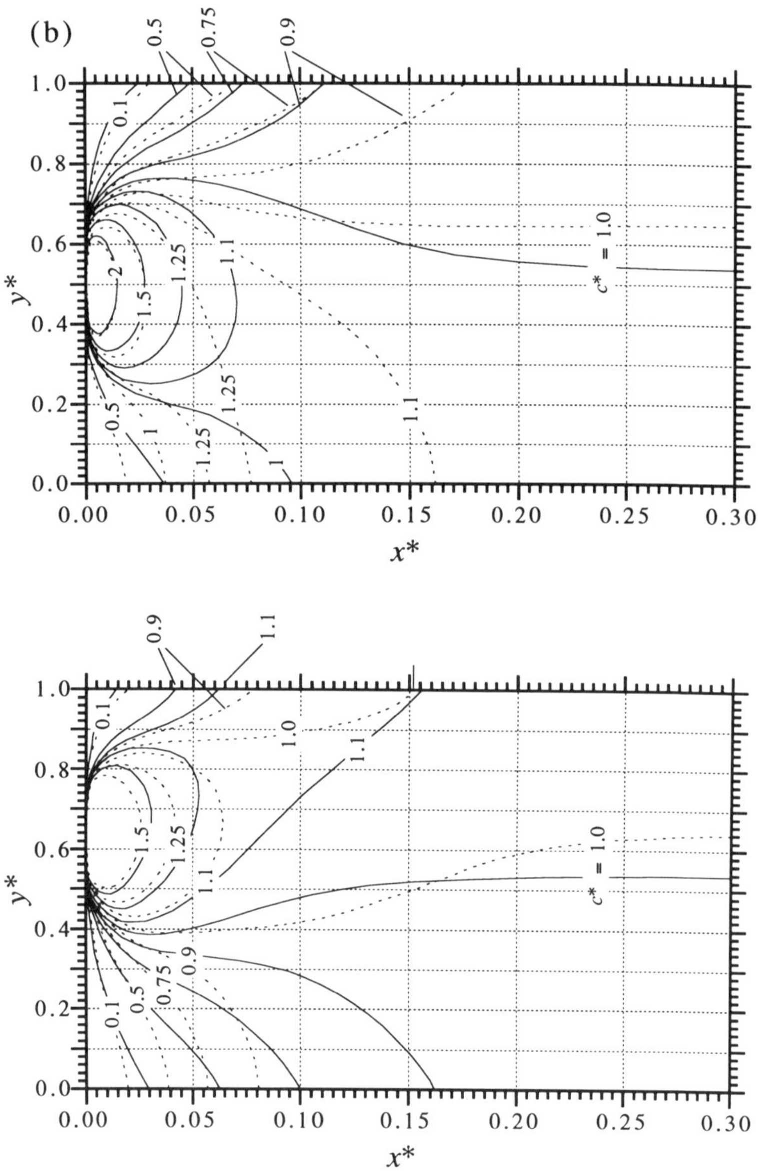

Figure 2.7. *(continued)*

(c)

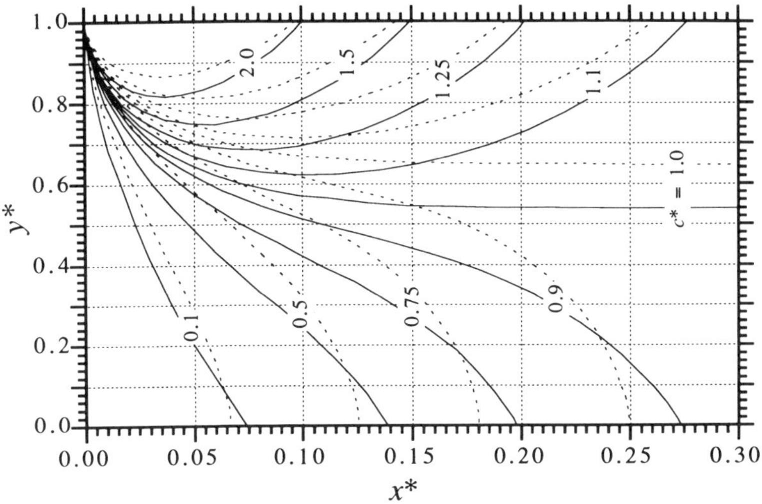

Figure 2.7. *(continued)*

of the channel has a high diffusivity and it turns out that this more than compensates for the low diffusivity near the surface.

In rough channels the variable-coefficient model predicts a shorter mixing distance than the constant-coefficient model for most source heights. The maximum discrepancy between the two models occurs for a source near $y^* = 0.65$ in a rough channel: the variable-coefficient model predicts a mixing distance of $x^* = 0.07$ whereas the constant-coefficient model predicts $x^* = 0.22$, larger by a factor of three. For source heights between $y^* = 0.35$ and 0.55, however, the variable-coefficient model predicts a longer mixing distance than the constant-coefficient model. For these source depths, the high local diffusivity moves tracer into the low velocity and low diffusivity region near the bed, thereby increasing the mixing distance. Thus it would be unwise to assume that the constant-coefficient model always gives a conservative estimate of the mixing distance. For a source at $x^* = 0.5$ (the optimum source height for the constant-coefficient model) the constant-coefficient model predicts $x^* = 0.075$ whereas the variable-coefficient model predicts $x^* = 0.17$, which is larger by a factor of over two.

2.6.4 Nomographs for the Variable-coefficient Model

Figure 2.7 shows concentration contours predicted by Nokes' variable-coefficient model for several different source heights and two different values of the friction factor which cover the range likely to be encountered in natural channels ($f = 0.01$ and 0.30). These nomographs can be used in the same way as Figure 2.2. They give more realistic predictions, especially close to sources located near the bed and the surface where the diffusivity and velocity differ most from the depth-averaged values used in the constant-coefficient model.

2.7 VERTICAL MIXING FROM A STEADY POINT SOURCE

Section 2.6 described vertical mixing downstream from a steady transverse line source in which transverse and longitudinal concentration gradients are negligibly small and the problem is one-dimensional. A commonly encountered problem is to predict concentrations downstream from a steady point source, such as a simple pipe discharging into a river. This is a more difficult problem because both transverse and vertical dispersion are important. We start by considering the three-dimensional mixing equation

$$\frac{\partial c}{\partial t} + u_x \frac{\partial c}{\partial x} + u_y \frac{\partial c}{\partial y} + u_z \frac{\partial c}{\partial z} = \frac{\partial}{\partial x}\left(e_x \frac{\partial c}{\partial x}\right) + \frac{\partial}{\partial y}\left(e_y \frac{\partial c}{\partial y}\right) + \frac{\partial}{\partial z}\left(e_z \frac{\partial c}{\partial z}\right) \quad (1.64)$$

Longitudinal dispersion can be neglected because the source is steady. If the channel is straight the x axis can be aligned with the direction of flow, vertical and transverse velocities can be assumed zero (i.e. $u_y = u_z = 0$) and equation (1.64) simplified to

$$u_x \frac{\partial c}{\partial x} = \frac{\partial}{\partial y}\left(e_y \frac{\partial c}{\partial y}\right) + \frac{\partial}{\partial z}\left(e_z \frac{\partial c}{\partial z}\right) \quad (2.42)$$

An approximate solution to this problem can be obtained if the channel is assumed to be uniform and rectangular and the velocity and diffusivities are assumed to be constant. Making these assumptions simplifies equation (2.42) to

$$u_x \frac{\partial c}{\partial x} = e_y \frac{\partial^2 c}{\partial y^2} + e_z \frac{\partial^2 c}{\partial z^2} \quad (2.43)$$

where u_x, e_y and e_z are constant. Equation (2.43) has the analytical solution in an unbounded channel

$$c(x, y, z) = m \frac{\exp\left(-\frac{u_x(y - y_o)^2}{4e_y x}\right)}{\sqrt{4\pi e_y x}} \frac{\exp\left(-\frac{u_x(z - z_o)^2}{4e_z x}\right)}{\sqrt{4\pi e_z x}} \quad (2.44)$$

where m = tracer mass inflow rate and the source is located at $x = 0$, $z = z_o$ and $y = y_o$. The boundary conditions in a natural channel are zero flux at the water surface, river bed and banks which requires that

$$e_z \frac{\partial c}{\partial z} = 0 \qquad \text{at } z = 0 \text{ and } z = b \tag{2.45}$$

and

$$e_y \frac{\partial c}{\partial y} = 0 \qquad \text{at } y = 0 \text{ and } y = h \tag{2.27}$$

As e_y and e_z are constant, equations (2.27) and (2.45) require that the concentration gradients vanish at the boundaries. This can be accommodated using the method of images described earlier. The unbounded solution (2.44) is rewritten

$$c(x, y, z) = \frac{m}{4\pi x\sqrt{e_y e_z}} \, \mathrm{f}(x, y)\mathrm{g}(x, z) \tag{2.46}$$

where

$$\mathrm{f}(x, y) = \exp\left(- \frac{u_x(y - y_o)^2}{4e_y x} \right) \tag{2.47}$$

$$\mathrm{g}(x, z) = \exp\left(- \frac{u_x(z - z_o)^2}{4e_z x} \right) \tag{2.48}$$

Then the solution in a bounded channel is

$$c(x, y, z) = \frac{m}{4\pi x\sqrt{e_y e_z}} \, \mathrm{F}(x, y)\mathrm{G}(x, z) \tag{2.49a}$$

where

$$\mathrm{F}(x, y) = \mathrm{f}(x, y \pm y_o) + \sum_{n=1}^{\infty} \mathrm{f}(x, 2nh \pm y \pm y_o) \tag{2.49b}$$

and

$$\mathrm{G}(x, z) = \mathrm{g}(x, z \pm z_o) + \sum_{m=1}^{\infty} \mathrm{g}(x, 2mb \pm z \pm z_o) \tag{2.49c}$$

The first expressions on the right-hand side of equations (2.49b) and (2.49c) contain two terms whereas the second expression contains four terms for each value of n or m.

Figure 2.8 shows contours of equal concentration predicted using equation (2.49) for a point source at mid-depth. Concentrations are shown on a vertical

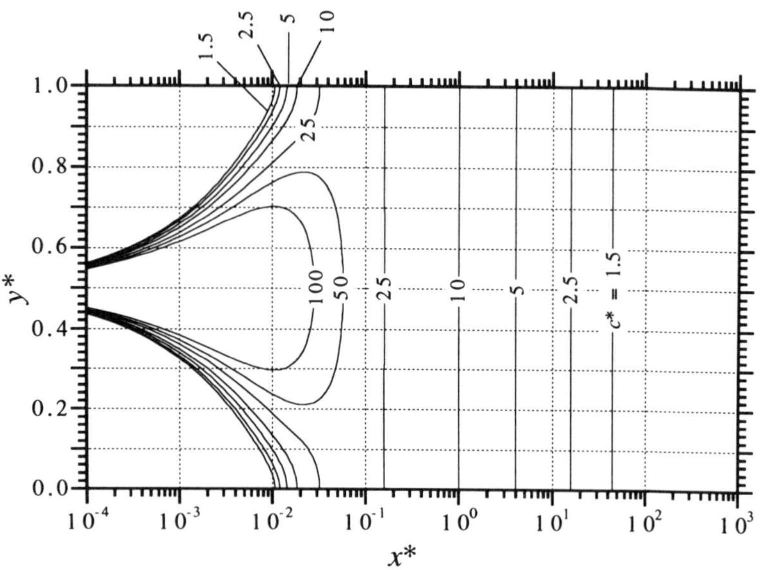

Figure 2.8. Concentrations downstream from a steady point source predicted using the constant-coefficient mixing model (equation 4.49). The source is located at $y/h = 0.5$ and $z/b = 0.5$ in a channel where $b/h = 50$ and $e_z/e_y = 2$. Concentrations are shown on a vertical plane located at $z/b = 0.5$ which passes through the source

plane passing through the tracer source, which is located at the mid-point of a channel with a width/depth ratio $b/h = 50$. The ratio of transverse/vertical diffusion coefficient is taken as $e_z/e_y = 2$. Clearly higher concentrations are found immediately downstream from a point source than below a transverse line source with the same mass flow-rate. The vertical mixing distance is, however, the same as that given in equation (2.32) although at $x^* = 0.137$, $c^* = 25$ in the centre of the plume because transverse mixing is incomplete. At distances greater than the vertical mixing distance, concentrations are well-mixed vertically but continue to be affected by transverse mixing until they eventually become well-mixed across the channel. Figure 2.8 reinforces the point made previously in Section 2.4 that vertical mixing occurs more rapidly than transverse mixing.

It is possible to draw a whole family of contour plots like Figure 2.8 for sources at different points across the channel and/or at different depths. These plots differ slightly depending on the channel aspect ratio and the ratio of transverse and vertical diffusivity specified. The same general pattern emerges, however, with significant vertical concentration gradients only occurring at $x^* < 0.5$.

2.8 VERTICAL MIXING FROM A NON-STEADY POINT SOURCE

We now discuss briefly the difficult problem of vertical mixing downstream from a non-steady discharge. The three-dimensional mixing equation is

$$\frac{\partial c}{\partial t} + u_x \frac{\partial c}{\partial x} + u_y \frac{\partial c}{\partial y} + u_z \frac{\partial c}{\partial z} = \frac{\partial}{\partial x}\left(e_x \frac{\partial c}{\partial x}\right) + \frac{\partial}{\partial y}\left(e_y \frac{\partial c}{\partial y}\right) + \frac{\partial}{\partial z}\left(e_z \frac{\partial c}{\partial z}\right) \quad (1.64)$$

As we are dealing with only a short reach of the channel in the near-field we will align the x axis with the main flow direction and assume that the channel is straight and uniform. This allows us to set $u_y = u_z = 0$ leaving

$$\frac{\partial c}{\partial t} + u_x \frac{\partial c}{\partial x} = \frac{\partial}{\partial x}\left(e_x \frac{\partial c}{\partial x}\right) + \frac{\partial}{\partial y}\left(e_y \frac{\partial c}{\partial y}\right) + \frac{\partial}{\partial z}\left(e_z \frac{\partial c}{\partial z}\right) \quad (2.50)$$

Equation (2.50) indicates that downstream from a non-steady point source we must consider diffusion in all three coordinate directions. Even in a uniform channel the velocity $u_x(y)$ varies with depth (e.g. logarithmically). While the vertical and longitudinal diffusivities e_y and e_x vary parabolically with depth, as can be seen from equation (2.18). The transverse diffusivity e_z may also vary with depth although (as is discussed in more detail in Chapter 3) the form of its variation is not known precisely. A complete analysis of equation (2.50) in which the source is unsteady while the velocity and diffusivities all vary with depth has not been made. An approximate analysis can be made, however, if we assume that u_x, e_x, e_y and e_z are constant. We can then adapt the methods outlined in Section 2.6.

Assuming constant-coefficients, equation (2.50) has the analytical solution for instantaneous input of mass M occurring at time $t = 0$ from a point source located at $x = 0$, $y = y_o$, $z = z_o$ in an unbounded channel

$$c(x, y, z, t) = M \frac{\exp\left(-\frac{(x - u_x t)^2}{4e_x t}\right)}{\sqrt{4\pi e_x t}} \frac{\exp\left(-\frac{(y - y_o)^2}{4e_y t}\right)}{\sqrt{4\pi e_y t}} \frac{\exp\left(-\frac{(z - z_o)^2}{4e_z t}\right)}{\sqrt{4\pi e_z t}} \quad (2.51)$$

The no-flux boundary conditions at $y = 0$ and h and $z = 0$ and b can again be accommodated using the method of images. It is straightforward to adapt equation (2.49). Equation (2.51) gives a tolerably accurate prediction of concentrations provided the tracer cloud lies in mid-water near the centre of the channel. Once the tracer cloud approaches the bed our assumption of uniform velocity becomes invalid. Tracer near the bed is carried downstream more slowly than tracer near mid-depth and as a result the tracer cloud is distorted. This distortion increases the vertical concentration gradient, which in turn increases the rate of vertical diffusion. Predictions made downstream from a non-steady source using a constant-coefficient model and inserting realistic values of the vertical and longitudinal diffusivity would be significantly in error. (Note that if the source is steady this distortion has much less effect on concentrations and as a first approximation the variation in velocity can be neglected. We saw in Section 2.6 that for a steady transverse line source the constant-coefficient model is reasonably accurate.) The only way to match observed tracer concentrations is by augmenting the vertical and longitudinal diffusivities to account for the dispersion which arises from vertical velocity shear. (The same argument applies if there is significant tranverse velocity shear). A complete analysis of this problem can only be made using equation (2.50) and allowing the velocity and diffusivities to vary spatially. Nokes (1986) and McNulty (1983) have solved this problem when transverse concentration gradients are negligibly small as discussed in Section 2.6, but a complete analysis of both vertical and transverse mixing below a non-steady point source has not been made. One reason is that the spatial variation of the transverse diffusivity is not known theoretically (as is discussed in more detail in Chapter 3).

2.9 EFFECTS OF BUOYANCY ON VERTICAL MIXING

In this section we discuss the effects of buoyancy on vertical mixing in rivers. One situation where we need to consider buoyancy is when warm water (e.g. cooling water from a power station) is discharged to a river. In this problem buoyancy effects are important only in the near-field because bed friction promotes rapid vertical mixing. Another situation where buoyancy is important is in tidal rivers where density stratification occurs with saline (more dense) water underlying fresh (less dense) water. In both situations vertical mixing is inhibited because turbulence must perform additional work to lift dense fluid and depress light fluid. Stratification also damps the secondary currents which enhance vertical mixing. Thus the overall effect of stratification is to reduce the rate of vertical mixing and in some situations this reduction can be substantial (see, for example, Table 2.1).

Our understanding of vertical mixing in stratified flows is far from complete and it is impossible to make accurate predictions of vertical diffusivities. Methods are outlined in the following, however, which can be used to make approximate calculations. For a more complete exposition of the topic the reader is referred to texts such as Turner (1973) and Officer (1976).

2.9.1 Vertical Mixing in Stratified Flow

In the situation where the density structure and velocity distribution are known, the strength of the stratification can be quantified by the local gradient Richardson number Ri which is defined

$$Ri = -g \frac{\partial \rho}{\partial y} \Big/ \rho \left(\frac{\partial u}{\partial y} \right)^2 \tag{2.52}$$

where g = acceleration of gravity; $\rho(y)$ = density; $u(y)$ = longitudinal velocity; and y = vertical distance. The Richardson number is the ratio of the buoyancy forces to the inertial forces acting in the flow. $Ri = 0$ in non-stratified flows and increases with the strength of the stratification. In fast flowing river channels typically $Ri < 1$ but in stratified flows $Ri \approx 5$–10. Flows in which $Ri > 10$ exhibit a sharp density interface and vertical mixing is principally the result of instabilities in internal waves travelling along the interface rather than eddy diffusion associated with bottom shear. We will not consider strongly stratified flows as they are rarely encountered in rivers.

Both theoretical arguments and empirical analysis suggest that stratification reduces the vertical diffusion coefficient (Officer 1976, Bowden 1983). This reduction has been quantified by experimental studies and has the form

$$e_s = e_o(1 + aRi)^b \tag{2.53}$$

where e_o and e_s = vertical diffusion coefficient in non-stratified and stratified conditions, respectively, a and b = constants estimated variously as shown in Table 2.4. It is common practice to use tidally averaged values of $\partial \rho / \partial z$ and $(\partial u / \partial z)^2$ to calculate Ri (Fischer 1976). Figure 2.9 shows how the vertical diffusivity varies with Richardson number as predicted by equation (2.53). Weak stratification ($Ri = 0.20$) reduces the diffusion coefficient by about 40%, whereas strong stratification ($Ri = 1.0$) reduces it by about 80%.

The local Richardson number varies with depth and this variation is important in situations where the density changes abruptly (e.g. in stratified lakes). In most tidal rivers, however, bed friction is sufficient to mix the salt and fresh

Table 2.4. Coefficients describing the effects of stratification on the vertical diffusion coefficient (see equation 2.53)

Reference	a	b
Munk and Anderson (1948)	10/3	−3/2
Officer (1976)	1	−2
Schiller and Sayre (1975)	10/3	−5/4

Note: Richardson number is calculated using tidally averaged data.

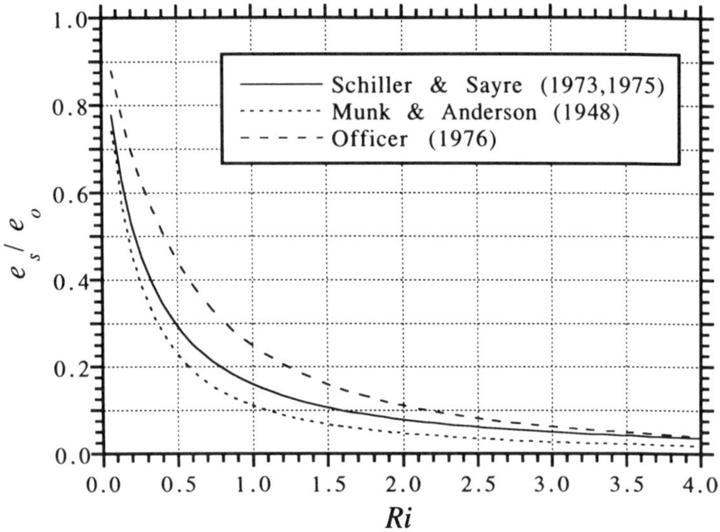

Figure 2.9. Variation of vertical diffusivity with Richardson number in stratified flow predicted using equation (2.53) for three sets of empirical coefficients

waters so that density increases monotonically with depth. It is common practice in such flows to use the bulk Richardson number, which assumes a linear variation of velocity and density with depth (Bowden 1983)

$$Ri_b = -gh\,\frac{\Delta\rho}{\rho(\Delta u)^2} \tag{2.54}$$

where Ri_b = bulk Richardson number; $\Delta\rho$ and Δu = difference between surface and bottom density and velocity, respectively; ρ = depth-averaged density; and h = depth.

Example 2.3

Lucas Creek is one of several small tidal rivers which drain into the Upper Waitemata Harbour, Auckland, New Zealand. Vertical velocity and salinity profiles were measured at a cross-section near the mouth of Lucas Creek over one tidal cycle during a field experiment to study tidal flushing. At high tide the salinity was fairly uniform vertically and close to that of adjacent coastal waters (29–30 ppt), but at low tide the fresh water inflow caused partial stratification as shown in Figure 2.10. At low tide vertical mixing is likely to be low and an estimate was required of how quickly tracer released at mid-depth was likely to mix vertically.

The channel has a mean depth of 2 m at low tide and the slope is

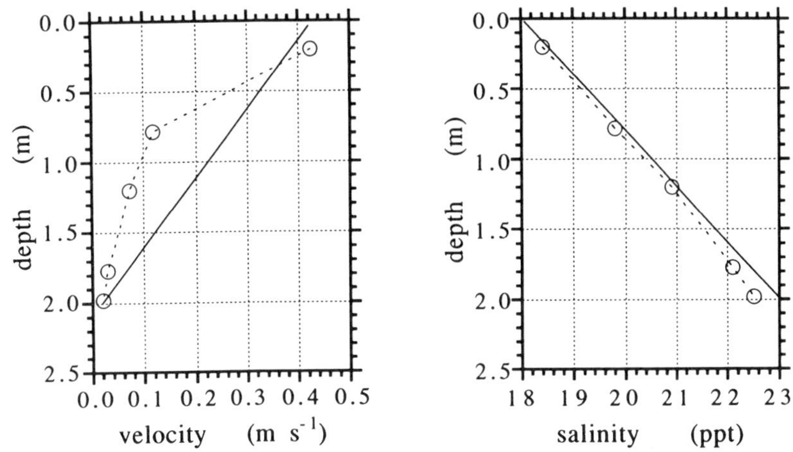

Figure 2.10. Velocity and salinity profile measured in Lucas Creek 30 minutes before low tide. The coastal water salinity at the time was $S = 29.7$ ppt. Solid lines were fitted by eye and used to estimate surface and bottom values

1.0×10^{-4}. Thus the shear velocity is $u^* = 0.044\ \mathrm{m\,s^{-1}}$. If the channel is unstratified, then from equation (2.19) $e_y/hu^* = 0.067$ and hence $e_y = 0.0058\ \mathrm{m^2\,s^{-1}}$ ($58\ \mathrm{cm^2\,s^{-1}}$).

The vertical salinity and velocity profiles measured 30 minutes either side of low water are shown in Figure 2.10. The salinity and velocity gradients were approximated by straight lines (as shown) and the surface and bottom values of velocity and salinity are summarised in Table 2.5. Assuming that the water temperature averaged 20°C the densities and density gradients

Table 2.5. Velocity and salinity measurements in Lucas Creek and the estimated Richardson number

Mean velocity (V_x)	$0.12\ \mathrm{m\,s^{-1}}$
Surface velocity (u_{top})	$0.42\ \mathrm{m\,s^{-1}}$
Bottom velocity (u_{bottom})	$0.02\ \mathrm{m\,s^{-1}}$
Mean water depth (h)	2.0 m
Velocity difference (Δu)	$0.40\ \mathrm{m\,s^{-1}}$
Surface salinity (S_{top})	18 ppt
Bottom salinity (S_{bottom})	23 ppt
Mean water temperature (T)	20°C
Surface density (ρ_{top})	$1011.9\ \mathrm{kg\,m^{-3}}$
Bottom density (ρ_{bottom})	$1015.7\ \mathrm{kg\,m^{-3}}$
Mean density ($\rho_{average}$)	$1013.8\ \mathrm{kg\,m^{-3}}$
Density difference ($\Delta\rho$)	-3.8
Bulk Richardson number (Ri_b)	0.46

Table 2.6. Variation of density with salinity and temperature. Reproduced with the permission of the author from Millero *et al* (1976)

Salinity (ppt)	Temperature (°C)									
	0	2	4	6	8	10	15	20	25	30
0	−13	−3	−0	−5	−16	−32	−87	−177	−293	−433
5	397	403	402	394	381	362	301	207	87	−57
10	801	804	799	788	772	750	685	586	462	315
15	1204	1204	1195	1181	1162	1138	1067	964	836	686
20	1607	1603	1589	1573	1551	1532	1450	1342	1210	1057
25	2008	2001	1988	1970	1947	1920	1832	1720	1585	1428
30	2410	2400	2384	2363	2340	2308	2215	2098	1960	1801
32	2571	2560	2543	2521	2494	2464	2364	2250	2110	1950
34	2732	2719	2701	2678	2651	2619	2522	2402	2261	2100
36	2893	2879	2860	2836	2808	2775	2676	2554	2412	2250
38	3055	3040	3019	2994	2965	2931	2830	2707	2563	2400
40	3216	3200	3179	3153	3122	3088	2985	2860	2714	2550
42	3377	3361	3337	3310	3279	3243	3138	3011	2864	2700

Note: Values are density anomoly σ. Density is given by $\rho = 1000 + \sigma/100\ \mathrm{kg\,m^{-3}}$.

can be estimated from Table 2.6. From the calculations summarised in Table 2.5, the bulk Richardson number is 0.46. Equation (2.53) and Table 2.4 indicate that the effects of stratification are to reduce the vertical diffusivity to

$$e_s = 15\ \mathrm{cm^2\,s^{-1}}\ \text{(Munk and Anderson 1948)}$$
$$e_s = 27\ \mathrm{cm^2\,s^{-1}}\ \text{(Officer 1976)}$$
$$e_s = 18\ \mathrm{cm^2\,s^{-1}}\ \text{(Schiller and Sayre 1973, 1975)}$$

From equation (2.32) the vertical mixing time for a source at mid-depth is:

$$L_y/u_x = 184\ \mathrm{s}\ \text{(Munk and Anderson 1948)}$$
$$L_y/u_x = 98\ \mathrm{s}\ \text{(Officer 1976)}$$
$$L_y/u_x = 148\ \mathrm{s}\ \text{(Schiller and Sayre 1973, 1975)}$$

Note that if the channel is unstratified

$$L_y/u_x = 46\ \mathrm{s}$$

In this example stratification reduces the vertical diffusivity and increases the vertical mixing time by a factor of 2–4. Lucas Creek is only stratified for about 30 minutes either side of low water and these calculations suggest that tracer is still likely to become well-mixed within this period.

2.9.2 Non-neutrally Buoyant Tracer

When a large volume of buoyant effluent is released into a channel it has a significant impact on the velocity and density structure. The most common example of this problem is the discharge of cooling water into a river from a thermal power station. In these circumstances the Richardson number is not known in advance but is determined by the inflow itself. Schiller investigated this problem in the laboratory (Schiller and Sayre 1973, 1975) and found that the vertical mixing distance (i.e. the distance required to achieve complete vertical mixing) was related to the initial densimetric Froude number

$$Fd_o = \frac{V_x}{\sqrt{gH \dfrac{(\rho_a - \rho_e)}{\rho_a}}} \tag{2.55}$$

where V_x = mean (cross-sectional averaged) river velocity; H = mean water depth; ρ_e = density of the cooling water; and ρ_a = ambient river water density (measured above the hot water inflow). Where $Fd_o < 1$ stable stratification can be expected whereas for $Fd_o > 15$ tracer is effectively of neutral buoyancy. The densimetric Froude number is akin to the square root of the reciprocal of the bulk Richardson number.

The vertical mixing distance varies inversely with the initial densimetric Froude number. Thus for small values of Fd_o (highly stratified flows) the mixing distance is large, but it decreases with increasing Fd_o to approach that for neutrally buoyant tracer (Figure 2.11). Because of the difficulties of measuring very small temperature differences, Schiller and Sayre found it more convenient to define the vertical mixing distance as the distance at which

$$\frac{c_{max} - c_{min}}{C_m} = 0.2 \tag{2.56}$$

where c_{max} and c_{min} = maximum and minimum concentrations over the depth and C_m = fully mixed concentration. This is very close to

$$P_m = \frac{c_{min}}{c_{max}} = 0.80 \tag{2.57}$$

The results in Figure 2.11 are strictly applicable only to situations where the rate of heat loss from the channel is similar to that in the experimental set-up (surface heat loss rate of the order of 40 $W\,m^{-2}\,{}^\circ C^{-1}$). Where the heat loss is significantly higher, as may be the case in natural channels, stratification will decrease faster and hence the mixing distance will be smaller than indicated.

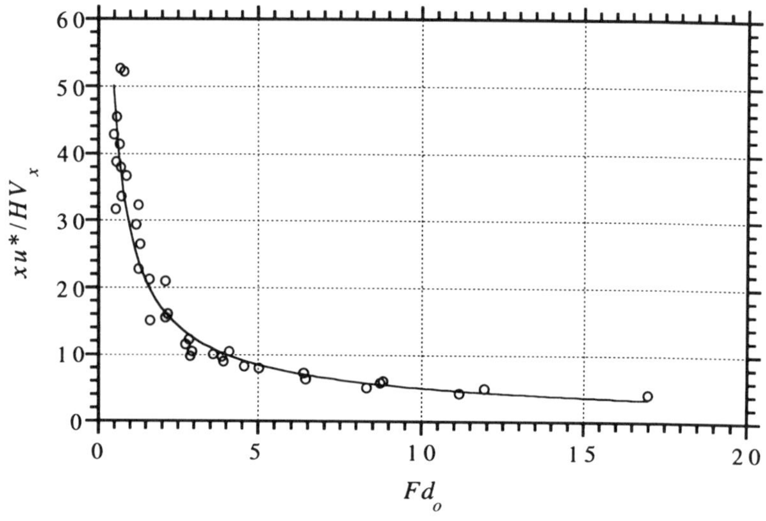

Figure 2.11. Dependence of the mixing distance of a buoyant tracer with the initial densimetric Froude number. Reproduced with the permission of the Iowa Institute of Hydraulic Research from Schiller and Sayre (1973)

Example 2.4

The Waikato River in the vicinity of the Huntly power station, 35 km downstream from Hamilton, New Zealand, has a mean depth of 2.0 m, a mean velocity of 0.65 $\mathrm{m\,s^{-1}}$ and a shear velocity of 0.05 $\mathrm{m\,s^{-1}}$. Cooling water is discharged from a bankside outfall near the surface. The average cooling water flow rate and excess temperature are 32 $\mathrm{m^3\,s^{-1}}$ and 8.4°C, respectively. The ambient river temperature in winter is 15°C. An estimate of the vertical mixing distance can be made using Figure 2.11.

For an unstratified flow the depth-averaged diffusivity from equation (2.19) is $e_y = 0.0067\ \mathrm{m^2\,s^{-1}}$. The density anomaly at the ambient river water temperature of 15°C for zero salinity is from Table 2.6 $\sigma = -87$ giving $\rho_a = 999.13\ \mathrm{kg\,m^{-3}}$. At a temperature of 23.4°C and salinity zero by interpolation $\sigma = -256$ giving $\rho_e = 997.44\ \mathrm{kg\,m^{-3}}$. At the point of discharge $\Delta\rho = \rho_a - \rho_e = +1.69\ \mathrm{kg\,m^{-3}}$.

From equation (2.48) the initial densimetric Froude number is $Fd_o = 3.57$. As $Fd_o > 1$ we would not expect a well-defined stratified flow to occur (Schiller and Sayre 1975). From Figure 2.11 for $Fd_o = 3.6$, $xu^*/hu_x = 10$. Thus $x = 260$ m. By comparison, for a neutrally buoyant effluent Figure 2.2 indicates that for a source at the surface $xe_y/u_xh^2 = 0.3$, $x = 110$ m.

According to these calculations stratification roughly doubles the vertical mixing distance. This analysis is only approximate because turbulence is strong near the Huntly outfall as a result of jet momentum and this may well promote more rapid vertical mixing than was present in Schiller and Sayre's laboratory channel. In addition, Schiller and Sayre's experiments used a transverse line source whereas at Huntly the outfall is at the bank. Field tests below the Huntly outfall have in fact shown that there are detectable vertical temperature gradients within 300 m of the outfall, but that at a site 450 m downstream vertical temperature gradients are small. This is in reasonable agreement with the calculations presented here.

NOTATION

a, b	empirical coefficients
b	channel width
c	tracer concentration
$\langle c \rangle$	ensemble mean tracer concentration
c_{min}, c_{max}	minimum and maximum concentration over the depth
c_u	predicted tracer concentration in an unbounded channel
c^*	non-dimensional tracer concentration, c/C_m
C_m	fully mixed tracer concentration
d	port diameter
e	eddy diffusivity
e_o	eddy diffusivity in non-stratified flow
e_s	eddy diffusivity in stratified flow
e_x, e_y, e_z	eddy diffusivities in the x, y, z directions
$\overline{e_y}$	depth averaged value of e_y
f	Darcy friction coefficient
Fd_o	initial densimetric Froude number
g	gravitational acceleration
h	local water depth
H	mean water depth
L, L_y, L_z	length scales for mixing
L_m	Prandtl's mixing length
L_o	outfall length
L_y	distances for complete vertical mixing
m	tracer mass inflow rate
M	total tracer mass injected
n, m	number of terms
n	number of ports
P_m	degree of mixing, c_{min}/c_{max}
q_e	effluent flow rate
Q	total river flow
Ri	gradient Richardson number
Ri_b	bulk Richardson number

S	salinity
Sc	Schmidt number
T, T_z, T_y	time-scales for mixing
T	water temperature
u_t	a turbulent velocity
u_x, u_y, u_z	velocities in the x, y and z direction
u'_x, u'_y, u'_z	turbulent velocity fluctuations
$\langle u_x \rangle$	ensemble mean velocity in the x direction
u^*	shear velocity
Δu	velocity difference over the depth
V_x	cross-sectional averaged velocity
v_x	depth-averaged velocity
$\langle v_x \rangle$	ensemble mean depth-averaged velocity
x, y, z	longitudinal, vertical and transverse distance
x^*, x'	non-dimensional distances
x_o, y_o, z_o	source location
y^*, y'	non-dimensional depths
y	height above bed
y_0	height at which velocity is zero
κ	von Karman's constant
ν	kinematic viscosity
ν_t	eddy viscosity
ρ	density of water
$\bar{\rho}$	depth averaged density
ρ_a	density of ambient river water
ρ_e	density of cooling water
$\Delta\rho$	density difference over the depth
σ	density anomaly
τ	viscous shear stress
τ_o	bed shear stress
τ_t	turbulent shear stress (the Reynolds stress)
χ	non-dimensional velocity variation with depth
ψ	non-dimensional diffusivity variation with depth

Chapter 3
Transverse Mixing

3.1 INTRODUCTION

In Chapter 2 it was shown that in most rivers neutrally buoyant tracer becomes well-mixed vertically long before it becomes well-mixed transversely. In this chapter we focus attention on the mid-field and the problem of transverse mixing. In the mid-field it is convenient to neglect any depth variations in tracer concentration and to consider only transverse and longitudinal changes of the depth-averaged tracer concentration. The three-dimensional mixing equation is integrated over the depth as outlined in Chapter 1 so that the problem becomes two-dimensional. Figure 1.9 and Table 1.2 illustrate these points.

Transverse mixing is arguably more important in water quality management than either vertical or longitudinal mixing, especially when dealing with the discharge of wastes from point sources or the mixing of tributary inflows. In such problems vertical mixing occurs rapidly and is only important very close to the source, whereas longitudinal dispersion is only important if the source is rapidly varying. It is surprising, therefore, to find that there is much less published research on transverse mixing than on longitudinal dispersion. Part of the reason is that the theory of transverse mixing is far from complete, in marked contrast with vertical diffusion for which there is a sound theoretical basis as discussed in Chapter 2. When faced with transverse mixing problems the practitioner is forced to use semi-empirical techniques.

3.2 SUMMARY

A summary of the main conclusions reached in this chapter is given at the outset to help the reader appreciate where the ensuing discussions are leading and to act as a checklist subsequently.

1. Prandtl's mixing length hypothesis (which provides estimates of the vertical and longitudinal eddy diffusivity in plane shear flow) is inadequate for predicting the transverse eddy diffusivity e_z.

2. Laboratory studies indicate that $e_z/hu^* \approx 0.13$ in plane shear flow. Plane shear flow only occurs where the channel is straight, uniform and nearly rectangular.
3. There is some controversy over the relative importance for transverse mixing of turbulent diffusion and secondary currents in straight channels. Further theoretical and experimental work on this subject is needed.
4. Experimental data from irregular but only gently meandering natural channels indicate that $0.3 < (e_z + k_z)/hu^* < 0.9$, where k_z is the transverse dispersion coefficient arising from depth variations of velocity and secondary currents.
5. Channel curvature increases the rate of transverse mixing dramatically because it induces strong secondary currents and typically at bends $1 < (e_z + k_z)/hu^* < 3$. As yet a relationship has not been found between transverse mixing rates at bends and bulk channel parameters which fits both laboratory and field data.
6. An approximate constant-coefficient model is a valuable tool for understanding the behaviour of tracer plumes and for making initial calculations of tracer concentrations, the degree of mixing and the transverse mixing distance. Nomographs are presented to illustrate plume behaviour and to facilitate problem solving. A useful rule of thumb is that tracer becomes mixed across the channel 100–300 channel widths downstream from a point source near mid-stream.
7. The constant-coefficient model has a number of deficiencies but a heuristic modification can be used to make improved localised predictions. At present there is insufficient understanding of velocity, secondary currents and eddy diffusivities in natural channels to develop a variable-coefficient model for transverse mixing.
8. Streamtube models are effective tools for investigating transverse mixing in steady river flow, especially if the tracer source is also steady. A simplified streamtube model (which makes use of earlier nomographs) can be used to calculate tracer concentrations in uniform channels.
9. Transverse mixing downstream from a source of finite length, or when the tracer concentration profile across the river is known at the upstream boundary of the study reach, can be addressed with the constant-coefficient model using the superposition principle.
10. Two-dimensional (depth-averaged) numerical models are powerful tools which allow for non-steady sources and flow and non-uniform depth, width, velocity and dispersion coefficients. They are time consuming to calibrate, test and use properly and they require extensive bathymetric, flow and tracer data to do so accurately. The user may encounter numerical problems in some situations and find it difficult to achieve a unique model calibration.
11. Buoyancy can affect the rate of transverse mixing in river channels especially close to the source. In the mid-field, where vertical concentration

and temperature gradients are small, the effects are only second order and can be handled by making small changes to the transverse dispersion coefficient. Buoyant tracer released from a bankside outfall (e.g. a discharge canal) moves away from the bank because of buoyant spreading and then reattaches to the bank. Experimental results can be used to estimate the initial dilution.

12. Close to strongly jetted outfalls, momentum affects the rate of mixing. Two analytical models can be used to predict the amount of entrainment and initial dilution in the near-field.

3.3 MECHANISMS CAUSING TRANSVERSE MIXING

3.3.1 The Depth-averaged Mixing Equation

In Chapter 1 a detailed derivation was given of the depth-averaged mixing equation which is restated here in rectangular Cartesian coordinates

$$\frac{\partial s}{\partial t}+\frac{\partial}{\partial x}(hv_x s)+\frac{\partial}{\partial z}(hv_z s)=\frac{\partial}{\partial x}\left[h(e_x+k_x)\frac{\partial s}{\partial x}\right]+\frac{\partial}{\partial z}\left[h(e_z+k_z)\frac{\partial s}{\partial z}\right] \quad (3.1)$$

where x and z = longitudinal and transverse distance; s = depth-averaged tracer concentration; h = local depth; v_x and v_z = depth-averaged velocities in the x and z directions; e_x and e_z = turbulent diffusion coefficients (eddy diffusivities) in the x and z directions; k_x = longitudinal dispersion coefficient (which accounts for the effects of vertical variations of longitudinal velocity); and k_z = transverse dispersion coefficient (which accounts for the effects of vertical variations of transverse velocity).

In equation (3.1) the rate of transverse mixing is determined by two processes: turbulent diffusion (quantified by the diffusivity e_z) and vertical variations in the transverse velocity (quantified by the dispersion coefficient k_z). It is important to remember that these two coefficients arise from distinct processes, even though when we come to solve equation (3.1) they are usually combined into a single mixing coefficient for convenience. In the remainder of this chapter we shall use e_z where transverse dispersion is negligibly small, $(e_z + k_z)$ where both processes are important and k_z where transverse dispersion is significantly larger than diffusion.

3.3.2 Transverse Turbulent Diffusion

In plane shear flow turbulence is generated by vertical velocity shear caused by bed friction. As discussed in Chapter 2, Prandtl's mixing length hypothesis provides theoretical predictions of the vertical eddy diffusivity in plane shear flow which closely match experimental results. Unfortunately, there is as yet no theoretical basis for predicting the transverse diffusivity. Prandtl's mixing length hypothesis relates the vertical diffusivity to the vertical velocity gradient

$$e_y = L_{\mathrm{m}}^2 \left| \frac{\partial \langle u_x \rangle}{\partial y} \right| \tag{3.2}$$

where y = vertical distance; e_y = vertical diffusivity; $\langle u_x \rangle$ = ensemble mean longitudinal velocity; and L_{m} = Prandtl's mixing length (a turbulence length scale). By considering the balance of shear stress on an element of fluid it can be shown (Hinze 1972) that

$$e_x = e_y \tag{3.3}$$

where e_x = longitudinal diffusivity. This leaves only the transverse eddy diffusivity e_z to determine. It is tempting to draw an analogy with equation (3.2) and write

$$e_z = L_{\mathrm{t}}^2 \left| \frac{\partial \langle u_x \rangle}{\partial z} \right| \tag{3.4}$$

where z = transverse distance and L_{t} = a transverse turbulent length scale. This equation is inadequate because in plane shear flow the mean velocity $\langle u_x \rangle$ does not vary across the channel which implies that the diffusivity vanishes. This is not what is found experimentally (Rodi 1980). Equation (3.4) could perhaps be used near the banks where there is strong transverse velocity shear, but this would still leave the problem of estimating e_z in the central portion of the flow. Prandtl's mixing length hypothesis cannot do this.

Although equation (3.4) is inadequate, the first part of Prandtl's mixing length hypothesis relates the diffusivity to a turbulent velocity and a length scale (Tennekes and Lumley 1972). Applying this to transverse mixing

$$e_z = L_{\mathrm{t}} u_{\mathrm{t}} \tag{3.5}$$

A convenient turbulent velocity is

$$u_{\mathrm{t}} = \sqrt{\langle u_x'^2 \rangle} \tag{3.6a}$$

where u_x' = turbulent velocity fluctuation

$$u_x = \langle u_x \rangle + u_x' \tag{3.6b}$$

u_x = velocity; and $\langle u_x \rangle$ = ensemble mean velocity. We must still determine the length scale L_{t}. This cannot be done theoretically because we do not know how vertical velocity shear drives transverse eddies and further theoretical work is needed to address this problem. At present, therefore, it is not possible to predict theoretically the value of the transverse eddy diffusivity and it must be determined experimentally.

We saw in Chapter 2 that it was appropriate to scale e_y using depth h and shear velocity u^*. This is sensible in plane shear flow because turbulence is generated by vertical velocity shear which arises as a result of bed friction. The shear velocity is a measure of the bed friction which generates the turbulent eddies and the depth determines the length scale of the largest vertical eddies which transport most of the tracer mass. The same length and velocity scales are commonly used to non-dimensionalise the tranverse diffusivity. Thus in equation (3.5) it is commonly assumed that $L_t = h$ and $u_t = u^*$. In Section 3.4 experimental results are summarised using this assumption. We do not have a quantitative understanding of how vertical turbulent eddies generated by bed friction undergo rotation and become transverse eddies. In particular, the maximum length scale of transverse eddies is not known, although it is no longer restricted to the water depth. The upper bound is clearly the channel width, but in a shallow river channel friction ensures that transverse eddies are rarely as large as this. In Section 3.4 we also use the channel width as a length scale when examining published transverse mixing coefficients.

3.3.3 Transverse Dispersion

In equation (3.1) the transverse dispersion coefficient k_z quantifies the transverse mixing which occurs as a result of depth variations in the transverse velocity. It is important to distinguish between the effects of two-dimensional meander currents and three-dimensional secondary currents.

3.3.3.1 Meander currents

In natural channels the main flow (the thalweg) tends to meander within the channel. When such meander currents are uniform with depth, they move tracer bodily within the channel but do not cause any concentration changes. Figure 3.1 shows three cross-sections in a non-uniform river channel in which there is a steady release of tracer near the left bank. At section A tracer is contained within the flow q and water on either side of the plume contains no tracer. The plume is near the true left bank (i.e. the bank to the left of an observer looking downstream) at section A. At section B the channel is shallow near the left bank and because of bed friction the velocity is lower. As a result the main flow moves towards the right bank, carrying the plume bodily across the channel. At section C the main channel returns close to the left bank carrying the plume with it. Refering to equation (3.1) there is a positive depth-averaged transverse velocity v_z between sections A and B and a negative velocity between sections B and C. Some transverse mixing occurs as the tracer cloud is carried between A and C, which is shown as an exchange flow of e between each section. If the transverse velocity is uniform over the depth then this mixing would arise solely from turbulent diffusion and would

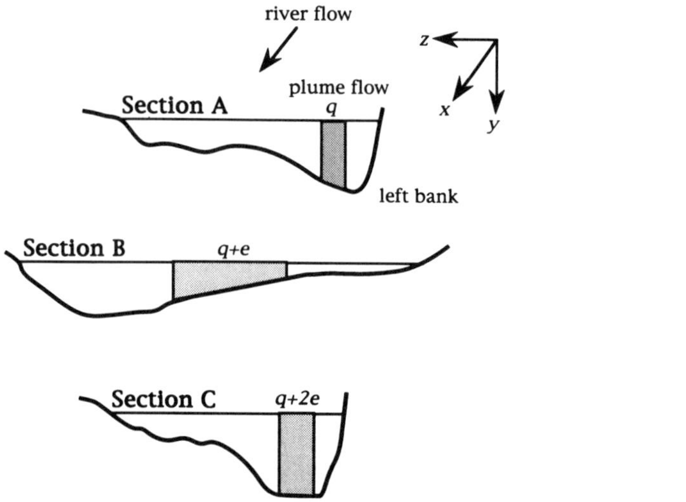

Figure 3.1. Schematic representation of three cross-sections in a non-uniform channel. Meander currents carry the tracer cloud bodily across the channel but do not cause any changes of concentration. Transverse mixing occurs as a result of turbulent diffusion and depth variation in the meander currents

be very small. In practice the transverse velocity has a similar depth variation to the longitudinal velocity (viz. a logarithmic variation) and is lower near the bed. Consequently there is a tendency for tracer to move across the channel faster near the surface than near the bed. This increases the vertical concentration gradients and promotes vertical mixing. The net result is an accelerated rate of transverse spreading of the tracer cloud termed transverse dispersion.

It is also worth noting that changes of bathymetry cause the width of the plume to vary cyclically along the channel. In Figure 3.1 the plume expands between sections A and B because the channel becomes shallower and wider while it contracts between sections B and C because the channel becomes deeper and narrower. Thus plume width has an underlying increasing trend resulting from transverse mixing, but in a non-uniform channel there is a cyclical variation superimposed by changes in channel bathymetry. By comparison, plume flow (i.e. the total flow in which there are significant tracer concentrations) increases monotonically with distance downstream.

Although two-dimensional meander currents do not contribute to transverse mixing it is very important to take account of them when analysing tracer data in an irregular natural channel. If equation (3.1) is used to analyse tracer data in an irregular channel and the transverse velocity v_z is erroneously neglected, then the resulting estimates of the transverse dispersion coefficient may be seriously in error. This point was first elucidated by Holley *et al* (1972). In extreme cases estimated values of k_z can be negative in regions of converging flow (El-Hadi *et al* 1984) and very large in regions of diverging flow.

3.3.3.2 *Secondary currents*

At some places in the channel the transverse velocity is highly non-uniform over the depth. For example, helical secondary currents occur at bends with water moving towards the outside of the bend at the surface and towards the inside of the bend along the bed. The reasons why secondary currents occur are explained by Henderson (1966). Secondary currents cause tracers to move in opposite directions at different depths and this behaviour greatly increases the rate of tranverse mixing. The mixing caused by secondary currents is distinct from transverse turbulent diffusion and is termed transverse dispersion. An analogy can be drawn between the effects of vertical variations in transverse velocity on transverse mixing and the effects of transverse variations of longitudinal velocity on longitudinal mixing. The latter were analysed by Taylor (1954), who showed that longitudinal dispersion obeys a gradient-diffusion law (i.e. Fick's law) provided sufficient time has elapsed since the introduction of the tracer into the fluid (Section 1.8.4 outlines Taylor's analysis of dispersion in turbulent shear flow). Taylor argued that sufficient time has elapsed once a tracer particle has sampled the entire flow so that an equilibrium becomes established between the effects of velocity shear and turbulent diffusion. The time taken to sample the entire flow is closely related to the time taken for tracer to become well-mixed vertically. We saw in Chapter 2 that a useful rule of thumb is that tracer becomes well-mixed vertically within a longitudinal distance of 50 times the water depth.

Taylor's arguments have been applied (in a much less rigorous way) to transverse mixing and the assumption made that transverse dispersion can also be modelled by a gradient-diffusion law provided sufficient time has elapsed since the introduction of the tracer for an equilibrium to become established between vertical velocity shear in the transverse velocity and vertical turbulent diffusion. Holly (1985) points out, however, that there are difficulties in applying Taylor's analysis to transverse mixing in natural river channels. Secondary currents are continuously varying along an irregular channel so that sufficient time may not elapse for an equilibrium to become established between velocity shear and turbulent diffusion before the tracer moves into a different hydraulic regime. Certainly one can expect changes in bathymetry over distances along the channel of 50 times the water depth. Intuitively, however, one feels that during its passage along a river reach each tracer particle will sample the entire range of conditions associated with secondary currents. If so, then it is not unreasonable to expect that an equilibrium of sorts will become established between turbulent diffusion and velocity shear, as is required by Taylor's analysis. Certainly there is experimental evidence that the transverse distribution of tracer in natural channels is often approximately Gaussian and this supports the use of the gradient-diffusion model (Yotsukura and Cobb 1972, Yotsukura and Sayre 1976).

3.4 PREDICTING THE RATE OF TRANSVERSE MIXING

3.4.1 Transverse Turbulent Diffusion Coefficient

There have been several experimental studies of transverse mixing in straight rectangular laboratory channels. These provide estimates of the transverse diffusivity e_z provided the flow does not depart significantly from plane shear flow. Figure 3.2 and Table 3.1 summarise the results from a number of studies and it can be seen that the transverse diffusivity lies in the range

$$0.10 < \frac{e_z}{hu^*} < 0.26 \tag{3.7}$$

It is likely that secondary currents were present in some of these experiments, which would increase the rate of transverse mixing in channels. A lower-bound

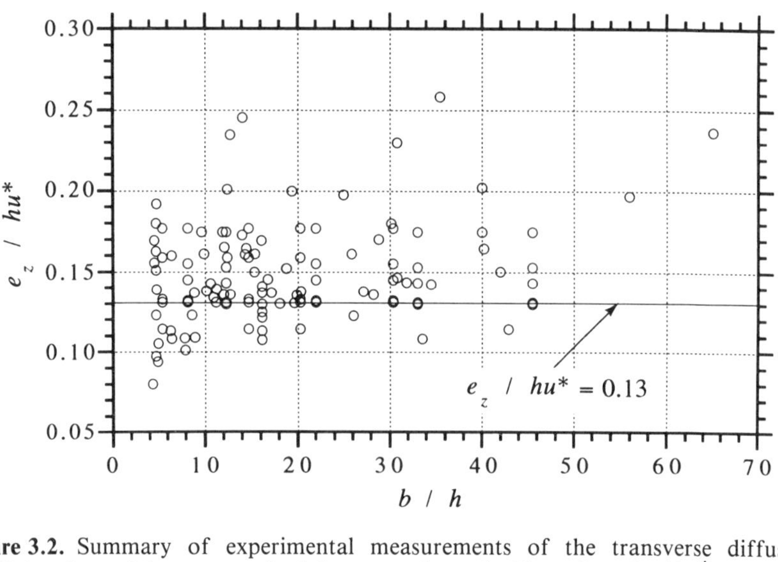

Figure 3.2. Summary of experimental measurements of the transverse diffusion coefficient in straight, rectangular laboratory channels. The value of e_z/hu^* appears to have a constant value of about 0.13 (see text for details). Data are listed in Table 3.1

Table 3.1. Reported transverse diffusion coefficients in straight laboratory channels

Reference*	b (cm)	h (cm)	u ($cm\,s^{-1}$)	u^* ($cm\,s^{-1}$)	e_z ($cm^2\,s^{-1}$)	Reference*	b (cm)	h (cm)	u ($cm\,s^{-1}$)	u^* ($cm\,s^{-1}$)	e_z ($cm^2\,s^{-1}$)
1	60	3.9	20	1.88	1.11	10	49	9.0	14	1.88	2.22
	60	3.3	19	2.03	0.88		49	6.0	10	1.53	1.21
	60	1.7	20	2.39	0.59		49	4.0	6.9	1.25	0.65
	60	1.4	20	2.80	0.60		49	9.0	17.1	1.88	2.25
	45	3.8	19	1.82	0.93		49	6.0	12.6	1.53	1.22
	45	3.2	20	2.09	1.16		49	4.0	9.2	1.25	0.66
	45	1.8	20	2.39	0.85		49	9.0	13.5	1.49	1.78
	45	1.3	20	2.69	0.88		49	6.0	10.0	1.21	0.96
	60	3.1	20	1.54	0.62		49	4.0	7.2	0.99	0.52
	60	2.3	20	1.78	0.50		49	9.0	10.5	1.15	1.38
	60	1.8	20	1.87	0.36		49	6.0	7.7	0.94	0.74
	60	1.4	20	2.09	0.34		49	4.0	5.6	0.77	0.40
	45	4.0	20	1.58	0.88		49	9.0	14.2	1.51	1.55
	45	2.8	20	1.76	0.67		49	6.0	10.8	0.94	0.82
	30	3.4	20	1.91	0.88		49	4.0	8.1	0.77	0.44
	30	2.5	20	2.08	0.92		49	9.0	16.7	1.15	1.65
	30	1.6	20	2.39	0.74		49	6.0	12.8	0.94	0.87
	60	4.2	34	1.96	1.34		49	4.0	9.8	0.77	0.47
	60	3.9	31	1.82	1.14		49	9.0	12.9	0.73	1.16
	60	4.2	18	1.10	0.74		49	6.0	10.0	0.59	0.63
	60	4.9	15	0.93	0.91		49	4.0	7.5	0.49	0.34
	60	5.0	30	1.72	1.41		132	9.0	13.9	1.88	2.22
							132	6.0	10.0	1.53	1.21
2	85	1.5	31	2.12	0.64		132	4.0	6.9	1.25	0.65
	85	3.0	27	1.57	0.79		132	9.0	17.1	1.88	2.25
	85	5.3	43	2.23	1.33		132	6.0	12.6	1.53	1.22
	85	5.3	43	2.09	1.33		132	4.0	9.2	1.25	0.66
	85	5.3	43	2.18	1.43		132	9.0	13.5	1.49	1.78

(continued)

Table 3.1. (*Continued*)

Reference*	b (cm)	h (cm)	u ($cm\,s^{-1}$)	u^* ($cm\,s^{-1}$)	e_z ($cm^2\,s^{-1}$)	Reference*	b (cm)	h (cm)	u ($cm\,s^{-1}$)	u^* ($cm\,s^{-1}$)	e_z ($cm^2\,s^{-1}$)
	85	5.3	42	2.14	1.21		132	6.0	10.0	1.21	0.96
	85	5.3	42	2.12	1.45		132	4.0	7.2	0.99	0.52
	85	10.7	42	1.93	2.09		132	9.0	10.5	1.15	1.38
	85	17.3	37	1.75	2.86		132	6.0	7.7	0.94	0.74
	110	1.7	33	1.98	0.79		132	4.0	5.6	0.77	0.40
	110	2.7	50	2.64	1.19		132	9.0	14.2	1.51	1.55
	110	2.8	30	1.64	0.79		132	6.0	10.8	0.94	0.82
	110	3.5	32	1.75	0.87		132	4.0	8.1	0.77	0.44
	110	5.4	44	2.18	1.62		132	9.0	16.7	1.15	1.65
	110	5.5	42	2.20	1.65		132	6.0	12.8	0.94	0.87
	110	10.8	39	1.86	2.78		132	4.0	9.8	0.77	0.47
	110	17.3	35	1.66	3.25		132	9.0	12.9	0.73	1.16
	110	22.0	31	1.41	3.26		132	6.0	10.0	0.59	0.63
	110	6.8	36	5.01	4.81		132	4.0	7.5	0.49	0.34
	110	8.7	41	5.13	6.03		182	9.0	13.9	1.88	2.22
	110	10.4	43	5.06	7.48		182	6.0	10.0	1.53	1.21
	110	17.1	35	3.64	6.74		182	4.0	6.9	1.25	0.65
							182	9.0	17.1	1.88	2.25
3	110	4.1	35	1.90	1.06		182	6.0	12.6	1.53	1.22
	110	6.6	45	2.14	2.04		182	4.0	9.2	1.25	0.66
	110	11.1	46	1.99	3.56		182	9.0	13.5	1.49	1.78
	110	3.9	37	3.73	1.98		182	6.0	10.0	1.21	0.96
	110	6.4	46	4.01	3.52		182	4.0	7.2	0.99	0.52
							182	9.0	10.5	1.15	1.38
4	238	14.8	23	3.81	9.57		182	6.0	7.7	0.94	0.74
	238	24.8	34	4.93	21.4		182	4.0	5.6	0.77	0.40
	238	37.1	48	6.03	35.8		182	9.0	14.2	1.51	1.55
							182	6.0	10.8	0.94	0.82
5	80	7.3	23	1.20	1.18		182	4.0	8.1	0.77	0.44
	80	9.0	19	0.99	0.97		182	9.0	16.67	1.15	1.65
	80	10.2	15	0.82	0.91		182	6.0	12.8	0.94	0.87

6	60	12.7	30	3.01	3.72		182	4.0	9.8	0.77	0.47
	60	12.7	53	5.30	8.27		182	9.0	12.9	0.73	1.16
	60	12.5	81	7.86	18.9		182	6.0	10.0	0.59	0.63
	60	12.5	31	5.41	9.38		182	4.0	7.5	0.49	0.34
	60	12.6	53	8.14	15.5						
	60	13.2	77	11.1	22.7	11	36	1.2	25.4	1.86	1.89
	60	12.7	31	6.46	14.8						
	60	12.6	53	11.6	23.8	12	122	4.0	26.4	2.14	1.24
	60	13.1	77	16.3	36.2		122	4.1	22.9	1.81	1.32
							122	4.7	12.3	1.13	0.86
7	244	17.4	50	3.51	15.0		122	6.5	19.7	1.57	1.55
8	220	5.5	30	3.60	4.0	13	120	9.7	11.2	0.60	0.92
	220	17.3	31	1.60	6.5						
						14	69	15.8	94.5	5.18	6.65
9	56	6.5	28	1.55	1.24						
	56	5.0	23	1.40	0.92						

* 1 = Lau and Krishnappan (1977); 2 = Okoye (1970); 3 = Prych (1970); 4 = Sayre and Chang (1968); 5 = Sullivan (1968); 6 = Miller and Richardson (1974); 7 = Sayre and Chamberlin (1964); 8 = Engelund (1969); 9 = Nokes (1986); 10 = Webel and Schatzmann (1984); 11 = Elder (1959); 12 = Engmann (1974); 13 = Holley and Abraham (1973b); and 14 = Kalinske and Pien (1944).

estimate of the transverse eddy diffusivity appears to be

$$\frac{e_z}{hu^*} = 0.13 \tag{3.8}$$

This is the value suggested by Webel and Schatzmann (1984) and Nokes and Wood (1988) but is lower than other published values. Elder (1959) reported $e_z/hu^* = 0.23$, which is one of the earliest estimates and has been quoted extensively. Fischer also measured a value of $e_z/hu^* = 0.23$ in a straight irrigation canal, but he noted that the thalweg meandered within the channel and suggested that secondary currents associated with this meandering were responsible for increasing the rate of transverse mixing (Fischer 1973). Sayre (1968a) measured an average value of $e_z/hu^* = 0.17$ in a laboratory channel, which is identical to the value given by Orlob (1959). Fischer *et al* (1979) reviewed several studies and suggested that $e_z/hu^* \approx 0.15$. Equation (3.8) is arguably the most likely value of the transverse diffusivity in plane shear flow.

By comparing equations (3.8) and (2.19) it can be seen that

$$\frac{e_z}{e_y} \approx 2 \tag{3.9}$$

3.4.1.1 Spatial variability

We have only a limited understanding of how the transverse diffusivity varies spatially. The vertical diffusivity e_y is known from both theory and experiment to vary parabolically with depth [see equation (2.18)]. Experimental studies suggest that the transverse diffusivity varies with depth in much the same manner as the longitudinal velocity (i.e. logarithmically: Okoye 1970, Nokes 1986) although there is no body of theory to support this observation. The depth variation of e_z is seldom modelled and in the two-dimensional mixing equation (3.1) the depth-averaged value is all that can be accommodated.

Of more practical importance is the transverse variation of e_z and it is particularly important to know whether the rate of transverse mixing is significantly reduced near the banks. It is not possible to predict theoretically the transverse variation in e_z and there are few published studies in which e_z has been measured systematically in different places across the channel. Such a study would be complicated by the presence of secondary currents near the banks.

It is important to remember that equation (3.8) quantifies the transverse eddy diffusivity and should only be used to predict mixing rates in plane shear flow where it is reasonable to neglect transverse dispersion (i.e. to assume that $k_z = 0$). In practice such flows occur near mid-stream in straight rectangular channels.

3.4.1.2 *Role of secondary currents*

Okoye (1970) found that in laboratory channels e_z/hu^* increased with aspect ratio (b/h). His explanation was that as the channel width increases, so does the length scale of the transverse eddies and hence the transverse diffusivity. Lau and Krishnappan (1977), however, reviewed several studies but could find no consistent relationship between e_z/hu^* and b/h. Figure 3.2 illustrates this point. It seems reasonable to assume that the non-dimensional diffusivity e_z/hu^* is independent of the channel aspect ratio.

There is something of a controversy about the relative importance of turbulent diffusion and secondary currents in determining the rate of transverse mixing in straight laboratory channels. Webel and Schatzmann (1984) found that in smooth laboratory channels e_z/hu^* showed some dependence on the Darcy friction factor f, but that in fully rough channels ($f > 10$) it was remarkably constant at $e_z/hu^* = 0.13$. They inferred from this that turbulent diffusion dominates the transverse mixing process. Lau and Krishnappan (1977), on the other hand, collated data from a number of earlier studies and carried out their own experiments. They argued that secondary circulations are more important than turbulent diffusion in causing transverse mixing even in straight laboratory channels. Based on this premise they advocated using the channel width as the length scale for non-dimensionalising the diffusivity and found that e_z/bu^* varies inversely with the aspect ratio of the channel (Figure 3.3). It is a little worrying, however, that channel width appears on

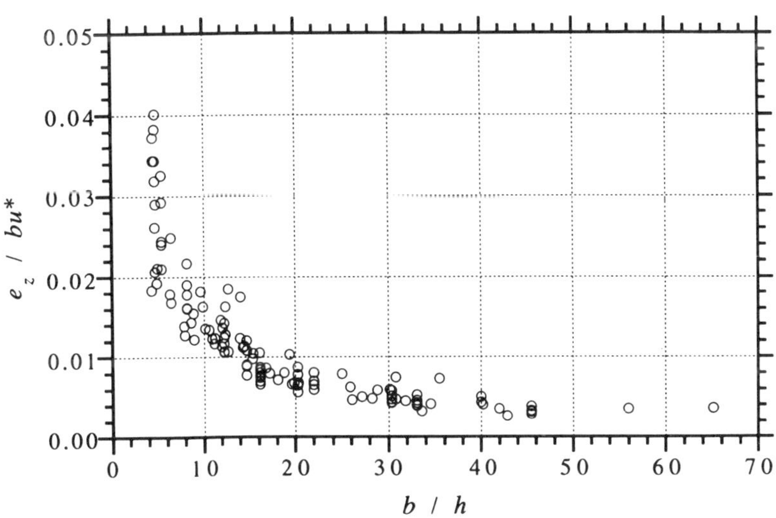

Figure 3.3. Summary of experimental measurements in laboratory channels showing the relationship between e_z/bu^* and b/h. Data are the same as in Figure 3.2

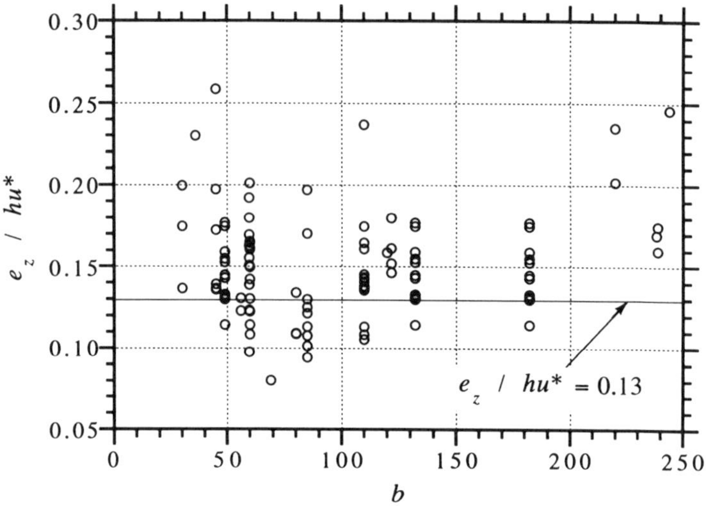

Figure 3.4. Summary of experimental measurements in laboratory channels showing that e_z/hu^* does not vary systematically with b. Data are the same as in Figure 3.2

both axes of this figure, thereby introducing a degree of self-correlation which is entirely spurious. Figure 3.4 shows that when the non-dimensional diffusivity e_z/hu^* is plotted against the channel width b no dependence is apparent. This is at variance with the suggestion of Lau and Krishnappan that the channel width is the dominant length scale.

Clearly there is a need for a better theoretical understanding of the processes which contribute to transverse mixing in straight rectangular channels to help resolve these apparent conflicts. There can be no doubt, however, that secondary currents are very important in irregular and meandering natural channels, as is discussed in more detail in the following.

3.4.2 Transverse Dispersion Coefficient

Published data from a number of sources on transverse dispersion in natural channels have been collated in Table 3.2. Where possible the original references have been checked in an effort to minimise transcription errors. Several databases have been analysed more than once using different approaches (notably on the Missouri River) and these are treated as independent estimates of k_z in Table 3.2. Figures 3.5 and 3.6 summarise the data from Table 3.2 graphically. As the channels are irregular, dispersion coefficients are scaled by the mean depth H and the average shear velocity U^* defined

$$U^* = \sqrt{gRs} \approx \sqrt{gHs} \tag{3.10}$$

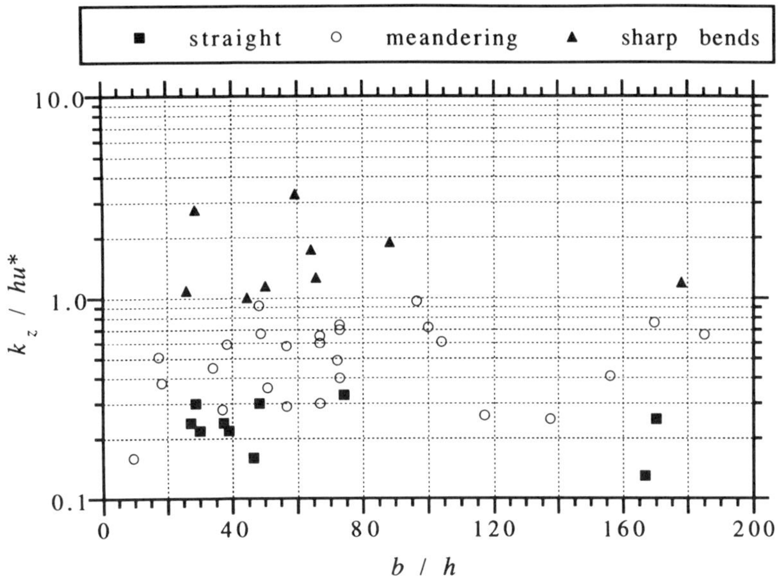

Figure 3.5. Summary of field measurements of the transverse dispersion coefficient in rivers. Data are given in Table 3.2

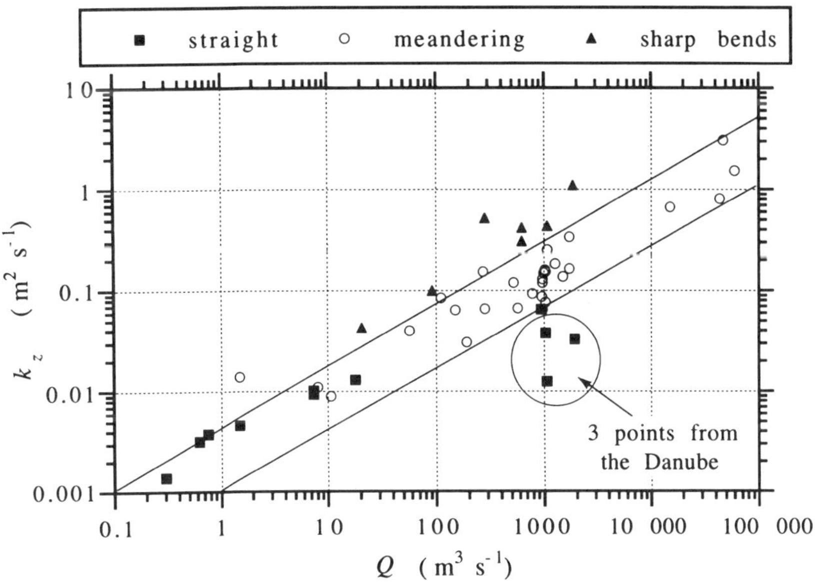

Figure 3.6. Summary of field measurements of the transverse dispersion coefficient in rivers showing the relationship between k_z and Q. Data are given in Table 3.2

Table 3.2. Reported transverse dispersion coefficients in straight, ice-covered, sinuous and meandering natural channels

Reference*	River	b (m)	Q ($m^3 s^{-1}$)	H (m)	V_x ($m s^{-1}$)	U^* ($m s^{-1} \times 10^2$)	k_z ($m^2 s^{-1} \times 10^4$)	k_z/HU^*
Straight								
1	Atrisco	18.3	7.4	0.68	0.63	6.3	102	0.24
1	Atrisco	18.3	7.4	0.67	0.66	6.1	102	0.25
1	Atrisco	18.3	7.4	0.67	0.67	6.2	93	0.22
1	Bernardo	20	17.8	0.70	1.25	6.2	130	0.30
1	South	18.3	1.6	0.37	0.24	4.0	140	0.95
1	South	18.3	1.5	0.44	0.18	4.0	47	0.26
2	Athabasca	373	776	2.20	0.95	5.6	930	0.76
Ice covered								
3	Athabasca	252	240	1.9	0.49	5.2	410	0.42
3	Beaver	38.7	6.5	0.61	0.28	3.5	200	0.94
3	Athabasca	276	105	0.96	0.40	5.4	100	0.19
4	Slave	36	—	2.0	0.37	4.3	105	0.12
Bends								
5	Missouri	201	623	3.06	1.01	7.9	3100	1.27
5	Missouri	206	1063	4.11	1.26	9.2	4350	1.15
5	Missouri	210	1700	5.49	1.47	10.6	3420	0.59
5	Missouri	195	623	3.04	1.05	7.9	4200	1.74
5	Missouri	201	1063	4.13	1.28	9.2	2550	0.67
5	Missouri	204	1700	5.54	1.50	10.7	1650	0.28
6	Missouri	240		4.0	1.98	8.5	11 110	3.40
3	Beaver	42.7	20.5	0.96	0.50	4.5	430	1.01
7	Isere	60–70	250	2.25	1.40	—	—	0.50–1.60
7	Isere	145–160	400	1.7–2.0	1.60	—	—	2.0–2.5
8	Mississippi	178	92–120	1.0	0.52–0.67	8.1	1000	1.2
Meandering								
9	Columbia	305	—	3.05	1.35	8.8	1860	0.72

2	Missouri	183	966	2.74	1.74	7.3	1010	0.50
10	Missouri	183	966	2.74	1.75	7.4	1210	0.60
5	Missouri	214	1020	2.94	1.58	7.4	1600	0.74
5	Missouri	214	992	2.94	1.58	7.4	1500	0.70
5	Missouri	214	949	2.94	1.58	7.4	870	0.40
5	Missouri	192	524	1.99	1.39	6.10	1200	0.97
11	Potomac	350	—	0.73–1.74	0.29–0.58	3.3–5.1	—	0.52–0.65
12	Grand	59.2	10.6	0.51	0.35	6.9	90	0.26
13	Grand	59.1	12.5	0.59	0.39	—	72	—
14	Waal	266	1028	4.70	0.82	5.7	770–1550	0.29–0.58
14	Waal	266	1480	5.25	1.06	7.4	1390	0.36
14	IJssel	69.5	270	4.0	0.97	7.6	1550	0.51
4	Slave	42	—	2.3	0.59	4.6	402	0.38
3	Athabasca	320	556	2.05	0.86	7.9	670	0.41
12	Saskatchewan	213	—	1.55	0.58	8.0	310	0.25
12	Bow	104	—	1.0	1.05	13.9	850	0.61
15	Danube	415	1030	2.9	0.87	5.2	380	0.25
16	Danube	418	1050	2.9	0.87	3.7	140	0.13
16	Danube	475	1900	4.2	0.95	6.0	320	0.13
16	Kris-Raba	10	8.0	1.06	0.80	6.9	110	0.16
17	Rea	5.1	0.30	0.11	0.47	6.5	14	0.16
17	Rea	6.6	0.62	0.17	0.70	7.2	32	0.22
17	Rea	6.7	0.75	0.18	0.73	7.5	38	0.24
18	Orinoco	1340	43 600	1.79	1.83	—	8 100	—
18	Orinoco	3580	47 000	1.19	1.11	—	31 000	—
18	Orinoco	2560	59 600	1.65	1.41	—	15 500	—
19	Waikato	85	150	2.5	0.70	5.7	640	0.45
20	Waikato	210	232	2.2	0.63	4.0	950	1.1
20	Waikato	300	232	1.7	0.63	4.6	100	0.31
11	MacKenzie	1240		6.7	1.77	15.2	6700	0.66

* 1 = Fischer (1973); 2 = Yotsukura and Cobb (1972); 3 = Beltaos (1980b); 4 = Engmann and Kellerhals (1974); 5 = Sayre (1979); 6 = Sayre and Yeh (1973); 7 = Holly and Nerat (1984); 8 = Demetracopoulos and Stefan (1983); 9 = Glover (1964); 10 = Yotsukura *et al* (1970); 11 = Fischer *et al* (1979); 12 = Lau and Krishnappan (1981); 13 = Gowda (1983); 14 = Holley and Abraham (1973a); 15 = Somlyody (1982); 16 = Somlyody (1977); 17 = Cotton and West (1980); 18 = Stallard (1987); 19 = Rutherford *et al* (1980); 20 = Rutherford and Williams (1992).

where R = hydraulic radius (cross-sectional area/wetted perimeter). For shallow channels the hydraulic radius is approximately equal to the mean depth.

In straight channels (denoted by squares in Figures 3.5 and 3.6) it appears that

$$0.15 < \frac{e_z + k_z}{HU^*} < 0.30 \tag{3.11}$$

A similar range has been suggested by Fischer *et al* (1979), Cotton and West (1980) and Holly (1985). Most workers who report mixing coefficients do not distinguish between the effects of diffusion and dispersion, which explains the appearance of both e_z and k_z in equation (3.11).

Weak secondary currents are found close to the banks in straight rectangular channels where their presence can be detected by the fact that the velocity maximum is depressed below the water surface (Henderson 1966). These secondary currents have been predicted successfully using k–ε turbulence models (Naot and Rodi 1982), but their effects on transverse mixing have not been quantified precisely. In artificial channels (e.g. canals) there are slight irregularities in the bed and the bank which also promote weak secondary currents and cause the thalweg to meander across the channel. Comparing equations (3.8) and (3.11) it can be inferred that weak secondary currents increase the coefficient of transverse mixing by a factor of about two.

The transverse dispersion coefficient k_z depends on secondary currents rather than turbulence generated by bed friction so that H and U^* may no longer be the appropriate scaling variables for k_z. At present, however, there is no general agreement about suitable alternatives.

Figure 3.5 indicates that in irregular but only gently meandering natural channels (shown as open circles)

$$0.3 < \frac{k_z}{HU^*} < 0.9 \tag{3.12}$$

This is similar to the range of values suggested by Fischer *et al* (1979) and Lau and Krishnappan (1981). Comparing equations (3.8) and (3.12) it is clear that turbulent diffusion is relatively unimportant in a curved channel and we can drop e_z.

3.4.3 Effects of Channel Curvature

Channel curvature increases the rate of transverse mixing because at bends there are strong secondary currents which carry tracer sideways in different directions at different depths. Experimental results (shown as triangles in

Figure 3.5) suggest that in curved channels

$$1 < \frac{k_z}{HU^*} < 3 \tag{3.13}$$

A local value of $k_z/HU^* = 10$ has been reported for a sharp bend on the Missouri River (Sayre 1973).

Several attempts have been made to establish the relationship between k_z and bulk channel parameters (such as width, depth, shear velocity, curvature and sinuosity). Fischer (1969) derived a theoretical relationship which described adequately tracer dispersion rates in a circular flume

$$\frac{k_z}{HU^*} = \frac{0.25}{\varkappa^5}\left(\frac{V_x H}{U^* r_c}\right)^2 \tag{3.14}$$

where V_x = cross-sectional averaged velocity; r_c = radius of curvature and $\varkappa$ = von Karman's constant. Equation (3.14) also fits the laboratory measurements made by Chang (1971) but it seriously underestimates the value of k_z in natural rivers. Yotsukura and Sayre (1976) reported that in the Missouri River the average value around a bend was

$$\frac{k_z}{HU^*} = 0.4\left(\frac{V_x b}{U^* r_c}\right)^2 \tag{3.15}$$

where b = channel width and the values of b, H, U^*, V_x and k_z are averaged around the bend. Sayre (1979) reported a further set of experiments in the Missouri River which he analysed using the streamtube model (discussed later in this chapter). His results can be rewritten

$$\frac{k_z}{HU^*} = 0.3 - 0.9\left(\frac{V_x b}{U^* r_c}\right)^2 \tag{3.16}$$

The constant in equation (3.16) decreased from 0.9 to 0.3 as the flow increased from 565 to 1700 $m^3 s^{-1}$.

It must be stressed that equations (3.15–3.16) do not fit the observed laboratory flume data (Yotsukura and Sayre 1976, Sayre 1979). The reasons for the mis-match are unclear but probably relate to differences in the structure of the secondary flows. Clearly there is a need for further research on this topic. If estimates of k_z are made using equations (3.14–3.16) they should be afforded a high uncertainty as is discussed in Example 3.1.

3.4.4 Spatial Variations in Mixing Coefficients

3.4.4.1 Transverse variations

Equations (3.11)–(3.13) suggest that

$$\frac{k_z}{HU^*} = \text{constant} \tag{3.17}$$

in a particular channel. If this relationship holds locally then

$$\frac{k_z}{hu^*} = \text{constant} \tag{3.18}$$

where h = local depth and u^* = local shear velocity defined by

$$u^* = \sqrt{ghs} \tag{3.19}$$

Substituting equation (3.19) into (3.11)–(3.13) indicates that k_z varies across the channel as $h^{3/2}$ and hence is low in shallow water near the banks.

This analysis is complicated by the fact that where the depth is changing plane shear flow does not occur. Changes of depth promote secondary currents which enhance transverse mixing and increase the value of k_z. It is difficult to predict the strength of secondary currents in an irregular channel and to deduce their effect on k_z. Surprisingly there have been few attempts to measure systematically the transverse distribution of k_z in natural channels. Several attempts have been made, however, to infer the distribution by matching observed tracer concentration data (Holley *et al* 1972, Lau and Krishnappan 1981) and these results are discussed in Section 3.5.4. As yet there is, however, no general consensus on the transverse distributions of k_z in natural channels.

3.4.4.2 Longitudinal variations

In a straight, uniform channel the transverse dispersion coefficient is unlikely to vary significantly along the channel, but in a natural river channel the situation is different. Chang (1971) measured local values of the transverse dispersion coefficient in a sinuous flume using soluble tracers. He found that in the upstream part of a bend k_z was about one-half the reach-averaged value, whereas at the downstream end of bends it was twice the reach-averaged value. Thus along a bend the local value of k_z varied by a factor of four. Sayre (1973) found a similar pattern in a meandering six-mile reach of the Missouri River. Sayre (1979) also analysed temperature profiles measured at several cross-sections below the Fort Calhoun power station on the Missouri River and found that at a given flow values of k_z for adjacent sub-reaches in the same locality varied by at least a factor of six. By comparison two other river studies using dye tracers report a local variation along the channel in k_z of only two: Yotsukura and Cobb (1972) in the South and Missouri Rivers and Lau and Krishnappan (1981) in the Grand River, Ontario.

It is clear that the structure and effects of secondary currents vary along the length of a bend, but to date there has been little success in relating the flow structure to the rate of transverse mixing in a quantitative fashion. The situation may be further complicated by sudden changes of bathymetry (e.g. sand banks), obstacles (e.g. bridge piers) and islands which induce wakes and further influence the local transverse mixing rate.

Over a long reach which includes several complete bends we would expect some averaging out of localised variations in the transverse dispersion coefficient. If the reach consists of a number of uniform sub-reaches then

$$\overline{k_z} = \frac{1}{L} \sum_{i=1}^{N} k_i \Delta x_i \tag{3.20}$$

where the overbar denotes a reach-average; k_i = dispersion coefficient in sub-reach i; Δx_i = length of sub-reach i; N = total number of sub-reaches; and L = total reach length.

Example 3.1

The Waikato River below Huntly, New Zealand, has the average hydraulic characteristics shown in Table 3.3. As part of a modelling study of temperature a preliminary estimate was made of the transverse dispersion coefficient in this reach.

The river channel is gently meandering for most of this reach, except that 3.5 km below Huntly there is a sharp bend and a constriction in the channel. The length of the bend (measured along the centreline of the channel) is 400 m and the radius of curvature is 350 m. There are four surveyed cross-sections around the bend from which the bend-averaged depth and width were calculated (see Table 3.3). The reach was divided into three sub-reaches, the second of which included the bend. The top and bottom sub-reaches were assumed to have identical channels. The mean depth and width of each sub-reach are shown in Table 3.4. Mean velocity was estimated from Q/Hb, the water slope was assumed to be uniform and the shear velocity was estimated from equation (3.18).

For sub-reaches 1 and 3 it seems reasonable, based on equation (3.12), to assume that $k_z/HU^* = 0.4$. This gives $k_z = 0.043\ \mathrm{m^2\,s^{-1}}$. In sub-reach 2 equations (3.14–3.16) provide estimates which differ by two orders of magnitude (Table 3.5). Equation (3.14) was derived from laboratory flume

Table 3.3. Hydraulic parameters of the Waikato River near Huntly

River flow (Q)	$300\ \mathrm{m^3\,s^{-1}}$
Reach length (L)	5.5 km
Mean depth* (H)	2.2 m
Mean width* (b)	210 m
Slope (s)	120×10^{-4}

* Mean denotes the average along the entire reach.

Table 3.4. Hydraulic parameters of each sub-reach

	Sub-reach 1	Sub-reach 2	Sub-reach 3
Sub-reach length (Δx_i) (km)	3.5	0.4	1.6
Mean depth* (H) (m)	2.13	3.15	2.13
Average width* (b) (m)	226	140	226
Mean velocity* (V_x) ($\mathrm{m\,s^{-1}}$)	0.62	0.63	0.62
Shear velocity (U^*) ($\mathrm{m\,s^{-1}}$)	0.050	0.0609	0.050

* Averaged along the sub-reach.

data and gives a very low estimate of k_z. Equations (3.15–3.16) yield higher values of k_z/HU^* than typically reported (see Table 3.2). It might be wise to discount the highest and lowest estimates and to assume that $k_z/HU^* = 5–7$. This gives $k_z = 1.0–1.3\ \mathrm{m^2\,s^{-1}}$ in sub-reach 2. Subsequent to this analysis a dye test was carried out at this bend which yielded a value of $k_z = 0.52\ \mathrm{m^2\,s^{-1}}$.

The reach averaged value from equation (3.20) is then $k_z = 0.11–0.13\ \mathrm{m^2\,s^{-1}}$. This is similar to the value of $k_z = 0.095\ \mathrm{m^2\,s^{-1}}$, which was estimated by calibrating a two-dimensional numerical model to measured water temperature distributions (Rutherford and Williams 1992).

3.4.4.3 Flow variations

Data from Table 3.2 are shown plotted against river flow in Figure 3.6. Most of the data for straight and meandering channels fall between the two lines drawn by eye. Somlyody's three estimates for the River Danube are outliers for some unknown reason and neglecting them the variation in k_z at a given flow is about a factor of five. At sharp bends k_z is consistently higher than in a comparable straight or meandering channel. It is clear that the larger the river, the larger is k_z. In these data the mean depth increases with the size of the river (as would be expected), but the shear velocity shows no consistent pattern and can be assumed constant. If k_z/HU^* remains constant (as is suggested by the data collated in Table 3.2), then as the size of the river increases

Table 3.5. Three estimates of the transverse dispersion coefficient at a sharp bend in the Waikato River

	k_z/HU^*	k_z ($\mathrm{m^2\,s^{-1}}$)
Equation (3.14)	0.21	0.04
Equation (3.15)	7	1.3
Equation (3.16)	5–15	1.0–3.0

so too does the dispersion coefficient. Figure 3.6 adds little to our understanding of transverse mixing but is a convenient way of summarising the available data and is a potentially useful tool for making an initial estimate of the transverse dispersion coefficient.

It would be unwise to interpret Figure 3.6 as saying that in a particular river the dispersion coefficient increases with river flow. Rather it says that big rivers have large values of k_z. Sayre (1979) measured the dispersion coefficient around the same bend in the Missouri River at three different flows and found that k_z decreased by a factor of three as the flow increased (see equation 3.16). Presumably for the bend Sayre studied, the strength of the secondary flow decreased as the flow increased. It is not clear whether this result holds true at all river bends and further investigations are desirable.

It can be argued that in a straight channel the transverse mixing coefficient should increase with flow. In a straight channel transverse mixing is principally the result of turbulence generated by bed friction which scales with the depth and shear flow. Both H and U^* increase with flow and so if k_z/HU^* remains constant then k_z showed an increase with flow. There is evidence to support this argument. Somlyody (1977) found that in a straight reach of the River Danube k_z increased from 0.0125 to 0.0330 $m^2 s^{-1}$ as flow increased from 1050 to 1900 $m^3 s^{-1}$. Rutherford and Williams (1992) found in the Waikato River that k_z increased from 0.09 to 0.11 $m^2 s^{-1}$ as the river flow increased from 220 to 330 $m^3 s^{-1}$.

3.4.4.4 *Sensitivity analysis*

Given the large variation in published values of the transverse dispersion coefficient, it is advisable when making predictions to choose a range of values for k_z and to undertake a careful analysis of the sensitivity of predictions to changes in this coefficient. If these preliminary calculations indicate that a more refined estimate of k_z is required, then it is necessary either to conduct field tracer tests or to carry out physical modelling studies. A potentially fruitful area of research is to improve our quantitative understanding of the structure of secondary flows and bed-generated turbulence on transverse mixing so that we can reduce the uncertainty inherent in estimating values of e_z and k_z. These two processes are best studied separately.

3.5 APPROXIMATE CONSTANT-COEFFICIENT MODEL

3.5.1 Model Derivation

An important practical problem is to predict the rate of transverse mixing downstream from a steady point source. Although very close to the source both vertical and transverse dispersion are important, tracer becomes vertically well mixed close to the source and in the mid-field the problem becomes two-dimensional (see Table 1.2 and Figure 1.9). In the case of a steady source the longitudinal dispersion term becomes negligible and the depth-averaged mixing equation (equation 3.1) can be simplified to

$$\frac{\partial}{\partial x}(hv_x s) + \frac{\partial}{\partial z}(hv_z s) = \frac{\partial}{\partial z}\left(hk_z \frac{\partial s}{\partial z}\right) \tag{3.21}$$

In equation (3.21) the transverse eddy diffusivity e_z has been incorporated into the transverse dispersion coefficient k_z. A simplified analysis of this problem can be made by assuming that the channel is straight, rectangular and uniform so that $v_z = 0$. The further assumption is made that k_z, h and v_x are constant. Equation (3.21) then simplifies to

$$v_x \frac{\partial s}{\partial x} = k_z \frac{\partial^2 s}{\partial z^2} \tag{3.22}$$

This can be termed the constant-coefficient model of transverse mixing. Few natural channels comply exactly with these assumptions but equation (3.22) provides a useful model for preliminary investigations of transverse mixing.

At both banks the transverse tracer flux is zero and as k_z is constant and non-zero this requires that

$$\frac{\partial s}{\partial z} = 0 \qquad \text{at } z = 0 \text{ and } z = b \tag{3.23}$$

At the upstream boundary it is necessary to specify the tracer concentration distribution across the channel. At the downstream boundary it can be assumed that

$$\frac{\partial s}{\partial x} = 0 \tag{3.24}$$

although in practice this boundary condition has little influence on the solution because of the large advective velocity v_x.

For a steady vertical line source at $x = 0$, $z = z_o$ the well-known solution to equation (3.22) in an unbounded flow is

$$s_u(x, z) = \frac{m}{H\sqrt{4\pi k_z x v_x}} \exp\left[-\frac{v_x(z - z_o)^2}{4k_z x}\right] \tag{3.25}$$

where m = tracer mass inflow rate. The boundary condition given in equation (3.23) can be accommodated using the method of images (Carslaw and Jaegar 1959) as outlined in Section 2.6.1. This gives

$$s(x, z) = s_u(x, z \pm z_o) + \sum_{n=1}^{N} s_u(x, 2nb \pm z \pm z_o) \tag{3.26}$$

where the banks are at $z = 0$ and $z = b$ and $s_u(x, z)$ is the unbounded solution given by equation (3.25). The shorthand notation used in equation (3.26) is the same as that used in equations (2.28) and (2.29).

3.5.2 Predicted Concentration Distributions

Non-dimensional variables can be defined

$$c^* = \frac{s}{C_m} = \frac{sQ}{m} \tag{3.27a}$$

$$z^* = \frac{z}{b} \tag{3.27b}$$

$$x^* = \frac{xk_z}{v_x b^2} \tag{3.27c}$$

where c^*, z^* and x^* = non-dimensional concentration, transverse distance and longitudinal distance, respectively; C_m = fully-mixed tracer concentration; Q = river flow. Equations (3.27) have the same form as equations (2.30) and consequently Figure 2.2 can be used to solve transverse mixing problems provided that z^* is used in place of y^*. Figure 3.7 shows results from Figure 2.2 replotted using a logarithmic scale for x^*. This gives a better resolution when working with large values of c^*.

3.5.3 Transverse Mixing Distance

An important parameter in transverse mixing problems is the longitudinal distance required for tracer to become well mixed across the channel. This can be termed the transverse mixing distance and denoted L_z. The constant-coefficient model predicts that the transverse concentration gradient decreases monotonically with distance from the source and vanishes asymptotically. Tracer can be assumed to be fully mixed when the ratio of the minimum to maximum concentration across the channel $P_m = s_{min}/s_{max}$ reaches a certain limit (commonly 0.98, 0.95 or 0.90). For $P_m = 0.98$ the mixing distance is

$$L_z = 0.536 \frac{v_x b^2}{k_z} \tag{3.28}$$

for an outfall at either bank and

$$L_z = 0.134 \frac{v_x b^2}{k_z} \tag{3.29}$$

for a source in mid-channel.

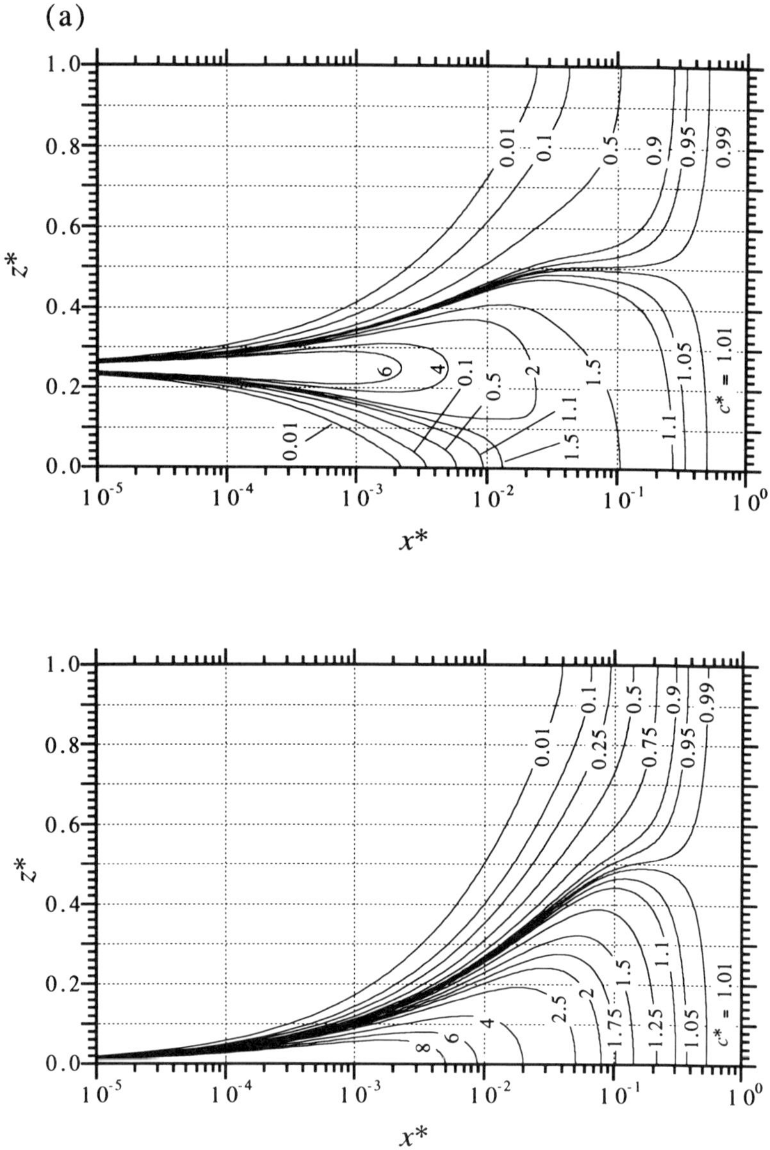

Figure 3.7. (a) Concentration contours downstream from a vertical line source at $z^* = 0.00$ and 0.25 predicted using the constant-coefficient model. (b) Concentration contours downstream from a vertical line source at $z^* = 0.33$ and 0.50 predicted using the constant-coefficient model

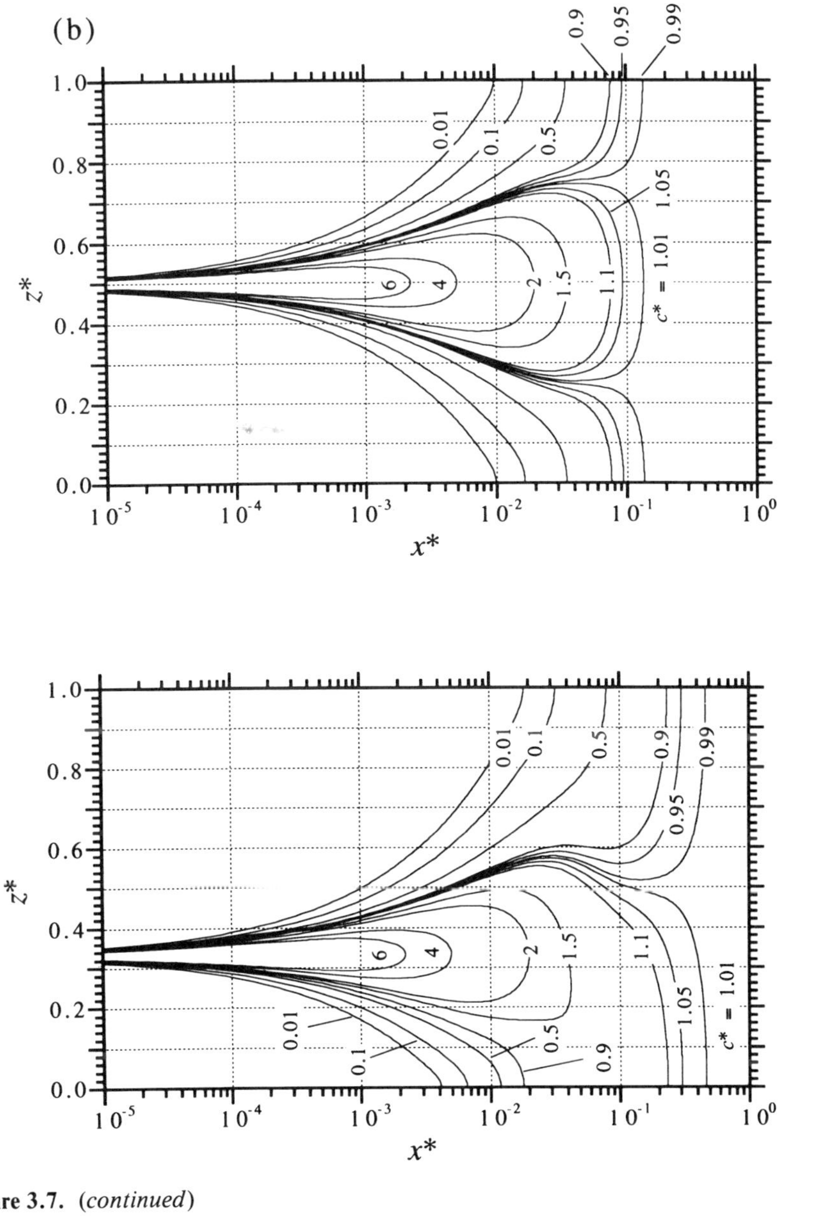

Figure 3.7. *(continued)*

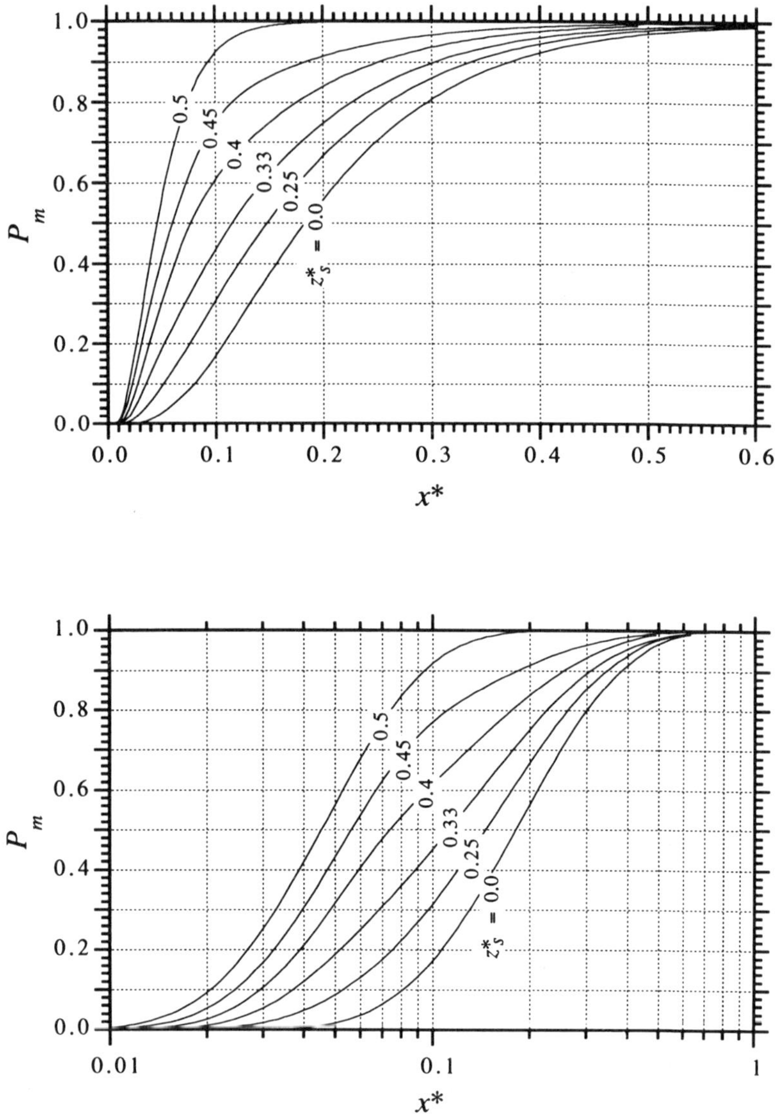

Figure 3.8. Plots showing how the degree of transverse mixing ($P_m = s_{min}/s_{max}$) increases with distance downstream from a vertical line source at six different transverse locations

In most laboratory and field studies it is difficult to measure the tracer accurately enough to resolve 2% concentration differences and larger differences (e.g. 5 or 10%) are more convenient. Figure 3.8 shows the value of P_m plotted against non-dimensional longitudinal distance for several source locations. Yotsukura and Cobb (1972) give similar graphs for diffusers of finite length. Figures 2.2 and 3.7 can be used to estimate the transverse mixing distance for other values of P_m.

For a source located in mid-channel, equation (3.29) can be rewritten

$$\frac{L_z}{b} = 0.134 \frac{v_x b}{k_z} \tag{3.30}$$

Data in Table 3.2 were used to evaluate the right-hand side of equation (3.30) and it was found that for straight and meandering channels tracer becomes well mixed across the channel a distance of about 100–300 channel widths downstream from the source. This is a useful rule of thumb for a source in mid-channel. For a source near either bank the mixing distance is four times this value.

The constant-coefficient model can also be used to estimate the point at which tracer from a non-steady source becomes well mixed across the channel. In the case of a non-steady source it is a slight over-simplification to talk of complete mixing because transverse velocity shear results in transverse concentration gradients which are attenuated, but never entirely eliminated by transverse dispersion. This point is discussed more fully in Chapter 4 in connection with longitudinal dispersion. In practice the transverse concentration gradients encountered in the far-field below a non-steady source are of little significance and the concept of complete mixing is useful. Consequently equations (3.28)–(3.29) and Figure 3.8 can be used to estimate the transverse mixing distance below a non-steady source.

A parameter of particular interest below a bankside source is the distance required for tracer to first reach the opposite bank. This can be termed the crossing distance (Holley *et al* 1972) and denoted L_c. For a bankside outfall the far-bank concentration reaches 2% of the near-bank concentration (i.e. $P_m = 0.02$) at a distance

$$L_c = 0.054 \frac{v_x b^2}{k_z} \tag{3.31}$$

Note that the crossing distance is only one-tenth of the mixing distance given by equation (3.29). Figure 3.7 can be used to estimate the crossing distance for other source locations and concentration ratios.

Example 3.2

The Ohinemuri is a small river in the Hauraki Plains, New Zealand, which supports a locally important trout fishery. The channel is deeply incised, the banks are lined with willows and the channel contains numerous fallen trees. Channel characteristics are summarised in Table 3.6. The river receives treated effluent from a mining operation whose water right requires that the effluent mixes rapidly across the entire channel. An assessment was made of the likelihood of a point source located 1 m from the right bank achieving the target of complete mixing within five minutes of the time of release.

Slope, depth and width were surveyed during low flow (similar to the design low flow) at five sections in the study reach. The mean velocity was estimated using $V_x = Q/Hb$ and the shear velocity from equation (3.10). As a check on the consistency of these data, the average Manning friction coefficient for the study reach was estimated to be

$$n = \frac{R^{2/3} s^{1/2}}{V_x} \approx 0.15$$

This value is high but plausible given that the channel contains fallen trees, stumps and tree roots (Chow 1959).

The channel is straight and uniform and from equation (3.11) the likely range of the transverse dispersion coefficient is

$$0.15 < \frac{e_z + k_z}{HU^*} < 0.30$$

From Table 3.6 this gives $0.005 < e_z + k_z < 0.010\ \mathrm{m^2\,s^{-1}}$.

For a point source located at three-quarter width, Figure 3.7a indicates that mixing is 90% complete (i.e. $P_m = s_{min}/s_{max} = 0.90$) at $x^* = 0.35$. The same number can be derived from Figure 3.8. Substituting in equation (3.27c) values for V_x and b from Table 3.6 $100 < x < 300$ m. The uncertainty in the estimated transverse mixing distance is high, but this is typical for calculations of this nature. The time taken to achieve complete transverse mixing is $x/V_x = 17–33$ minutes. Even the lower bound estimate of mixing time is longer than the target of five minutes and these preliminary

Table 3.6. Hydraulic parameters of the Ohinemuri River

Slope (s)	5.5×10^{-4}
Depth* (H)	0.62 ± 0.28 m
Width* (b)	5.60 ± 0.86 m
Hydraulic radius* (R)	0.50 ± 0.20 m
River flow (Q)	$0.30\ \mathrm{m^3\,s^{-1}}$
Mean velocity (V_x)	$0.10 \pm 0.05\ \mathrm{m^3\,s^{-1}}$
Shear velocity (U^*)	$0.058\ \mathrm{m^3\,s^{-1}}$

* Mean ± standard deviation of five measurements along a 1 km reach.

calculations indicate that a diffuser of some kind is required to achieve the rapid initial dilution required.

Dye tests were carried out to corroborate these preliminary calculations. An instantaneous release of dye was made 1 m from the right bank and dye concentrations were measured at five sites across each of several cross-sections. Dye was found to become well mixed ($P_m = 0.88$) at $x = 185$ m. This lies within the range of the mixing distance predicted above. The mean velocity determined from the time of travel of the dye peak was found to be $V_x = 0.12 \text{ m s}^{-1}$ (slightly higher than the $V_x = 0.10 \text{ m s}^{-1}$ used in the previous calculations). This implies that the transverse dispersion coefficient was $0.007 \text{ m}^2 \text{s}^{-1}$, which is within the range of values estimated earlier.

3.5.4 Deficiencies of the Constant-coefficient Model

The constant-coefficient model assumes that the depth, velocity and transverse dispersion coefficient are constant along and across the channel. This allows the mixing equation to be solved analytically and greatly facilitates problem solving. When first tackling a practical mixing problem there are usually rather sparse data available on river bathymetry and velocity distributions, while the transverse diffusion and dispersion coefficients are rarely known accurately. In this situation there seems to be little point in using a sophisticated mixing model and the constant-coefficient model is arguably the most appropriate for making an initial analysis. In a natural channel, however, the depth and velocity are not constant but vary with transverse location. This can be expected to have some effect on the rate of transverse mixing. It is also likely that the transverse diffusion and dispersion coefficients vary transversely, although a satisfactory body of theory has not yet been developed to predict this variation. The question arises of how best to handle the non-uniformity encountered in natural channels. It is always possible to calculate an average depth, velocity and dispersion coefficient and to use these average values in the constant–coefficient model. The following sections discuss the problems that arise from this approach.

3.5.4.1 Effects of non-uniform depth

Few natural channels are rectangular and most are markedly shallower near the banks than in mid-channel. It is natural to ask whether transverse variations in depth affect the rate of transverse mixing. Holley *et al* (1972) addressed this question by conducting numerical experiments and comparing predicted tracer concentrations in a trapezoidal and a rectangular channel. The reach-averaged velocity, depth and dispersion coefficient were the same so that the differences in predicted tracer concentration reflected only the different depth distributions. They modelled a steady tracer source located at one bank and solved equation (3.1) using a finite difference scheme. Bankside concentrations were found to be markedly higher in the shallow water of the trapezoidal channel close to the source. The transverse diffusive tracer flux from equation (3.1), assuming that $v_z = 0$, is

$$F_z = \int_{y=0}^{h} k_z \frac{\partial c}{\partial z} \, \mathrm{d}y \tag{3.32}$$

where it is assumed that $k_z \gg e_z$. For constant k_z the flux depends on the depth and the transverse concentration gradient. In the trapezoidal channel the depth is small close to the bank and hence the transverse flux is small. This causes tracer to accumulate near the bank, which increases the concentration gradient. Holley's results indicate, however, that the increased concentration gradient is insufficient to compensate for the small depth. As the tracer is carried downstream differences between channels decrease, which is easily understood because as the tracer spreads out the differences in depth experienced by the cloud diminish.

3.5.4.2 *Effects of non-uniform diffusivity*

Holley *et al* (1972) also addressed the question of whether transverse variations in k_z affect the rate of transverse mixing. Holley considered a trapezoidal channel and assumed three distributions for k_z

$$k_z = \text{constant} = k_1 H U^* \tag{3.33}$$

$$k_z = k_2 h v_x \tag{3.34}$$

$$k_z = k_3 h u^* \tag{3.35}$$

where h = local depth; v_x = local depth-averaged velocity; u^* = local shear velocity (see equation 3.19); H = mean depth; and U^* = cross-sectional averaged shear velocity (see equation 3.10). The coefficients k_1, k_2 and k_3 were chosen so that the mean values of k_z were the same. For a steady source at the bank, bankside concentrations near the source were significantly higher for equations (3.34) and (3.35) than for equation (3.33). Differences decreased with distance below the source as would be expected. The velocity and depth distributions were identical so the concentration differences were solely the result of different values of k_z. The differences can be explained by the fact that for the trapezoidal channel considered, equations (3.33) and (3.35) predicted lower values of k_z near the banks than equation (3.33). Equations (3.33) and (3.35) are only suggestions about how the diffusivity might vary across the channel, but these simulations indicate that if the diffusivity is low near the bank then tracer is likely to be retained in that area.

We can conclude this section by saying that Holley's numerical experiments provide evidence that non-uniformities in the depth and dispersion coefficient affect the rate of transverse mixing. It is surprising that in the 20 years since this paper was published there has been so little progress in quantifying the variability in the transverse diffusion and dispersion coefficients. A probable explanation is the lack of a reliable theory of transverse mixing to help guide and interpret experimental work.

3.5.4.3 Transverse mixing in the near-field

Equation (3.22) and Figures 3.7 and 3.8 are based on the assumption that the tracer originates from a vertical line source and that vertical concentration gradients are negligible. This introduces a degree of approximation when applied to the case of a point source because tracer takes a finite time to become fully mixed vertically (the distance required for tracer to become vertically well mixed can be estimated using the methods outlined in Chapter 2). If transverse mixing calculations are being made well below this point then equation (3.22) can be used with some confidence. If not, then both vertical and transverse mixing will need to be considered in which case equations (2.42)–(2.49) and figures like 2.8 are required.

In the near-field where both vertical and transverse mixing proceed simultaneously the transverse diffusivity has been found to be smaller than it is in the mid-field (Okoye 1970, Nokes 1986). The reason is that close to the source large transverse eddies (which effect most of the mass transport) are unable to break up the tracer plume and simply cause it to move bodily backwards and forwards across the flow. As discussed in Section 3.3 such meandering does not affect tracer concentrations. It is well known from ocean dispersion studies that as the size of the tracer patch increases the rate of dispersion increases (Okubo and Ozmidov 1970) as larger and larger eddies become involved in the mixing process. If we consider relative diffusion (i.e. diffusion of the tracer cloud relative to its centroid, which may wander within the flow), then in the near-field the lateral diffusivity should be reduced. Nokes (1986) suggests that the reduction is of the order of 50% within about 10 depths of the source. When the principal concern is with predicting tracer concentrations in the mid-field then this local reduction of transverse diffusivity close to the source can safely be neglected.

3.5.5 Heuristic Modifications to the Constant-coefficient Model

The discussion in the preceding section shows that the constant-coefficient model is only approximate as a description of transverse mixing in natural channels because it ignores transverse variations in the depth, velocity and dispersion coefficient. If, however, we restrict our attention to just one small region of the channel then it is reasonable to assume that the coefficients in that region are constant. Provided we choose the appropriate coefficients then it can be argued that the constant-coefficient model should provide reasonable predictions. This approach is best illustrated with a practical example.

Example 3.3

During the 1970s a point-source of dairy factory waste with a mass flow of suspended solids of 200 $g\,s^{-1}$ was located close to the left bank of the Waikato River near Hamilton, New Zealand. Milk solids have a strong visual impact and the problem was to estimate how far downstream, and over how much of the channel width, the plume extended in which concentrations exceeded 5 $g\,m^{-3}$. The channel parameters at low flow are shown in Table 3.7.

The fully mixed concentration is $C_m = m/Q = 0.80\ g\,m^{-3}$. For $s = 5\ g\,m^{-3}$, equation (3.27a) gives $c^* = 6.25$. In Figure 3.7a the $c^* = 6.25$ contour for a source at the bank extends downstream a maximum distance of about $xk_z/V_xb^2 = 0.0075$. In the absence of other information we can assume that $k_z/HU^* \approx 0.4$. For the width-averaged parameters in Table 3.7, $k_z \approx 0.047\ m^2\,s^{-1}$. (Table 3.1 shows that the measured value of k_z about 3 km upstream was $0.064\ m^2\,s^{-1}$, which is remarkably similar.) Adopting $k_z \approx 0.047\ m^2\,s^{-1}$, $V_x = 0.74\ m\,s^{-1}$ and $b = 140$ m then the plume length is $x \approx 2300$ m. The plume width for the $c^* = 6.25$ contour is (from Figure 3.7a) $z/b \approx 0.08$ and hence $z \approx 11$ m.

The last calculation shows that the plume is confined very close to the left bank. The depth, velocity and probably the dispersion coefficient are low in this near-bank region. This is likely to reduce the rate of transverse mixing near the outfall. Measurements of hydraulic parameters within 10 m of the bank are summarised in Table 3.8.

It remains to estimate the transverse dispersion coefficient near the bank and here considerable conjecture enters the calculations. One possibility is that the non-dimensional dispersion coefficient remains constant at $k_z/hu^* = 0.4$. Assuming that the Darcy friction coefficient remains

Table 3.7. Hydraulic parameters of the Waikato River near Hamilton

Flow (Q)	$250\ m^3\,s^{-1}$
Mean depth* (H)	2.4 m
Width (b)	140 m
Slope (s)	1.0×10^{-4}
Mean velocity* (V_x)	$0.74\ m\,s^{-1}$
Shear velocity (U^*)	$0.049\ m\,s^{-1}$

* Mean denotes width-averaged.

Table 3.8. Hydraulic parameters close to the bank

Local depth* (h)	1.0 m
Local velocity* (v_x)	0.35 m s^{-1}

* Local denotes depth-averaged value close to the bank

constant, where

$$f = 8\left(\frac{u^*}{v_x}\right)^2_{\text{bank}} = 8\left(\frac{U^*}{V_x}\right)_{\text{width-averaged}}$$

then

$$\left(\frac{u^*}{v_x}\right)_{\text{bank}} = \left(\frac{U^*}{V_x}\right)_{\text{width-averaged}}$$

and the shear velocity in the shallows can be estimated to be $u^* = 0.023\ \text{m s}^{-1}$. Taking $h = 1.0$ m and $u^* = 0.023\ \text{m s}^{-1}$ gives $k_z = 0.0092\ \text{m}^2\,\text{s}^{-1}$ which is only 20% of the width-averaged value. It must be stressed that we do not know how k_z varies across the channel and this is just one plausible estimate: only field tests would provide a reliable estimate.

Taking $xk_z/V_xb^2 = 0.0075$ and substituting the near-bank coefficients yields $x \approx 5600$ m. This latter estimate of plume length is more than twice the earlier estimate based on the width-averaged coefficients. Note that near the bank the velocity is low, which tends to reduce the length of the plume. This is more than offset by the estimated decrease in the dispersion coefficient so that overall the plume length increases substantially. The residence time of particles within the plume x/V_x increases from 52 minutes (width-averaged coefficients) to 267 minutes as a result of lower velocity and reduced transverse mixing.

There are two obvious weaknesses in this approximate method of analysis. Firstly, the values of velocity, depth and dispersion coefficient change as the size of the plume increases. In the iterative approach outlined here the likely extent of a given contour is estimated using a first guess of the coefficients (in this example the width-averaged values). The values of the model coefficients are then determined within that part of the channel containing the contour of interest and the calculation is updated. Were a different concentration contour selected (say $c^* = 2$) then the average coefficients within the plume would be different. It is obvious that this approach neglects the variations in model coefficients that occur within the plume, but it enables a refined estimate to be made of the gross plume behaviour. As the value of c^* approaches unity the need for this correction diminishes.

The second weakness is that the constant-coefficient model assumes that the model coefficients are constant across the whole channel. In Example 3.3 it is appropriate to reduce the dispersion coefficient and velocity near the bank, but near the centre of the channel both coefficients remain high. This means that the constant-coefficient model almost certainly underestimates the rate of transverse mixing along the outer edge of the $c^* = 6.25$ contour. One consequence is that in Example 3.3 the corrected tracer predictions are likely to be pessimistic (i.e. the predicted concentrations are likely to be higher than the actual concentrations). Taking this into account, Example 3.3 tells us that the $c^* = 6.25$ contour probably extends more than 2300 m but less than 5600 m downstream. Despite these shortcomings the outlined modification to the constant-coefficient model is useful, especially when assessing the impact of bankside sources.

3.6 TOWARDS A VARIABLE-COEFFICIENT MODEL

In Chapter 2 reference was made to the work of McNulty (1983) and Nokes (1986) who solved the vertical mixing equation allowing the velocity and diffusivity to vary with depth. The velocity is known to vary logarithmically with depth (although a power law approximation is often used) and the vertical diffusivity is known to vary parabolically with depth. The mixing equations are difficult to solve when the coefficients are variable but McNulty (1983) developed a method for predicting the moments of the concentration distributions while Nokes (1986) found an elegant eigenfunction expansion solution for the concentrations themselves.

It is tempting to adopt the same approach for transverse mixing and to model transverse variations in velocity and the transverse dispersion coefficient. The main impediment is that in the general case the transverse variation in these coefficients is not known. In a particular channel it is easy to measure the transverse distribution of velocity and depth but determining the transverse distribution of the dispersion coefficient presents more of a problem. At present we do not know enough about the fluid mechanics of natural channels to estimate the distributions of e_z and k_z either theoretically or from the measured velocity distributions. It might be possible to conduct a large number of small-scale tracer tests to measure the transverse distributions of e_z and k_z although this presents logistical problems. If information were available on the distribution of depth, velocity and dispersion coefficient then it would be profitable to seek solutions to equation (3.1) using numerical methods.

There have been several theoretical studies which have sought solutions to equation (3.1) (Cleary and Adrian 1973, Wang *et al* 1978, Smith 1982). Each study has made different assumptions about the transverse variation of depth, velocity and transverse dispersion coefficient. Our poor understanding of natural river channels means that these results cannot be applied generally. They do, however, provide valuable qualitative insights into the effects of non-uniform coefficients. The most important differences between the approximate constant-coefficient model and the more realistic variable-coefficient models relate to the behaviour of the concentration maximum, the mixing distance and the optimum source location.

Consider first the constant-coefficient model. It predicts that in the absence of side boundaries the concentration maximum remains on the streamline which passes through the source. As tracer reaches the banks it is reflected and accumulates near the boundary thereby causing the concentration maximum to migrate towards the bank nearest the source. For a source in mid-channel the banks are equidistant and the constant-coefficient model predicts that the concentration maximum will remain in mid-channel. This behaviour is apparent in Figure 3.7. It is important to appreciate that in the constant-coefficient model migration of the tracer maximum (which occurs for sources displaced from mid-channel) is entirely due to the presence of the banks.

In the variable-coefficient model the behaviour of the concentration maximum is rather different. If either the velocity or the dispersion coefficient is non-uniform, then the concentration maximum migrates even in the absence of boundaries. This behaviour is apparent in the results of Wang *et al* (1978), who modelled a straight rectangular channel in which the longitudinal velocity V_x varied across the channel, the transverse velocity was everywhere zero and the dispersion coefficient was constant. In this situation the concentration maximum moved towards the region of low velocity. Wang concluded that the relative position of the source and the velocity maximum had a profound influence on the concentration distribution. Nokes (1986) and Nokes and Wood (1988) described the mechanisms which cause the concentration maximum to migrate vertically and their variable-coefficient model predicts that the tracer maximum migrates towards regions of low diffusivity and/or low velocity. In a natural channel the velocity is lower near the bank than in mid-channel and there are indications that the transverse dispersion coefficient may also be lower near the bank. Consequently, for a source displaced from the channel centreline, tracer is likely to impinge more rapidly on the near bank than is predicted by the constant-coefficient model. Unfortunately, it is not possible to quantify how much more quickly tracer reaches the bank in the general case because the distribution of the dispersion coefficient is unknown. In a practical problem, if bank impingment is a critical issue then field tests should be conducted.

Wastes are commonly discharged steadily from point sources and it is important to choose the correct location for the outfall. The littoral zone in rivers contains more abundant plant and animal communities than the main channel and a commonly encountered problem is to choose the optimum source location which minimises bankside concentrations. It so happens that this is also the source location which minimises the transverse mixing distance. For a straight and symmetrical channel it is intuitively obvious that the optimum source location is in mid-channel. Smith (1982) identified the optimum source location for two asymmetrical straight channels (semi-parabolic and triangular) assuming that the velocity and diffusivity varied as $h^{1/2}$ and $h^{2/3}$, respectively (where h = local depth). In each case the optimum source location was displaced from mid-channel towards deeper water by approximately 10% of the channel width, giving the optimum source location as approximately $0.6b$ from the shallow side of the channel (where b = channel width). In a meandering channel momentum effects also become important in determining the optimum source location. Smith considered an idealised meandering channel and showed that the optimum source location is displaced from the centreline towards the inside of the bends, although he was unable to quantify the extent of the displacement in the general case. The displacement may vary with the flow rate and can be expected to depend on the channel curvature and bathymetry. If the source is slightly displaced from the optimum location (by up to 10% of the channel width), then mixing and bankside concentrations are not greatly affected, which means that in a practical problem the engineer has some leeway. Nevertheless, for selecting the optimum source location in a complex natural channel it is advisable to conduct physical model or field tracer tests.

McNulty (1983) and Nokes (1986) have both shown that the variable-coefficient model for vertical dispersion predicts a shorter vertical mixing distance (typically by 10–20%) than the constant-coefficient model for a source on the bed or at the water surface (see Figure 2.6). Nokes points out, however, that it is not intuitively obvious how the mixing distance behaves when both the velocity and diffusivity are variable. There are significant differences between the depth variation of velocity (which is logarithmic and hence monotonic) and the transverse variation (which is parabolic and hence has a local maximum). It would be unwise, therefore, to draw an analogy from Chapter 2 and to suggest that the transverse mixing distance predicted by the constant-coefficient model is likely to overestimate the true transverse mixing distance. This question must remain unanswered and is a fruitful topic for further research. It would be wise to ascribe a fairly high uncertainty to estimates of the transverse mixing distance made using the constant-coefficient model and where the mixing distance is critically important then these estimates should be confirmed by field tests.

It is worth restating that in most circumstances the major impediment to the accurate estimation of tracer concentrations is not so much the approximate nature of the constant-coefficient model but the practical difficulty (viz. cost) of measuring the depth and velocity, together with the theoretical difficulty of estimating the transverse dispersion coefficient. For many practical problems, therefore, the combination of the constant-coefficient model and field tests will provide adequate predictions.

3.7 STREAMTUBE MODEL FOR TRANSVERSE DISPERSION

3.7.1 Introduction

In an irregular river channel, variations in the bathymetry result in non-zero values of the depth-averaged transverse velocity v_z, which cause the tracer cloud to move backwards and forwards across the channel. If the transverse velocity does not vary with depth, then meandering of the thalweg does not contribute to transverse mixing and simply moves the tracer cloud bodily within the x–z coordinate framework. If the transverse velocity varies over the depth then there will be some transverse dispersion superimposed on the transverse advection of the tracer cloud.

In most practical problems we are more interested in changes of concentration (i.e. with dispersion) than with bodily movements of the tracer cloud within the channel. It is natural, therefore, to analyse tracer dispersion in terms of cumulative discharge rather than transverse distance. Cumulative discharge is defined in Section 1.9.3 as

$$q(\alpha, \beta) = \int_{\beta' = -\infty}^{\beta} m_\beta h v_\alpha \, \mathrm{d}\beta' \tag{3.36}$$

where α and β = longitudinal and transverse coordinate directions in an orthogonal curvilinear system; α is parallel with the local depth-averaged longitudinal velocity and β is transverse across the flow; q = cumulative discharge (i.e. the flow between the left bank and the transverse location β); m_β = a metric coefficient which accounts for channel curvature (typically between 0.8 and 1.2 in a meandering channel but 1.0 in a straight channel); v_α = local depth-averaged velocity; and h = local depth. By convention $q = 0$ at the true left bank and then $q = Q$ at the true right bank. For steady river flow and a steady input of tracer it was shown in Section 1.9.3 that

$$\frac{\partial s}{\partial \alpha} = \frac{\partial}{\partial q}\left(m_\alpha h^2 v_\alpha k_\beta \frac{\partial s}{\partial q}\right) \tag{3.37}$$

where s = depth-averaged tracer concentration; m_α = a metric coefficient [which is close to unity and similar to m_β in equation (3.36)] and k_β = local transverse dispersion coefficient. This is the so-called streamtube model described by Yotsukura and Cobb (1972). Equation (3.37) only applies to the situation where the tracer input and flow are both steady, although the streamtube model has since been extended to include the case of steady flow but non-steady tracer input (Luk *et al* 1990).

The group of terms

$$D(\alpha, q) = m_\alpha h^2 v_\alpha k_\beta \tag{3.38}$$

is called the factor of diffusion.

At a given cross-section we can align the curvilinear coordinate axes α and β with the more familiar Cartesian axes x and z. For convenience we will use x and z in place of α and β for the remainder of this section. It is important to remember, however, that the z axis is always pointing transversely across the channel and the x axis is always pointing downstream.

As has been discussed earlier in this chapter, it is not generally known how the transverse dispersion coefficient varies across the channel. Lau and Krishnappan (1981) compared numerical solutions of equation (3.37) with observed tracer concentrations for a bankside discharge in the Grand River, Ontario. They found that predictions were fairly insensitive to transverse variations of k_z provided transverse variations of depth and velocity were modelled. This corroborates the conclusion of Holley *et al* (1972) (see Section 3.5.4) that transverse variations of depth strongly influence transverse mixing. The two studies differ, however, in that Holley concluded that transverse variations of k_z strongly influence tracer concentrations. As Lau and Krishnappan base their conclusions on matching tracer profiles in a natural river they should perhaps be given more weight than the results of Holley's numerical experiments in a hypothetical channel. Thus, although there is some conjecture about the need to model the transverse variation in k_z, there is clear evidence of the need to model the transverse variation of depth and velocity to make accurate simulations of tracer concentration, especially close to a tracer source. If the factor of diffusion D varies across the channel then equation (3.37) must be solved numerically (Lau and Krishnappan 1981).

3.7.2 Simplified Streamtube Model

As a first approximation the factor of diffusion can be assumed to be constant across the channel. The main justification for this assumption is that in practice variations of D across natural channels are smaller than variations along the channel (Lau and Krishnappan 1981). For convenience D can be written

$$D = \psi H^2 V_x k_z \tag{3.39}$$

where H = mean depth; V_x = mean velocity; and

$$\psi = \frac{1}{b} \int_{z=0}^{b} \left(\frac{h}{H}\right)^2 \frac{v_x}{V_x} \, dz \tag{3.40}$$

ψ = a dimensionless shape factor found to lie in the range $\psi = 1.0$–3.6 (Sayre 1979, Beltaos 1980b).

Several workers state that it is also acceptable to assume that D is constant along the channel (Yotsukura and Cobb 1972, Sayre 1979, Holly 1985). Making this assumption, equation (3.37) becomes

$$\frac{\partial s}{\partial x} = D \frac{\partial^2 s}{\partial q^2} \tag{3.41}$$

Equation (3.41) is identical in form to equation (3.22) and can be called the simplified streamtube model. At the banks the tracer flux is zero so that

$$D \frac{\partial S}{\partial q} = 0 \qquad \text{at } q = 0 \text{ and } q = Q \tag{3.42}$$

As D is assumed constant equation (3.42) implies that the concentration gradient is zero at the banks. (Note that if either the depth or velocity vanishes at the banks then strictly D vanishes and there is no longer a requirement that the concentration gradient vanish. This situation only occurs in the variable-coefficient model.) By analogy with equation (3.25) the solution to equation (3.41) for a vertical line source at $x = 0$, $q = q_o$ in an unbounded flow is

$$s_u(x, q) = \frac{m}{\sqrt{4\pi D x}} \exp\left(-\frac{q^2}{4Dx}\right) \tag{3.43}$$

where m = mass inflow rate. The effects of boundaries can be accommodated using the method of images, as has been described earlier and

$$s(x, q) = s_u(x, q \pm q_o) + \sum_{n=1}^{N} s_u(x, 2nQ \pm q \pm q_o) \tag{3.44}$$

where the banks are at $q = 0$ and Q and $s_u(x, q)$ is the unbounded solution given by equation (3.43). We can form new dimensionless variables

$$c^* = \frac{s}{C_m} \tag{3.45a}$$

$$q^* = \frac{q}{Q} \tag{3.45b}$$

$$x^* = \frac{xD}{Q^2} \tag{3.45c}$$

where C_m = fully mixed concentration; and c^*, q^* and x^* = non-dimensional concentration, cumulative flow and longitudinal distance, respectively. The nomographs in Figures 2.2 and 3.7 can then be used to solve streamtube mixing problems provided y^* or z^* is replaced with q^*, and x^* is defined using equation (3.45c).

In many natural channels the bathymetry varies longitudinally and so too does the shape factor ψ. The transverse dispersion coefficient is known to vary longitudinally, especially if the channel is sinuous. Consequently, in these circumstances it is desirable to make several estimates of D within the study reach and to calculate the reach-averaged value

$$D = \frac{1}{L} \sum_{i=1}^{N} D_i \Delta x_i \qquad (3.46)$$

where L = total reach length; Δx_i = sub-reach length; D_i = value of the factor of diffusion in sub-reach i; and N = number of sub-reaches. In each sub-reach D_i, $i = 1, 2, \ldots, N$ may be determined from equation (3.39) provided the bathymetry and velocity distribution is known and k_z can be estimated.

The transverse mixing distance (i.e. the longitudinal distance below a point

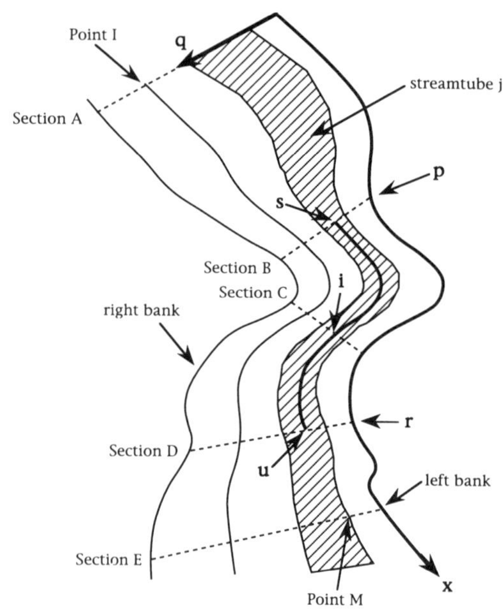

Figure 3.9. Sketch of a hypothetical irregular river channel showing four streamtubes and five cross-sections. The transverse and longitudinal scales are highly distorted. Tracer is injected at point I and measured at point M. The left bank is the line $q = 0$ (i.e. the x axis) and at section A $x = 0$ (i.e. the q axis). The heavy line s–u lies at the midpoint of streamtube j

source at which tracer becomes well mixed across the channel) can be estimated using the simplified streamtube model. By analogy with equations (3.28) and (3.29) the mixing distance to achieve 98% mixing (i.e. $P_m = s_{min}/s_{max} = 0.98$) is

$$L_z = 0.536 \frac{Q^2}{D} \tag{3.47}$$

for a source near either bank and

$$L_z = 0.134 \frac{Q^2}{D} \tag{3.48}$$

for a source at mid-flow. The mixing distance for other source locations and other values of P_m can be estimated from Figure 3.8 provided the non-dimensional variables are defined using equation (3.45).

Example 3.4

A hypothetical natural channel of non-uniform cross-section is shown in Figure 3.9. An estimate is required of the tracer concentration at point M for a steady point discharge of 100 g s^{-1} at point I. Depth and velocity distributions are available at five transects as summarised in Table 3.9. The channel is non-uniform and the problem is best tackled using the streamtube model.

As a first approximation we divide the channel into four uniform sub-reaches and use the simplified streamtube model. Depth, velocity and flow distributions are shown in Figure 3.10 for sections A, C and E: those for sections B and D are omitted for the sake of brevity. Several preparatory steps must be taken.

Firstly, at the five cross-sections where velocity and depth are known the cumulative discharge is calculated at 1 m intervals across the channel

$$q_k = \sum_{i=1}^{k} v_{xi} h_i \Delta z$$

Table 3.9. Hydraulic parameters at five sites in a hypothetical channel

Transect	Distance (m)	Width (b) (m)	Mean* velocity (V_x) (m s^{-1})	Mean* depth (H) (m)	Shape factor (ψ)	Dispersion coefficient (k_z) (m^2 s^{-1})	Factor of dispersion (D_j) (m^5 s^{-2})	Sub-reach length (Δx_j) (m)
A	0	25	0.60	1.0	1.7	0.018	0.0184	100
B	200	20	0.65	1.15	2.1	0.018	0.0325	250
C	500	12	0.85	1.47	1.3	0.018	0.0430	250
D	700	32	0.55	0.85	2.4	0.018	0.0793	250
E	1000	25	0.60	1.00	1.8	0.018	0.0194	150

*Mean denotes cross-sectional averaged.

where v_{xi} and h_i = local depth-averaged velocity and local depth measured at points Δz apart across the channel. In this example $\Delta z = 1$ m and the maximum value of k varies from 12 at section C to 32 at section D. The distributions of velocity, depth and cumulative discharge for sections A, C and E are sketched in Figure 3.10. The metric coefficient m_β does not appear in the summation because we have accounted for irregularities in channel width by allowing k_{max} to vary between sections.

Secondly, on a scaled map of the test reach, points of equal cumulative

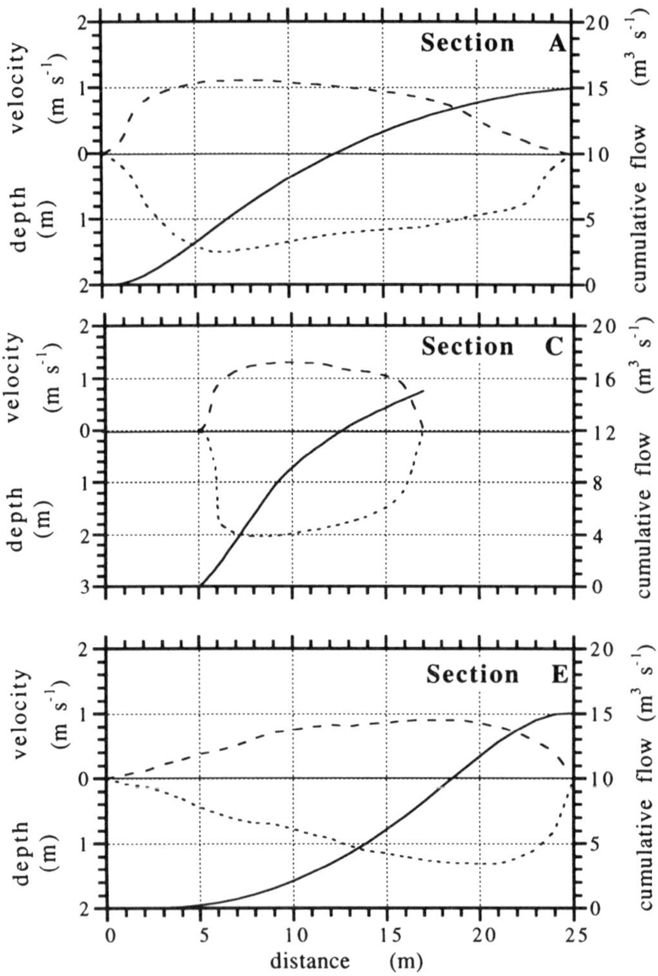

Figure 3.10. Velocity, depth and cumulative flow distributions at three cross-sections in a hypothetical channel. Distance is measured from the left bank

flow are marked on each cross-section and joined by eye to give the stream tubes. Only four streamtubes are shown in Figure 3.9 but in practice 10 or more are desirable. The metric coefficients m_α are then measured at the mid-point of each streamtube by scaling off the map. For example, at cross-section C for streamtube i

$$m_{\alpha i} = \frac{s-u}{p-r}$$

where $s-u$ and $p-r$ = distance between points s and u and p and r, respectively, measured along the streamline.

Thirdly, at each cross-section the shape factor is evaluated

$$\psi = \frac{1}{b} \sum_{i=1}^{N} m_{\alpha i} \frac{v_{xi}}{V_x}\left(\frac{h_i}{H}\right)^2 \Delta z_i$$

where b = channel width; v_{xi} = mean depth-averaged velocity in streamtube i; V_x = cross-sectional averaged velocity; h_i = mean depth in streamtube i; $m_{\alpha i}$ = metric coefficient in streamtube i; Δz_i = width of streamtube i; and N = total number of streamtubes. Values of ψ are given in Table 3.9.

Fourthly, the cross-sectional averaged factor of diffusion is estimated at each transect

$$D_j = (\psi k_z V_x H^2)_j$$

where j = transect number = 1, 2, ... 5. The mean depth, mean velocity and shape factor at each transect are known and it remains to estimate k_z. As a first approximation we can estimate an average k_z for the whole test reach. Knowing the slope $s = 10^{-4}$ and mean depth $H = 1.09$ m (Table 3.10), the shear velocity $U^* = 0.033$ m s^{-1}. Assuming that $k_z/HU^* = 0.50$ then $k_z = 0.018$ m^2 s^{-1} and D_j can be calculated for each transect (see Table 3.9).

Finally, the reach-averaged value of D is calculated

$$D = \frac{1}{L} \sum_{j=1}^{5} D_j \Delta x_j$$

which gives a value of $D = 0.0435$ m^5 s^{-2}.

Now the point of discharge I is located at $z = 14.5$ m from the left bank

Table 3.10. Reach-averaged parameters

Channel slope (s)	1.0×10^{-4}
Mean depth* (H)	1.09 m
Shear velocity (U^*)	0.033 m^3 s^{-1}
Discharge (Q)	15 m^3 s^{-1}
Dispersion (k_z/HU^*)	0.50
Factor of diffusion (D)	0.0435 m^3 s^{-1}

* Mean denotes cross-sectional averaged.

at section A. From Figure 3.10 at this point cumulative flow $q = 11.3\ \mathrm{m^3\,s^{-1}}$ and from equation (3.45) $q^* = 0.75$. We can use Figure 3.7a for a source at $q^* = 0.25$ as for the constant-coefficient model solutions are symmetrical about the centreline.

The measuring point M is located at $z = 8$ m from the left bank. From Figure 3.10 at point M, $q = 3\ \mathrm{m^3\,s^{-1}}$. Thus $q^* = 0.20$. At point M, $x = 1000$ m and from equation (3.45c), $x^* = 0.193$.

Finally from Figure 3.7a at $x^* = 0.193$ and $q^* = 0.20$, $c^* = 0.85$. Thus an initial estimate of tracer concentration at point M is $s = 5.7\ \mathrm{g\,m^{-3}}$.

3.7.3 Limitations of the Simplified Streamtube Model

The simplified streamtube model is a convenient tool for making preliminary estimates of tracer concentration in an irregular channel. It does assume, however, that the factor of diffusion is constant. In a natural channel we would expect the factor of diffusion to be low in shallow, slow moving water near the banks (see equation 3.38) and consequently the simplified streamtube model is likely to overestimate the rate of transverse mixing of tracer released into the near-bank region. Example 3.3 describes a modification to the constant-coefficient Cartesian mixing model which gives improved (although still only approximate) predictions near the banks. Similar modifications can be made to the streamtube model: the detailed implementation is straightforward and is left as an exercise for the reader.

The simplified streamtube model predicts that the optimum source location (which minimises both the transverse mixing distance and the bankside concentrations) lies at mid-flow. For straight channels this is a good choice, but at bends it is necessary to allow for momentum and move the source slightly towards deeper water as is discussed in Section 3.6.

The simplified streamtube model predicts that for all sources (except that at mid-flow) the concentration maximum moves slowly towards the nearest bank as a result of the no-flux boundary condition. In a natural channel the factor of diffusion is lower near the banks than in mid-channel and consequently the concentration maximum migrates towards the bank more rapidly than is predicted by the simplified streamtube model. We cannot say with confidence how the actual concentration contours will differ from those predicted by the constant-coefficient model. Because the factor of diffusion is not constant in natural channels, the transverse mixing distances estimated using the simplified streamtube model are only approximate and it would seem wise to ascribe them a high uncertainty. Where it is critically important to know L_z accurately then field tracer studies should be carried out.

3.8 TRANSVERSE MIXING FROM A DIFFUSER OF FINITE LENGTH

3.8.1 Graphical Method of Superposition

It is common for tracer to be released from a diffuser of finite length rather than from a point source. This may be necessary to increase the near-field dilution and to comply with a receiving water standard. As a first approximation we can use the constant-coefficient model to analyse this problem. If the momentum of the discharge is small so that it does not affect the flow pattern then the solutions given by equations (3.26) for a steady vertical line source can be adapted to determine the steady-state solution for a diffuser of finite length. Equation (3.22) is linear so that the superposition principle applies. The diffuser is assumed to consist of a number of point sources distributed evenly along the length of the pipe, each discharging a fraction of the total mass flow. Equation (3.26) gives the plume concentrations for each separate point source and these separate concentration distributions are added together. The superposition can be performed graphically, as is illustrated in the following example.

Example 3.5
For a hypothetical outfall with a mass flow rate of $20\ \mathrm{g\,s^{-1}}$, compare the length and width of the $4\ \mathrm{g\,m^{-3}}$ contour for a point source located at mid-channel and a mid-river diffuser whose ends are 4 m and 6 m from the bank. The channel is irregular and gently meandering with the average channel parameters given in Table 3.11.

The first step is to estimate the transverse dispersion coefficient. For a straight, non-uniform channel equation (3.12) gives $0.3 < k_z/HU^* < 0.9$ and hence $0.0093 < k_z < 0.028\ \mathrm{m^2\,s^{-1}}$. It is advisable to make predictions with a range of k_z values within this range, but in this illustration a single value of $k_z = 0.020\ \mathrm{m^2\,s^{-1}}$ will be used. The fully mixed concentration, $C_m = m/Q = 2\ \mathrm{g\,m^{-3}}$. From equation (3.27a) for $s = 4\ \mathrm{g\,m^{-3}}$, $c^* = 2$.

Table 3.11. Hydraulic parameters in a hypothetical channel

River flow (Q)	$10\ \mathrm{m\,s^{-3}}$
Mean depth (H)	1.0 m
Width (b)	10 m
Slope (s)	1.0×10^{-4}
Mean velocity (V_x)	$1.0\ \mathrm{m\,s^{-1}}$
Shear velocity (U^*)	$0.031\ \mathrm{m\,s^{-1}}$

First consider mixing below a point source in mid-channel. From Figure 3.7b for $c^* = 2$, $xk_z/V_x b^2 = 0.020$, $x = 100$ m, $z/b = 0.24$ and $z = 2.4$ m.

Now consider the case of the diffuser of finite length. As a first approximation, the plume contours can be obtained by assuming two point sources, each with a mass flow of 10 g s^{-1} located at $z = 4.5$ and 5.5 m. The displacement of these sources from the channel centreline is small and for much of the mid-field concentrations are not affected by the presence of the

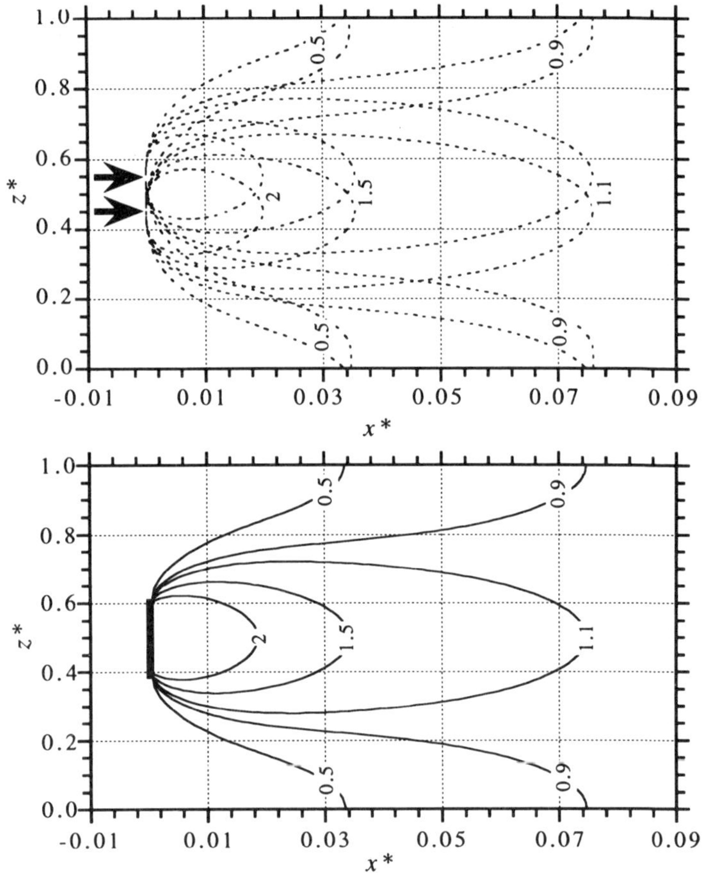

Figure 3.11. Concentration contours downstream from a diffuser of finite length. The diffuser stretches from $z^* = 0.4$ and 0.6 but is approximated by two point sources at $z^* = 0.55$ and 0.45. The top graph shows the concentration contours predicted using equation (3.26) for each source separately and the bottom graph shows the average concentration

banks. Consequently concentration contours below each of these sources can be assumed to be identical to those below a source at the centreline and hence can be estimated from Figure 3.7a. Two sets of contours from Figure 3.7a have been displaced by 0.1 m and superimposed as shown in Figure 3.11. Then for $c^* = 2$, $xk_z / V_x b^2 = 0.018$, $x = 90$ m, $z/b = 0.26$ and $z = 2.6$ m.

The length of the 4 g m^{-3} contour has been reduced by 10 m and its width increased by 0.2 m through the use of a short diffuser. Predictions could be improved by assuming three sources, each with a mass flow of 6.67 g s^{-1} located at $y = 4.3$, 5.0 and 5.7 m from the bank. In practice there is little point in combining more than five sets of contours.

3.8.2 Numerical Method of Superposition

Few people would attempt to solve problems like this graphically, but would use computer-based methods. The main reason for including Example 3.5 is to illustrate the superposition principle in a simple graphical way.

If the mass flux of tracer along a diffuser of length L is uniform and the total mass flux is m then an element of the diffuser of length $\Delta z'$ located at $x = 0$ and $z = z'$ has a mass flow of $m\Delta z'/L$. This elemental release causes the concentration at a point downstream to be (see equation 3.25)

$$\Delta s(x, z) = \frac{m\Delta z'}{LH\sqrt{4\pi k_z x v_x}} \exp\left(-\frac{v_x(z - z')^2}{4k_z x}\right) \tag{3.49}$$

Integrating along the length of the diffuser

$$s(x, z) = \int_{z'=0}^{L} \frac{m}{LH\sqrt{4\pi k_z x v_x}} \exp\left(-\frac{v_x(z - z')^2}{4k_z x}\right) dz' \tag{3.50}$$

Clearly if the mass flux of tracer is uneven, then numerical methods will be required to evaluate equation (3.50). Once the integral in equation (3.50) has been evaluated the effects of the banks can be accommodated using the method of images (see equation 3.26).

Example 3.6

The Waikato River receives cooling water from the Huntly Power Station. The bankside outfall is designed to mix cooling water rapidly with between 40–50% of the river flow. The near-field mixing region in which buoyancy and jet momentum are important has been shown to extend about 300 m downstream from the outfall. An initial assessment of mid-field plume temperatures was required for which the simplified streamtube model was deemed to be an appropriate tool.

Field surveys have shown that 450 m downstream from the outfall the excess temperature distribution is linearly distributed with respect to cumulative flow. Thus the appropriate boundary condition for equation (3.41) is

$$T(0, q) = T_o\left(1 - \frac{q}{q_1}\right) \qquad 0 < q < q_1$$
$$= 0 \qquad q_1 < q < Q$$

where T = depth-averaged excess temperature; q = cumulative discharge; q_1 = cumulative discharge at the edge of the plume

$$T_o = \frac{2I}{4.186 q_1}$$

where T_o = maximum plume temperature (which occurs at the left bank); I = waste heat inflow (MW); and q_1 = total plume flow. The value of q_1 varies with river flow, cooling water flow and river bathymetry, but averages $0.45Q$. Close to the outfall cooling can be neglected and the depth-averaged temperature approximated by equation (3.41) substituting T for s. There is no vertical stratification in the mid-field but there is a transverse temperature (and hence density) gradient which may slightly increase the transverse mixing rate. This can be accommodated in an augmented value of the factor of diffusion.

The solution to equation (3.41) for a point source at $x = 0$ and $q = q_o$ in an unbounded channel is

$$T_u(x, q) = \frac{T_e q_e}{\sqrt{4\pi D x}} \exp\left(-\frac{(q - q_0)^2}{4Dx}\right)$$

where T_e = cooling water excess temperature and q_e = cooling water flow. For a distributed source at $x = 0$ the superposition principle gives

$$T(x, q) = \int_{-\infty}^{\infty} \frac{T_u(0, q')}{\sqrt{4\pi D x}} \exp\left(-\frac{(q - q')^2}{4Dx}\right) dq'$$

For the excess temperature distribution given above this integral can be evaluated analytically taking account of the banks using the method of images (Rutherford and Williams 1992)

$$T(x, q) = \mathrm{f}(x, \pm q) + \sum_{n=1}^{n=\infty} \mathrm{f}(x, \pm 2nQ \pm q)$$

where

$$f(x, q) = \frac{T_o(q - q_1)}{2q_1}\left[\operatorname{erf}\left(\frac{q - q_1}{\sqrt{4Dx}}\right) - \operatorname{erf}\left(\frac{q}{\sqrt{4Dx}}\right)\right] + \frac{T_o}{q_1}\sqrt{\frac{Dx}{\pi}}\left[\exp\left(-\frac{(q - q_1)^2}{4Dx}\right) - \exp\left(-\frac{q^2}{4Dx}\right)\right]$$

and erf is the error function [see (for example) Spiegel (1968) for a definition and tabulation of values].

3.9 TWO-DIMENSIONAL NUMERICAL MIXING MODELS

3.9.1 Introduction

An initial assessment of transverse mixing in a non-uniform channel can be made using reach-averaged parameters and simplified analytical models. In many practical problems the resulting estimates of tracer concentration and mixing distance are sufficiently accurate for decision-making. In other problems, however, consistent mis-matches occur during model calibration between observed and predicted concentrations which indicate that the underlying model is inadequate. In this situation it may be necessary to allow the coefficients in equation (3.1) to vary spatially and to solve the equation using numerical methods. Numerical models based on the two-dimensional (depth-averaged) mixing equation provide a flexible and powerful tool for analysing mixing in the mid-field. Models of this type have been in use for about 15 years, their capabilities and shortcomings are well known (see, for example, Abbott and Basco 1989) and they are becoming a standard tool for analysing this important class of mixing problem.

Numerical models require considerably more effort and data to operate than the simpler analytical models and it is important to weigh up the potential advantages of numerical models against their high costs. It is seldom wise to use numerical models without first doing preliminary calculations using simple analytical models. The simple models can also be used to good effect to provide a rough check on numerical model results during the course of the study. There are several different types of numerical model and numerous computer programs in use around the world. A detailed discussion of currently available numerical models is beyond the scope of this book, but a brief discussion is included here to help give the reader an appreciation of the potential advantages and disadvantages of numerical models.

3.9.2 Hydrodynamic Sub-model

Equation (3.1) requires information about depth and velocity throughout the river channel. In a complex natural river it is not feasible to measure depth and velocity everywhere. These parameters can, however, be predicted using a hydrodynamic (HD) sub-model which solves the two-dimensional momentum and continuity equations (the so-called St Venant equations). An HD sub-model requires extensive input data on river bathymetry together with the bed friction coefficient (e.g. Manning's n), the upstream river flow, the flows of any tributaries and the downstream water level. Once the HD sub-model has been calibrated and verified, then predicted depths and velocities are available for use in the transport-dispersion (TD) sub-model which solves equation (3.1) numerically.

One important feature of all two-dimensional (depth-averaged) models is that they will not provide accurate flow or tracer predictions in the near-field if the source has significant momentum, induces strongly three-dimensional flows, or has an appreciable buoyancy flux. It is wise to place the upstream boundary of the TD sub-model below the point where the tracer becomes vertically well mixed and source momentum has become insignificant. The tracer distribution at this boundary must be specified either from analytical near-field mixing models (discussed later in this chapter), physical model studies, or field tests.

Many two-dimensional numerical models are formulated in rectangular Cartesian coordinates, notably those developed with oceanic and coastal applications in mind. This formulation is ideally suited to use with finite difference methods. In an irregular channel the grid points seldom correspond exactly with the land boundaries and a smooth boundary which is inclined at an angle to the computational grid must be approximated by a series of steps. As there can be no flow normal to the boundary, such steps cause slight inaccuracies in the flow near the boundary (Stelling 1984). To reduce the impact of these inaccuracies on flows in the main channel, a large number of grid points (at least 20 and preferably more) is required across the channel. If the channel is curved then a large number of grid points is required to accommodate channel meanders within the rectangular computational grid. Many of the grid points will be outside the channel and no calculations need be made at such points, but they do increase storage requirements and slow down input and output. There is an additional problem with using a rectangular grid in a meandering channel, namely that few codes are efficient at computing flows angled near 45° to the computational grid. It is not possible to align the flow with the coordinate axes everywhere in the channel and inaccuracies or numerical instabilities can occur when the flow is angled sharply across the computational grid. These numerical problems are most severe where the channel is narrow.

One simplification which can be made in a gently meandering channel which allieviates these two problems is to align the longitudinal grid lines with the main flow path and to set the transverse grid lines perpendicular to the main flow. Example 3.7 describes an application of this approach.

Example 3.7

A depth-averaged numerical model, SYSTEM21 (Anon. 1984) has been used to simulate the transverse mixing of cooling water in the Waikato River below the Huntly power station, New Zealand. SYSTEM 21 solves the two-dimensional (depth-averaged) momentum equation (S21HD sub-model) and the mass conservation equation (S21TD sub-model) using finite difference methods on a fixed rectangular grid. For the S21TD sub-model to give accurate temperature predictions, the S21HD sub-model must predict correctly not only the water level and mean velocity, but also the transverse velocity distribution. This example describes calibration and testing of the S21HD sub-model.

The Waikato River is gently meandering in the 40 km reach below the Huntly power station. A grid ($\Delta x = 50$ m, $\Delta y = 10$ m) is required to obtain satisfactory spatial resolution of temperatures. This grid is rectangular ($\Delta x/\Delta z = 5$) although SYSTEM 21 is generally used with square grids. A grid of 560 by 1200 mesh points would be required to encompass the channel meanders and computer storage and run times would be prohibitive. A grid was established in which longitudinal distance x was measured along the main flow path (identified from aerial photographs) and the transverse distance z was measured perpendicular to the centreline. This reduced the grid to 850 by 40 mesh points. Strictly the use of a curvilinear grid of this type requires the introduction of metric coefficients into the momentum and mass conservation equations, which take into account the fact that distances measured along the channel vary with transverse location as a result of channel curvature (see Section 1.9.2). As a first approximation, however, the metric coefficients were set to unity.

The S21HD sub-model was calibrated using water levels and flow gaugings made at a steady flow of 232 $m^3 s^{-1}$ and tested against data measured at a steady flow of 320 $m^3 s^{-1}$. The downstream boundary condition was the observed water level at Mercer (+38 km). Table 3.12 compares observed and predicted water levels during calibration and testing. Predicted water levels at the Rangiriri (+14.2 km) and Huntly (+0.0 km) recorders were within 3 cm of the rating curve values, which is within the accuracy of these curves. The predicted depths, velocities and cumulative flows matched those measured at two gauging sites fairly closely (Figure 3.12). Minor bed level changes near the banks would probably have improved the fit still further but are unlikely to have affected the predicted flow in the study reach as a whole. Thus despite neglecting channel curvature a good match between observed and predicted velocity distributions was achieved, giving

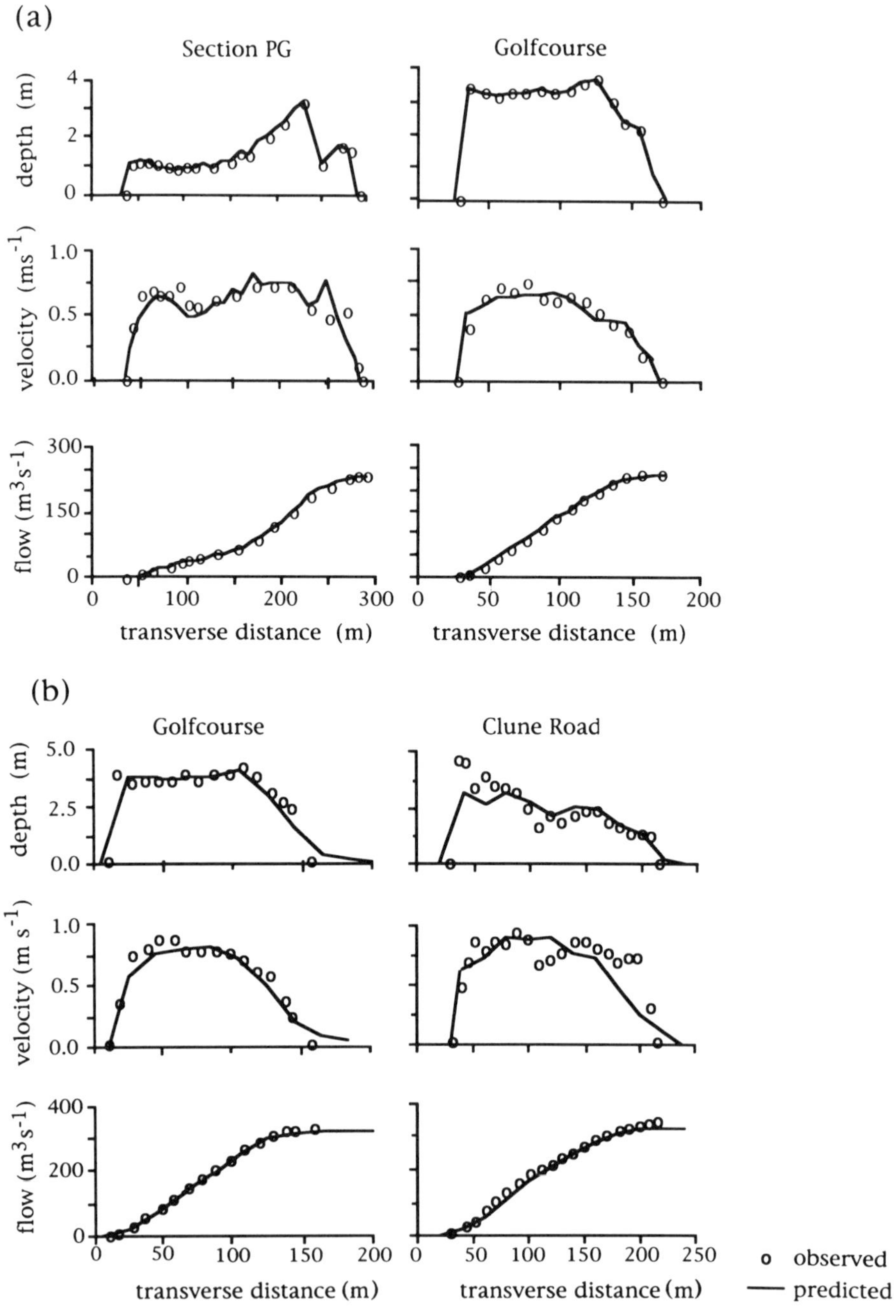

Figure 3.12. (a) Observed and predicted depth, velocity and flow during calibration of the System 21HD model in the Waikato River. (b) Observed and predicted depth, velocity and flow during verification of the System 21HD model in the Waikato River. Distances below Huntly are PG (0.5 km), golfcourse (6.1 km) and Clune Road (31.5 km)

Table 3.12. Observed and predicted water levels during S21HD model calibration and verification. The water level at Mercer provides the downstream boundary condition

	Calibration		Verification	
	Predicted (m)	Observed (m)	Predicted (m)	Observed (m)
Mercer (+38.7 km)	—	1.73	—	2.27
Rangiriri (+14.2 km)	5.63	5.64	6.05	6.06
Huntly (+0 km)	7.18	7.15	7.63	7.60

* Upstream flow: calibration 232 $m^3 s^{-1}$; verification 320 $m^3 s^{-1}$. Tributary inflow: calibration 0 $m^3 s^{-1}$; verification 10 $m^3 s^{-1}$.

some confidence about the model's ability to predict temperature. By neglecting curvature, momentum at the bends is not modelled accurately. The effect of channel curvature and momentum is to transport sediment and reshape the channel. Channel bathymetry is faithfully reproduced in the model even if curvature is not and in the gently meandering Waikato River it is bathymetry which largely dictates the flow distribution.

3.9.3 Transport–Dispersion Sub-model

The transport-dispersion (TD) sub-model solves equation (3.1) numerically allowing the coefficients v_x, v_z, h, k_x and k_z to vary both in time and space (Anon 1989). Information on v_x, v_z and h is passed from the HD to the TD sub-model and normally the same numerical grid is used for both. The HD sub-model does not provide any information about the transverse dispersion coefficients k_x and k_z which must be specified independently. It is common practice to infer values of the dispersion coefficients by calibrating the model. Calibration involves making trial and error adjustments of k_x and/or k_z until a satisfactory match is obtained between observed and predicted tracer profiles. This procedure is capable of identifying reach-averaged dispersion coefficients but it does not produce a unique calibration when the dispersion coefficients vary along or across the channel. A unique calibration could be achieved if the functional dependence of the dispersion coefficient on channel parameters (such as velocity, depth, width, curvature and slope) were known except for one or two constants whose values could be estimated by calibration. Unfortunately, we cannot yet specify this functional dependence in natural rivers.

There is an additional problem with rectangular Cartesian grid models in a meandering channel. The user may wish to specify different values for the longitudinal and transverse dispersion coefficients. Most Cartesian models only allow the dispersion coefficients to differ in the two coordinate directions, but in a meandering river these do not always correspond with the longitudinal and transverse directions. Few models include the cross-derivative terms which would allow for misalignment of the principal axes of the dispersion tensor with the coordinate axes (as is discussed in Chapter 1). In practice longitudinal dispersion is relatively unimportant, especially if the source is only slowly varying. Consequently, k_x and k_z can both be made equal to the transverse dispersion coefficient. This problem does not arise, of course, if the x axis is aligned with the centreline of the channel as in Example 3.7.

In mobile bed rivers the bathymetry is continuously changing. Normally a numerical model is developed, calibrated and tested using the bathymetry measured over a short period of time and the question then arises as to whether this model gives accurate predictions of mixing when the bathymetry changes. If the tracer concentration profile at a given cross-section is plotted against distance across the channel, then changes in bathymetry have a major impact. If, however, the tracer concentration is plotted against cumulative discharge, then the effects of bathymetry changes are eliminated. This approach works well in the situation where the bathymetry changes affect the distribution of water depth and velocity but not the transverse dispersion coefficient, as is illustrated in Example 3.8.

Example 3.8

The bed of the Waikato River is mobile and substantial changes in river bathymetry occur over periods of the order of months. This is illustrated by Figure 3.13 which shows measured bed level changes at a monitoring site

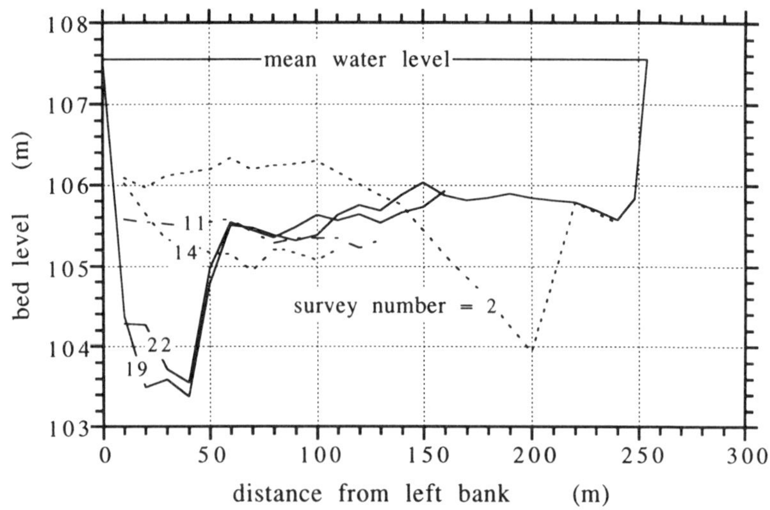

Figure 3.13. River bed levels measured over a 15-month period in the Waikato River 450 m downstream from the Huntly power station. Numbers denote survey number. Note the exaggerated vertical scale

450 m downstream from the Huntly power station. A modelling study was undertaken to determine whether changes in bathymetry are likely to have a significant effect on the transverse mixing of cooling water. A two-dimensional numerical model of the Waikato River below Huntly has been developed using SYSTEM21 (Anon. 1984, 1989). The S21HD sub-model was calibrated and tested at steady river flow (see Example 3.7) and the S21TD sub-model was calibrated against measured temperatures to determine the reach-averaged transverse dispersion and heat decay coefficients.

To determine the effects of bathymetry changes on the rate of mixing, temperatures were computed for two extreme bathymetries. The channel was surveyed during 1986–7 from which a model bathymetry (bathymetry 1) was generated. A hypothetical bathymetry (bathymetry 2) was generated by taking the mirror image of bathymetry 1. In practice the river bathymetry is unlikely to change to this extent because river bends and islands are largely fixed. In straight sections of the channel, however, the main flow path has been observed to cross from one side of the channel to the other (see, for example, Figure 3.13). The outfall was assumed to achieve mixing with 40% of the flow regardless of the bathymetry. Thus the initial temperature distributions were significantly different for the two bathymetries when plotted against transverse distance, but were identical when plotted against cumulative flow. The transverse dispersion coefficient was assumed constant everywhere in the study reach and was the same for both bathymetries ($k_z = 0.095\ \text{m}^2\,\text{s}^{-1}$). Temperature predictions were made over a 6.5 km reach using $\Delta x = 50$ m and $\Delta y = 10$ m.

The channel bathymetries and predicted temperature profiles at the downstream end of the study reach (6.5 km downstream from Huntly power station) are shown in Figure 3.14. As the plume is carried downstream it expands and contracts. Bathymetry changes have a significant effect on temperature distributions when plotted using Cartesian coordinates (i.e. transverse distance), as can be seen in Figure 3.14b. When plotted against cumulative discharge, however, the cooling water plume does not expand and contract as it moves downstream. There are only very small differences in predicted temperature for the different bathymetries when plotted using cumulative discharge. Indeed, the maximum (bankside) temperature decreases whereas the plume width (as a percentage of cumulative flow) increases monotonically with distance (details omitted here).

This study provides a strong indication that changes of bathymetry cause only small changes in predicted transverse temperature profiles when the results are plotted against cumulative flow, but large changes when plotted against transverse distance. This conclusion depends on bathymetry changes having little or no effect on either the transverse dispersion coefficient or the initial dilution in the near-field.

There is one aspect of channel curvature which is a source of difficulty

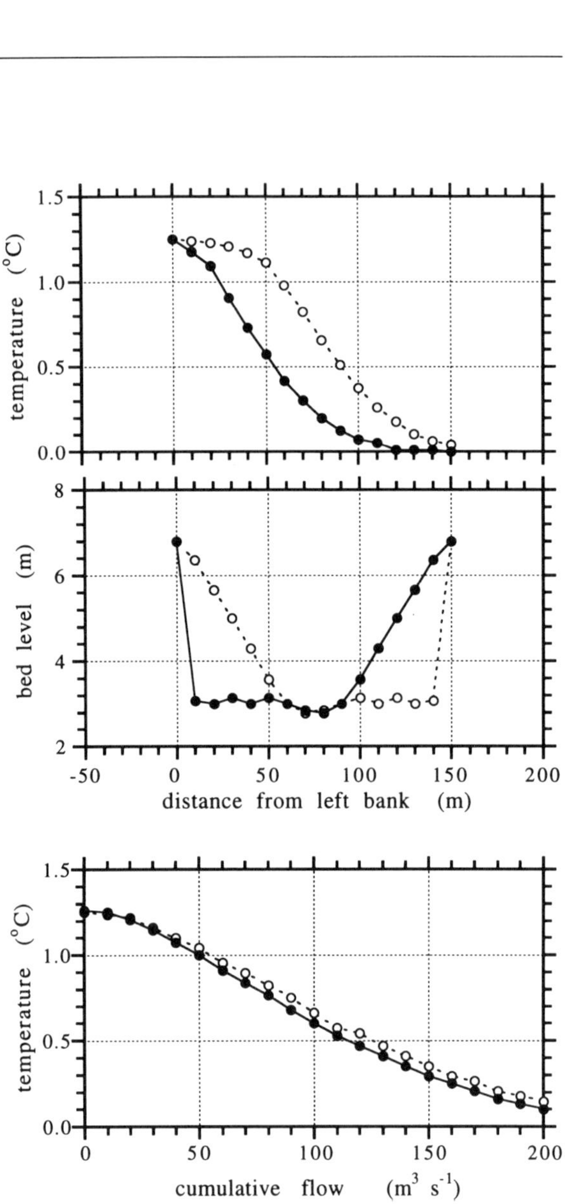

Figure 3.14. Comparison of predicted temperatures and bed levels 6.5 km downstream from Huntly power station. Bathymetry 1 was surveyed in 1986–7 and bathymetry 2 is the mirror image. Note that when compared using cumulative flow the two predicted temperature profiles are almost identical. The same initial dilution (40% mixing) and transverse dispersion coefficient ($k_z = 0.095\ \mathrm{m^2\,s^{-1}}$) were assumed for both bathymetries

in all two-dimensional models. At sharp bends secondary currents occur which greatly increase the rate of transverse mixing. Because these secondary currents are three dimensional, a depth-averaged HD model is not capable of reproducing their effects. In the TD model the transverse dispersion coefficient can be increased in the vicinity of the bend to account for the effects of secondary currents, but at present we cannot predict accurately the distribution of k_z around river bends.

3.9.4 Other Types of Numerical Model

One of the strengths of the better Cartesian finite difference models, such as SYSTEM 21, is that they have been tested and refined over many years. They will handle non-steady flow (including wetting and drying), multiple channels (e.g. islands) and inflows. There are, however, three other types of numerical model which can potentially overcome some of the shortcomings of the Cartesian finite difference model.

3.9.4.1 Numerical streamtube models

A small number of numerical models are formulated in terms of cumulative flow. There are great attractions to using streamtube models because transverse advection terms, which are difficult to handle, do not appear in the TD equation. The flow distribution must be known at regular intervals along the channel to define the streamtubes and this requires either extensive gauging data or a numerical HD model. The streamtube model is less attractive when the river flow is unsteady because the streamtubes change with time. Fischer (1966) developed a simplified but effective streamtube model for steady flows but unsteady tracer inputs in order to study longitudinal dispersion close to the tracer source. This model has also been used to predict transverse mixing rates in the Missouri River near Blair, Nebraska by Yotsukura *et al* (1970). Lau and Krishnappan (1981) developed a steady-state streamtube model called RIVMIX which solves equation (3.37) numerically and this model has been extended to the case of a non-steady tracer source (Luk *et al* 1990). RIVMIX does not consider non-steady river flows, but this is only a minor drawback in most river studies. A more serious shortcoming is that neither model contains an HD sub-model for predicting water depth and velocity and these data must be obtained either from another model or from field work.

3.9.4.2 Curvilinear models

In a meandering channel it is natural to write the HD and TD equations in curvilinear coordinates (see Section 1.9.2). In this way the banks can be made to coincide exactly with longitudinal grid lines, which overcomes the difficulties experienced with a Cartesian grid model when a curved boundary is represented by a series of steps. Another advantage of a curvilinear model is that all the grid points lie within the channel and computer storage requirements are reduced. Curvilinear models were originally developed for solving aeronautics problems but have been adapted for use in rivers and estuaries. There is much published work on methods to generate curvilinear grids and to transform the equations from rectangular Cartesian to curvilinear coordinates (Thompson *et al* 1985, Fletcher 1988). Delft Hydraulics has been closely involved with the development and testing of finite-difference numerical models for river flow and TD using both Cartesian and curvilinear grids (Stelling 1984). Their curvilinear model TRISULA has been extensively tested against data from laboratory flumes and Dutch rivers.

3.9.4.2 Finite element models

Finite element methods were originally developed for solving structural engineering problems but are finding increasing application in fluid mechanics. They offer two advantages. Firstly, element and land boundaries can be aligned. Secondly, elements can be smaller and more numerous in areas where the flow or concentration is changing rapidly to improve resolution and accuracy.

One disadvantage of both curvilinear and finite element models is that grid generation is complex, although this problem is diminishing as interactive graphics become widely available and grid generation software improves (Thompson *et al* 1985, Henry and Walters 1991).

3.9.5 Three-dimensional Numerical Models

Over the last 10 years improvements in computers have facilitated the development of three-dimensional numerical models. These offer a framework for studying difficult mixing problems which in rivers include the behaviour of buoyant and non-buoyant jets in the near-field mixing region, channel curvature and the effects of secondary circulations, saline intrusion and stratification of the ambient flow. Several studies have been published which shed light on these difficult problems and demonstrate the potential usefulness of three-dimensional models. At present, however, three-dimensional models are best viewed as research tools and further development work needs to be done before they become useful tools for the practitioner. One difficulty is that three-dimensional models require a powerful computer and long run times. A more important problem is that model predictions are very sensitive to the way in which turbulence is modelled. Turbulence modelling is a difficult area for all but the specialist and is rapidly changing (Rodi 1980, Abbott and Basco 1989). Finally, there are many opportunities for numerical difficulties with three-dimensional computer codes and a good understanding of numerical analysis is required to run them successfully. As models are refined and tested these difficulties are likely to diminish and three-dimensional models may soon become another of the tools used routinely in river mixing studies.

3.9.6 Conclusions

Several factors will dictate the choice of a numerical model. One important consideration is the nature of the channel; if it is only gently meandering then a rectangular Cartesian finite difference model may be adequate, whereas if it is sinuous then streamtube, curvilinear or finite element models are preferable. It is very important to select a model which has been extensively tested: numerical models are complex and many undesirable features do not show up until a particular set of circumstances is modelled. Benchmark tests against analytical solutions on simple problems are useful but do not guarantee adequate model performance in more complex problems. It is advisable to spend some time running test problems with a new model to ensure that it conserves flow and tracer mass, and that it correctly reproduces important features of tracer profiles such as time of travel. The simple analytical models described earlier can be very useful as a yardstick against which to test numerical model performance. Only when the user is confident about the model should it be applied to complex problems.

3.10 BUOYANCY EFFECTS ON TRANSVERSE MIXING

3.10.1 Introduction

Buoyant mixing is of considerable importance in oceans, estuaries, lakes and reservoirs and there is much published work on these water bodies. In rivers turbulence is strong, the depth is small and vertical mixing is rapid. Thus for most river mixing problems buoyancy forces can be neglected and the tracer can be assumed to be passive. There are, however, two types of river mixing problem in which buoyancy forces need to be considered: tracer mixing in stratified flows and mixing of non-neutrally buoyant tracers. The presence of buoyancy forces adds considerably to the complexity of the mixing problem; it is seldom possible to make accurate theoretical predictions from first principles and experimental studies are essential. In this section there is a brief discussion which is intended to give the reader an appreciation of the way buoyancy affects mixing and of the complexity of the problem.

3.10.2 Stratification of the Ambient Flow

In a tidal river the ambient flow is stratified with saline water underlying less dense fresh water. In this situation turbulence is suppressed and there is a reduction in the rate of vertical mixing. This point was discussed in Chapter 2 where an approximate formula was given relating the reduction in vertical diffusivity to the Richardson number [see equation (2.53)]. Meteorological observations suggest that stratification in the atmosphere reduces not only the vertical diffusivity but also the transverse diffusivity (Csanady 1973).

Sumer and Fischer (1977) studied the transverse mixing of a neutrally buoyant tracer in stably stratified tidal flow. They used a laboratory flume in which the channel was trapezoidal and the flow oscillatory. Their main conclusion was that stratification had a markedly different effect on transverse mixing in uniform and non-uniform channels. This difference highlights the complexity of mixing processes when buoyancy forces are important. In a straight, uniform channel the transverse diffusivity was reduced by stratification in approximately the same proportion as the vertical mixing coefficient. Stable stratification suppresses vertical turbulence because the eddies must do additional work to overcome buoyancy forces. In a uniform channel the reduction of vertical turbulence is accompanied by a reduction of transverse turbulence although, as discussed earlier, the processes linking vertical and transverse turbulence are not fully understood. Equation (2.53) and the coefficients of Table 2.4 can be used to make an initial estimate of the likely reduction in the transverse diffusivity of a straight, uniform channel.

Sumer and Fischer (1977) also conducted experiments in a trapezoidal channel where one bank was straight but the width varied sinusoidally. The rates of vertical mixing varied across the channel because of variations in depth and shear velocity. Consequently the vertical salinity structure varied across the channel. Typically the channel was well mixed in the shallows but stratified in the deeper parts, which resulted in transverse pressure gradients. Because of differences in depth and friction across the channel, the water level did not rise and fall uniformly. The resulting small differences in water level across the channel augmented the transverse pressure gradients arising from differences in salinity. The combination of salinity and water level differences induced transverse secondary currents which substantially increased the rate of transverse dispersion. It is not possible to generalise Sumer's results because it is not possible to separate the effects of stratification and oscillatory flow. For the particular channel configuration studied, the transverse dispersion coefficient was higher by a factor of three to five when the flow was stratified. It would be dangerous to assume that in all other irregular tidal channels stratification would cause the same increase in the transverse dispersion coefficient, but these results can at least be used to give a guide to the possible effects of stratification.

Ward (1974) made a laboratory study of transverse mixing in an oscillatory flow of constant density. Four experiments were conducted in a straight channel and it was found that during the peak flood flow the transverse dispersion coefficient lay in the range $k_z/HU^* = 0.2$–0.3, comparable with data in other straight channels [see equation (3.11)]. At slack tide, however, it was found that large-scale horizontal eddies accelerated the rate of transverse mixing. The tidally averaged transverse dispersion coefficient was found to lie in the range $k_z/HU^* = 0.4$–0.5, roughly twice the instantaneous values measured during peak flow. Ward's study also included measurements in curved channels which confirmed that helical secondary currents at bends significantly increase the rate of transverse mixing. In curved channels both the tidally averaged and the instantaneous peak flood values of k_z/HU^* were higher than in the straight channel by a factor of two to three.

3.10.3 Non-neutrally Buoyant Tracer

When a non-neutrally buoyant tracer is discharged into an unstratified channel, the rate of transverse mixing may be affected by buoyancy forces. A common example of a buoyant tracer is cooling water from a power station. If vertical mixing is very weak (as in lakes), then the hot water forms a thin surface layer and transverse buoyant spreading is very rapid (Jirka *et al* 1975). In most rivers, however, there is sufficient turbulence to mix cooling water vertically within a fairly short distance of the outfall as is discussed in Chapter 2. Once vertical mixing is complete the depth-averaged temperature may still vary across the channel. In this case there is a transverse density gradient which may accelerate transverse mixing. Jirka *et al* (1975) state that in most cases these forces are small and once vertical mixing has been achieved it is sufficient to assume that temperature behaves like a passive tracer. On the other hand, Fischer *et al* (1979) suggest that buoyancy has a significant effect on transverse mixing when

$$\frac{B}{hu^{*3}} \gg 5 \tag{3.51}$$

where

$$B = \frac{\Delta\rho}{\rho}\, g q_{\mathrm{e}} \tag{3.52}$$

in which B = buoyancy flux, ρ = density of river water, $\Delta\rho$ = density difference between cooling water and the river, and q_{e} = cooling water flow rate.

Equations (3.51) and (3.52) give no indication of how far downstream the buoyancy effects extend. Prych (1970) studied the transverse mixing of heated water in a laboratory flume and found that the transverse dispersion coefficient was two to four times the value for neutrally buoyant tracer over distances below the source of 10–20 times the channel width. He also studied heavy (sinking) tracers and found that they increased the transverse dispersion coefficient, but not by as much as light (rising) tracers. Smith (1979) made a theoretical investigation of this problem which provides qualitative support for Prych's experimental findings. Smith argued that once the heated water becomes mixed over the depth there is a tendency for it to spread sideways at the water surface. This induces weak secondary currents and promotes weak stratification, both of which promote further transverse mixing. There is also an additional longitudinal circulation which further accelerates transverse mixing. Prych's results are particular to the conditions of his experiments and have not yet been generalised. Similarly, Smith's descriptions are only semi-quantitative. Nevertheless, these two studies show that buoyancy can accelerate transverse mixing even after vertical mixing is complete. At present the only practical way to tackle this problem is to increase the value of the transverse dispersion coefficient close to the outfall, but it is not clear by how much or over what length of channel the dispersion coefficient is affected. For accurate predictions it is necessary to conduct either site-specific tracer tests or physical model studies.

3.11 TRANSVERSE MIXING IN THE NEAR-FIELD

3.11.1 Introduction

In discussing transverse mixing so far it has been assumed that the tracer distribution is known at the upstream boundary of the region where equation (3.1) applies. Commonly we assume that the tracer originates from a vertical line source, the discharge has no significant momentum and the tracer is passive (i.e. does not affect the flow). This assumption is valid for many discharges of treated sewage and industrial wastes. One commonly encountered problem where the assumption is invalid, however, is the discharge of large volumes of cooling water from steam-turbine power stations.

At full load a typical 1000 MW fossil-fuel plant using open-cycle cooling disposes of 40 $m^3 s^{-1}$ of cooling water heated by 9°C and containing 1500 MW of waste heat. There are two alternative strategies for cooling water disposal: bankside surface discharges (typically a discharge canal) and multi-port subsurface diffusers. The discharge canal usually gives low initial dilution and confines the plume to the surface waters near the outfall. In this region weakly stratified conditions are common, which accelerate buoyant plume spreading and surface heat loss. The multi-port diffuser, on the other hand, is strongly jetted and entrains a substantial fraction of the ambient river flow, giving a high initial dilution. Cooling water affects a larger proportion of the flow but near-field temperature increases are smaller. In rivers the combination of shallow depths and high entrainment mean that vertical mixing is achieved within a short distance of both types of diffuser.

3.11.2 Diffuser in a Strong Co-flow

This section discusses methods for calculating near-field dilution below a multi-port diffuser in a shallow river where there is a strong current. Figure 3.15 is a schematic diagram of flows in the vicinity of a diffuser. The channel is assumed to be uniform and rectangular. Upstream from the diffuser the approach velocity u_a is assumed uniform across the channel. The diffuser is of length L and contains a total of N ports each of area a_e discharging at velocity u_e. The total diffuser flow is therefore $q_e = Na_eu_e$. The velocity u_e is substantially higher than the river velocity u_a and the momentum of the discharge accelerates the flow. From Bernoulli's theorem the water level drops at the outfall and a converging flow is induced upstream of the diffuser (Figure 3.15). The point of maximum lateral contraction (the *vena contracta*) does not occur at the outfall itself but a distance downstream found by experiment (Lee 1985) to be about $L/2$. For temperature increases typical of power

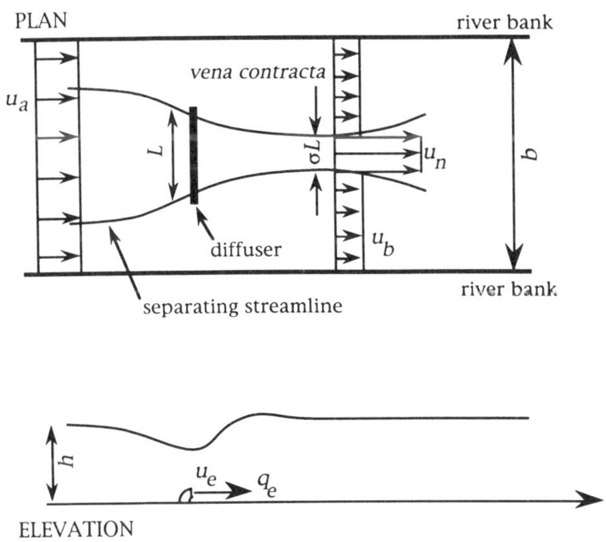

Figure 3.15. Diagram of a diffuser in a river channel and the variables used in the Parr and Melville (1981) and Adams (1982) models. The Adams model neglects the river banks and the velocity u_b. Key: u_a = approach velocity, u_n = slipstream velocity, σ = contraction factor, h = approach depth, q_e = effluent flow, u_b = by-pass velocity, L = diffuser length, b = channel width, u_e = effluent velocity

plants using open-cycle cooling (8–10°C) the buoyancy forces are small, vertical mixing is rapid and it has been shown that tracer becomes well mixed vertically within a short distance of the diffuser (Lee and Jirka 1980). Assuming inviscid flow, a dividing streamline can be postulated which separates water entrained by the diffuser from water which remains at ambient temperature. At the *vena contracta* the plume has accelerated to its maximum slipstream velocity u_n whereas outside the plume the flow has slowed to the bypass velocity u_b. Theoretical and experimental work has concentrated on predicting dilution at the *vena contracta* because near-field dilution occurs largely as a result of flow entrained into the slipstream. This problem is amenable to an approximate theoretical solution which provides a conservative (i.e. minimum) estimate of initial dilution.

3.11.2.1 Adams' model

Adams (1982) developed a theoretical model for shallow unbounded systems which has been tested using both laboratory and field data. This model assumes complete vertical mixing and is based on the classical analysis of flow induced by a propeller in inviscid and incompressible flow. When applied in rivers, the Adams model does not account for the presence of the river banks (which limit the amount of water available for entrainment) and overestimates dilution slightly. In spite of this inaccuracy, the Adams model is easy to use and is valuable for estimating the initial dilution. In non-dimensional form the model solutions are

dilution
$$d = \frac{q_n}{q_e} = \tfrac{1}{2}\left(\beta \nu_a + \sqrt{2\beta + \beta^2 \nu_a^2}\right) \tag{3.53}$$

slipstream velocity
$$\nu_n = \sqrt{\frac{2}{\beta} + \nu_a^2} \tag{3.54}$$

contraction
$$\sigma = \frac{1}{2}\left(1 + \frac{\nu_a}{\nu_n}\right) \tag{3.55}$$

where

$$\nu_a = \frac{u_a}{u_e} \tag{3.56}$$

$$\nu_n = \frac{u_n}{u_e} \tag{3.57}$$

and

$$\beta = \frac{LH}{Na_e} \tag{3.58}$$

The variables in these equations are defined in Figure 3.15, although note that the Adams' model neglects the banks and does not use u_b. When deriving these equations, Adams neglected the diffuser flow from the continuity equation and assumed that

$$d - 1 \approx d \tag{3.59}$$

This assumption enables the model equations to be solved directly but means that they are only accurate at high dilutions ($d > 5$). Adams (1982) does give a second-order correction which can be applied at low dilutions but which is rarely used.

One limiting case is that of a very strong cross-flow for which

$$\beta^2 \nu_a^2 \gg 2\beta \tag{3.60a}$$

Substituting in equation (3.53) gives

$$d = \beta \nu_a = \frac{u_a h L}{N a_e u_e} = \frac{u_a h L}{q_e} \tag{3.60b}$$

This is the ratio of the river flow passing over the diffuser at zero discharge divided by the diffuser flow and hence gives a minimum dilution estimate for the case where the momentum of the discharge is insufficient to induce any entrainment.

3.11.2.2 *Parr and Melville's model*

Parr and Melville (1981) extended Adams' analysis to the case of a channel of finite width and showed that the presence of the banks reduces the initial dilution. Parr tested the model against field data collected below the Quad Cities nuclear power plant and Lee (1985) conducted laboratory studies to corroborate the model over a wide range of conditions. The model equations in non-dimensional form are

dilution

$$d = \frac{q_n}{q_e} = \sigma\beta\sqrt{\frac{2}{\beta} + \nu_b^2} \tag{3.61}$$

bypass velocity

$$\nu_b = \nu_a\left(1 + \frac{1}{\xi}\right) - \frac{\xi\lambda}{4}\left(\frac{2}{\xi} + \frac{\nu_a^2 - \nu_b^2}{\nu_a}\right)\left(\sqrt{\frac{2\nu_a}{\xi\lambda} + \nu_b^2} - \nu_b\right) \tag{3.62}$$

slipstream velocity

$$\nu_n = \sqrt{\frac{2}{\beta} + \nu_b^2} \tag{3.63}$$

contraction

$$\sigma = \frac{1}{2}\left[1 + \frac{\xi}{2\nu_a}(\nu_a^2 - \nu_b^2)\right] \tag{3.64}$$

where in addition to equations (3.56) to (3.58)

$$\nu_b = \frac{u_b}{u_e} \tag{3.65}$$

$$\lambda = \frac{L}{b} \tag{3.66}$$

and

$$\xi = \frac{q_a}{q_e} \tag{3.67}$$

The Parr and Melville (1981) model represents an improvement on the Adams' model because it accounts for the presence of the banks. In a river there is a limited supply of water available to be entrained into the slipstream. The diffuser accelerates the flow within the slipstream (where the velocity increases from u_a to u_n) and the continuity equation requires that the flow outside the slipstream must decelerate (the velocity drops from u_a to u_b). This deceleration results in a back-pressure which opposes the diffuser momentum and reduces the dilution. The reduction in dilution is typically of the order of 10–20% (Lee 1985). One slight difficulty with Parr and Melville's model is that equation (3.62) must be solved iteratively for ν_b. By comparison the equations in the Adams' model can be solved directly. For most practical situations a rapidly converging solution comes from successive substitutions into the equation

$$\nu_b^{n+1} = \nu_a\left(1 + \frac{1}{\xi}\right) - \frac{\xi\lambda}{4}\left[\frac{2}{\xi} + \frac{\nu_a^2 - (\nu_b^n)^2}{\nu_a}\right]\left[\sqrt{\frac{2\nu_a}{\xi\lambda} + (\nu_b^n)^2} - \nu_b^n\right] \tag{3.68}$$

Of particular interest in river mixing problems is the determination of the diffuser length which entrains the entire river flow. This diffuser gives the maximum possible dilution

$$d = \frac{q_a + q_e}{q_e} \tag{3.69}$$

where q_a = river flow immediately upstream from the diffuser. The limiting diffuser length L_c estimated from the Parr and Melville equations is

$$\lambda_c = \frac{2\nu_a}{\xi}\left(\frac{1+\xi}{1 + \frac{\xi\nu_a}{2}}\right) \tag{3.70a}$$

where

$$\lambda_c = \frac{L_c}{b} \tag{3.70b}$$

Once the diffuser entrains the entire flow the separating streamline impinges on the bank just upstream from the diffuser. There can be no flow in the region outside the separating streamline, the bypass velocity $u_b = 0$ and recirculation is possible around the edges of the outfall. Recirculation may carry cooling water upstream as far as the intake, in which case plant performance will drop. A recirculating plume will also cause cooling water to impinge on the bank within a short distance of the outfall and hence will cause high temperatures in the littoral zone. As the littoral zone is often an important habitat for resident biota and is a fish migration route, it is desirable to avoid recirculation while still achieving a high initial dilution. This requires identifying the outfall length, which is slightly shorter than the critical length given by (3.69).

Example 3.9

The Huntly power station discharges cooling water to the Waikato River from a jetted bankside outfall. A feasibility study was carried out of a mid-river outfall and near-field dilution calculations were made using the Parr and Melville and Adams models. The channel parameters are given in Table 3.13.

Table 3.13. Channel parameters in the Waikato River near the Huntly power station at low flow

River flow (Q)	$150\ m^3 s^{-1}$
Cooling water flow (q_e)	$38\ m^3 s^{-1}$
Approach flow (q_a)	$112\ m^3 s^{-1}$
Port velocity (u_e)	$6\ m s^{-1}$
River mean depth (h)	2.10 m
River width (b)	180 m

Table 3.14. Initial dilution downstream from a 56 m long mid-river outfall estimated using the Adams (1982) and Parr and Melville (1981) models

	Parr and Melville	Adams
Approach velocity (u_a) ($m s^{-1}$)	0.296	0.296
Diffuser length (L) (m)	56	56
Bypass velocity (u_b) ($m s^{-1}$)	0.15	—
Slipstream velocity (u_n) ($m s^{-1}$)	1.61	1.63
Contraction (σ)	0.54	0.59
Dilution (d)	2.7	3.0*
Limiting diffuser length (L_c) (m)	115	—
Slipstream flow (q_n) ($m^3 s^{-1}$)	103	114
Approach flow (q_a) ($m^3 s^{-1}$)	112	112
Entrained flow ($q_c = q_n - q_e$) ($m^3 s^{-1}$)	65	76
Bypass flow (q_b) ($m^3 s^{-1}$)	47†	36‡
Total flow ($q_n + q_b = q_a + q_e$) ($m^3 s^{-1}$)	150	150

* Using the second-order correction (Adams 1982) $d = 3.2$.

† $q_b = u_b(b - \sigma L)h$.

‡ $q_b = q_a + q_e - q_n$.

From Table 3.14 it can be seen that the Parr and Melville model predicts a lower dilution ($d = 2.7$) than the Adams model ($d = 3.0$). This is because the Adams model does not account for the back-pressure induced by the deceleration of the flow outside the slipstream. Here the difference in estimated dilution is of the order of 10% which is typical of the error identified by Lee (1985) using laboratory data. Thus for many practical problems the rather simpler Adams model will provide satisfactory estimates of initial dilution.

Now consider the case where the length of the diffuser is set to $L_c = 115$ m as detailed in Table 3.15. The Parr and Melville model predicts that the bypass velocity drops to zero, the dilution reaches the maximum predicted by equation (3.69) and the entire approach flow is entrained into the plume. The Adams model predicts that the slipstream flow is greater than the approaching river flow and in this situation recirculation would be expected. Note, however, that there is nothing in the dilution calculations of the Adams model which alerts us to the likelihood of recirculation and it is only by comparing $q_n - q_e$ with q_a that recirculation can be identified. It should also be noted that the Parr and Melville model predicts that for $L > L_c$, the value of u_b is negative and the dilution exceeds the value given by equation (3.68). The negative value of u_b implies recirculation and the Parr and Melville model assumes that this recirculating flow contributes to initial dilution. In most practical problems, however, the upper limit to dilution is given by equation (3.68) because the recirculating flow does not dilute tracer in the effluent flow.

Table 3.15. Initial dilution downstream from a 115 m long mid-river outfall estimated using the Adams (1982) and Parr and Melville (1981) models

	Parr and Melville	Adams
Approach velocity (u_a) ($m\,s^{-1}$)	0.296	0.296
Diffuser length (L) (m)	115	115
Bypass velocity (u_b) ($m\,s^{-1}$)	0.00	—
Slipstream velocity (u_n) ($m\,s^{-1}$)	1.12	1.16
Contraction (σ)	0.56	0.63
Dilution (d)	3.95	4.63*
Limiting diffuser length (L_c) (m)	115	—
Slipstream flow (q_n) ($m^3\,s^{-1}$)	150	176
Approach flow (q_a) ($m^3\,s^{-1}$)	112	112
Entrained flow ($q_c = q_n - q_e$) ($m^3\,s^{-1}$)	112	138†
Bypass flow (q_b) ($m^3\,s^{-1}$)	0‡	−26§
Total flow ($q_n + q_b = q_a + q_o$) ($m^3\,s^{-1}$)	150	150

* Using the second-order correction (Adams 1982) $d = 4.8$.
† Entrained flow exceeds the approach flow, recirculation occurs.
‡ $q_b = u_b(b - \sigma L)h$.
§ $q_b = q_a + q_e - q_n$.

Both the Adams and Parr and Melville models assume that the flow is inviscid and that there is a discontinuity in the velocity across the dividing streamline (Figure 3.15). In real fluids, however, viscosity smears out this velocity discontinuity and the flow is accelerated for some distance beyond the ends of the diffuser. This results in some additional entrainment of fluid along the edges of the slipstream although Adams (1982) argues that the amount of flow entrained is small, being typically of the order 10% of the total entrainment. Indeed, model predictions are in substantial agreement with the results of several laboratory and field studies and can be used with some confidence to predict the initial dilution.

The Parr and Melville (1981) and Adams (1982) models both predict the initial dilution at the *vena contracta*. This point is located a distance downstream from the outfall of the order $L/2$ (Parr and Melville 1981). In the intermediate zone just downstream from the *vena contracta*, source momentum is still significant and the plume continues to entrain ambient river water. Mixing in the intermediate zone is not well understood. The practitioner has two options: either to neglect entrainment and assume passive mixing in the intermediate zone (which gives a conservative estimate of initial dilution), or to measure the initial dilution experimentally.

Eventually the plume velocity drops to the ambient river velocity and entrainment stops. Thereafter the plume lies in the mid-field where transverse mixing is entirely passive and can be modelled using the methods outlined earlier in this chapter.

Natural channels are not uniform and rectangular but Parr and Melville (1981) adapted their model to take this into account. They defined the effective depth, velocity and width just upstream from the diffuser as

$$h' = \frac{1}{L} \int_{-L/2}^{+L/2} h(z)\, \mathrm{d}z \tag{3.71}$$

$$u_{\mathrm{a}}' = \frac{1}{L} \int_{-L/2}^{+L/2} u_{\mathrm{a}}(z)\, \mathrm{d}z \tag{3.72}$$

$$b' = \frac{q_{\mathrm{a}}}{h'u'} \tag{3.73}$$

and used these values in place of h, u_{a} and b. As the diffuser causes the flow to contract, it is desirable to extend the averaging for some distance either side of the outfall. The actual distance over which the depth and velocity should be averaged (denoted L' where $L' > L$) can be determined by trial and error. An initial guess is made of L', the entrained flow ($q_{\mathrm{n}}-q_{\mathrm{e}}$) determined from the model and the flow distribution above the diffuser examined to see whether L' needs to be adjusted. This procedure is only necessary when the channel is highly variable in cross-section.

When the diffuser is attached to one bank the method of images can be used to calculate the initial dilution as shown in Figure 3.16. The model equations are solved assuming that the approach flow, effluent flow and diffuser length are doubled. It is left as an exercise for the reader to apply this modification.

3.11.3 Bankside Discharge Canal

The bankside discharge canal is a common method of discharging large volumes of cooling water into a river. The main features of the problem are shown in Figure 3.17. There is a strong ambient current u_a which deflects the cooling water plume downstream. The bank opposite the discharge canal causes a transverse pressure gradient which also helps to deflect the plume back towards the near bank. A buoyant effluent has the tendency to remain near the surface and to spread across the channel quickly. Most cooling water discharges are only weakly buoyant and in rivers vertical mixing is rapid so that temperature becomes well-mixed vertically within a short distance of the discharge canal (Figure 3.17). Because the effluent is slightly buoyant and has appreciable transverse momentum the plume has a tendency to detach from the bank for some distance below the source. The adverse pressure gradient caused by the far bank causes the plume to reattach a distance L downstream from the source. Between the source and the reattachment point there is a recirculation zone. The outer edge of the plume can be identified

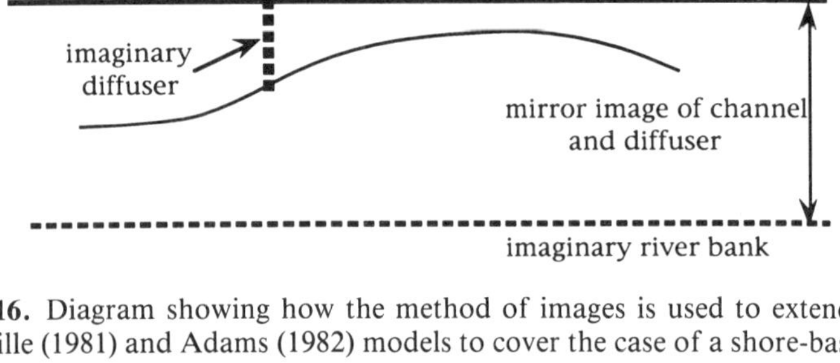

Figure 3.16. Diagram showing how the method of images is used to extend the Parr and Melville (1981) and Adams (1982) models to cover the case of a shore-based outfall

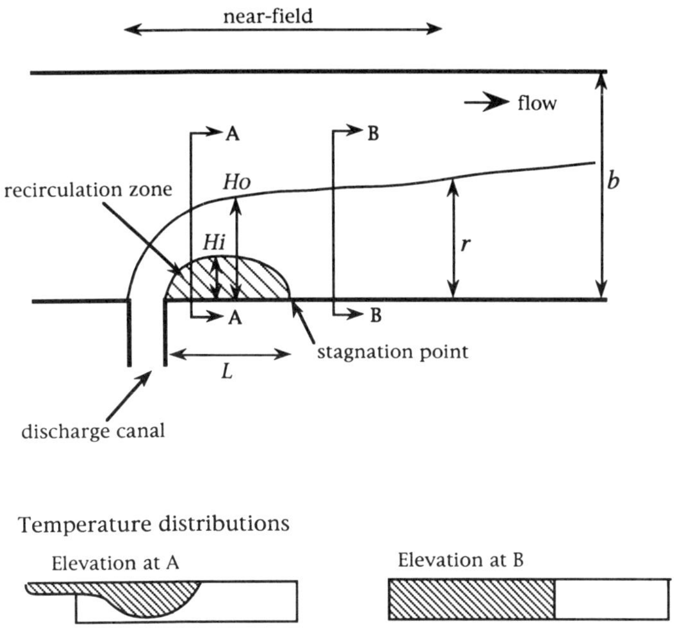

Figure 3.17. Sketch showing the cooling water plume downstream from a discharge canal, the recirculation zone and the variables used in Chu and Abdelwahed's model (1990)

experimentally as the point at which the temperature is not significantly different from ambient. Alternatively McQuirk and Rodi (1978) define the outer edge of the plume as the location of the streamline which originates at the upstream edge of the discharge canal. The inner part of the recirculation zone can be identified as the region in which tracer remains trapped against the bank. In experimental studies the maximum inner and outer widths of the recirculation zone are measured and denoted H_i and H_o, respectively. Both H_i and H_o attain their maximum value at about $L/2$.

The problem of a surface jet discharging into a deep and unbounded cross-flow has received considerable attention (Jirka *et al* 1975). Mathematical models are available for predicting the near-field dilution and these models have been verified by laboratory and field studies (Stolzenbach and Harleman 1971, Adams *et al* 1975). There are difficulties in applying these models to rivers where the cross-flow is very strong, the bed has a strong influence on the plume and the flow is confined by the far bank (Sayre 1979).

Chu and Abdelwahed (1990) carried out a laboratory study of plume reattachment in a river channel. They found that in a strong cross-flow (such as occurs in a river) buoyancy can be neglected and the initial dilution is determined by the balance between the momentum flux of the effluent and the receiving water.

The momentum flux of the effluent is

$$M_e = q_e u_e = u_e^2 b_e h_e \tag{3.74}$$

where q_e = effluent flow rate; u_e = discharge velocity; b_e and h_e = width and depth of the discharge canal. The momentum flux of the receiving water is

$$M_a = q_a u_a = u_a^2 bh \tag{3.75}$$

where q_a = approaching river flow; u_a = mean velocity; h = mean depth and b = channel width. Note that q_a, u_a, h and b are all measured just upstream from the source.

Abdelwahed and Chu (1981) defined a momentum length scale

$$l_m = \frac{M_e}{M_a} b \tag{3.76}$$

and showed experimentally that

$$\frac{H_o}{l_m} = \left(0.192 + \frac{l_m}{b}\right)^{-1} \tag{3.77}$$

$$\frac{H_i}{l_m} = \left(0.556 + 1.7 \frac{l_m}{b}\right)^{-1} \tag{3.78}$$

$$L = 7H_i \tag{3.79}$$

Downstream from the recirculation zone there is a turbulent wake where ambient river water is entrained into the plume. Chu's work does not give an estimate of the length of the turbulent wake, but in numerical simulations of this problem McQuirk and Rodi (1978) found appreciable entrainment downstream at least as far as

$$L_w \approx 2L \tag{3.80}$$

Equation (3.80) provides a conservative estimate of the length of the near-field (i.e. a likely underestimate). Equations (3.77–3.80) define the geometry of the problem but not the initial dilution.

In their experimental study Chu and Abdelwahed (1990) found that the width of the plume increased as

$$\frac{r}{H_o} = 1.0\left(\frac{x}{L}\right)^{1/2} \tag{3.81}$$

where r = width of the plume. On the basis of McQuirk and Rodi's work [see equation (3.80)] it seems likely that equation (3.81) is valid in the region $0 < x < 2L$.

Within the near-field the rate of transverse mixing is likely to be high largely because the momentum of the effluent causes entrainment. It is possible to infer from equation (3.81) the effective transverse dispersion coefficient within the near-field. The dispersion coefficient is defined

$$k_z = \frac{u_a}{2}\frac{\mathrm{d}\sigma_z^2}{\mathrm{d}x} \tag{3.82}$$

where σ_z^2 = transverse variance of the plume and x = distance downstream. The width of the plume can be closely approximated by

$$r = 4\sigma_z \tag{3.83}$$

Substituting equations (3.81) and (3.83) into (3.82) gives

$$k_z = \frac{u_\mathrm{a} H_\mathrm{o}^2}{32L} \tag{3.84}$$

which implies that the transverse dispersion coefficient in the wake is constant. The initial dilution is defined

$$d = \frac{q_\mathrm{e} + q_\mathrm{c}}{q_\mathrm{e}} \tag{3.85}$$

where q_c = flow entrained from the river into the plume. From Figure 3.17

$$q_\mathrm{c} \approx rhu_\mathrm{a} \tag{3.86}$$

Chu and Abdelwahed (1990) found, however, that recirculation reduces entrainment to

$$q_\mathrm{c} \approx 0.83rhu_\mathrm{a} \tag{3.87}$$

If the effluent tracer concentration is s_e then the average plume concentration at the end of the near-field is

$$\bar{\bar{s}} = \frac{s_\mathrm{c}}{d} \tag{3.88}$$

where the overbar denotes an average within the plume. Concentration varies within the plume from a maximum near the bank to a minimum at the outer edge of the plume. In the absence of any other information it is not unreasonable to assume a linear variation

$$s(z) \approx \begin{cases} \dfrac{2s_\mathrm{e}}{d}\left(1 - \dfrac{z}{r}\right) & 0 < z < r \\ 0 & z > r \end{cases} \tag{3.89}$$

Equation (3.89) establishes the boundary condition at the beginning of the mid-field.

Example 3.10

Cooling water is discharged from the Huntly power station from a bankside outfall. An estimate can be made of the initial dilution at low flow using Chu and Abdelwahed's results. The outfall extends 40 m along the left bank of the Waikato River and consists of 20 circular ports of diameter 0.70 m. There is strong jetting and a scour hole (typically 5 m deep) develops in front of the outfall (Table 3.16).

Chu's experiments were performed using a discharge canal which extended over the entire depth. We shall assume that the Huntly outfall behaves like a discharge canal 40 m long and 5 m deep.

The approach and effluent momentum fluxes are [from equations (3.74) and (3.75)]

$$M_e = 7.2 \text{ m}^4 \text{ s}^{-2}$$
$$M_a = 33.2 \text{ m}^4 \text{ s}^{-2}$$

Hence from equation (3.76) the momentum length scale is

$$l_m = 39.1 \text{ m}$$

and from equations (3.77)–(3.79)

$$H_o = 96 \text{ m}$$
$$H_i = 42 \text{ m}$$
$$L = 296 \text{ m}$$

These estimates are in reasonable agreement with field observations that the plume extends about halfway across the channel (which is about 180 m wide at this point) and that the near-field extends about 300 m downstream from the outfall.

At $x = 2L$ equation (3.81) indicates that

$$r = 136 \text{ m}$$

Table 3.16. Channel and outfall parameters at the Huntly power station during low flows

River flow (Q)	150 $\text{m}^3 \text{s}^{-1}$
Cooling water flow (q_e)	38 $\text{m}^3 \text{s}^{-1}$
Aproach flow (q_a)	112 $\text{m}^3 \text{s}^{-1}$
Approach depth* (h)	2.1 m
Channel width (b)	180 m
Approach velocity (u_a)	0.296 m s^{-1}
Outfall area† (a_e)	200 m^2
Outfall velocity† (u_e)	0.19 m s^{-1}

* The scour hole is about 5 m deep.
† Assuming the outfall behaves like a discharge canal 40 m long and 5 m deep.

and equation (3.87) gives

$$q_c = 70 \text{ m}^3 \text{s}^{-1}$$

Thus from equation (3.85) the initial dilution is

$$d = 2.84$$

The average cooling water excess temperature is 8.4°C and so the average plume temperature [from equation (3.88)] is about 3°C. Assuming a linear variation across the plume the maximum bankside temperature is about 6°C. Again this is in reasonable agreement with field observations under low flow conditions.

Equation (3.84) indicates that within the near-field the effective dispersion coefficient is $k_z = 0.3 \text{ m}^2 \text{s}^{-1}$. This is substantially higher than values of $k_z = 0.06 \text{ m}^2 \text{s}^{-1}$ measured outside the plume using dye and indicates that outfall momentum (and possibly buoyancy) significantly increase the transverse mixing rate in the near-field.

NOTATION

a_e	area of port
b	channel width
b'	effective channel width
b_e	width of the discharge canal
B	buoyancy flux
C_m	fully-mixed concentration
c^*	non-dimensional concentration
d	dilution $= q_n/q_e$
D	factor of diffusion
D_i	factor of diffusion in sub-reach i
e	flow entrained into the plume
e_x, e_y, e_z	turbulent diffusion coefficients in the x, y, z direction
f	Darcy friction coefficient $= 8(U^*/V_x)^2$
$f(x, q)$	excess temperature distribution in an unbounded channel
F_z	transverse tracer flux
g	acceleration of gravity
h	local depth
h'	effective depth
h_e	depth of the discharge canal
h_i	local depth at point i across the channel
H	mean depth
H_o	outer length of the recirculating eddy
H_i	inner length of the recirculating eddy

I	waste heat flux entering the river
k_1, k_2, k_3	empirical coefficients
k_x, k_z	longitudinal and transverse dispersion coefficients
k_β	transverse dispersion coefficient in curvilinear coordinates
k_i	value of k_z in sub-reach i
$\overline{k_z}$	reach-averaged value of k_z
l_{m}	momentum length scale
L	total reach length
L	length of the recirculating eddy
L	diffuser length
L_{w}	length of the near-field
L_{c}	crossing distance
L_{c}	limiting diffuser length
L_{m}	Prandtl's mixing length
L_{t}	transverse turbulent length scale
L_{z}	transverse mixing distance
m	mass inflow rate
m_α, m_β	metric coefficients
M_{a}	momentum flux of the approaching river flow
M_{e}	momentum flux of the effluent
n	Manning's friction coefficient
N	number of ports
N	number of sub-reaches
P_{m}	degree of mixing $= s_{\mathrm{min}}/s_{\mathrm{max}}$
q	cumulative discharge
q^*	non-dimensional cumulative discharge
q_{c}	entrained flow $= q_n - q_{\mathrm{e}}$
q_{e}	effluent flow
q_{n}	slipstream flow
q_{l}	cumulative discharge at the edge of the cooling water plume
q_k	cumulative discharge at point K across the channel
Q	total river flow
r	plume width
r_{c}	radius of curvature of the channel
R	hydraulic radius
s	water slope
s	depth-averaged concentration
s_{e}	effluent tracer concentration
$\bar{s}$	average plume tracer concentration
Δs	concentration resulting from an incremental release
$s - u$	distance between points s and u measured along curvilinear axes
s_{u}	predicted concentration in an unbounded channel
Δt	time step

t	time
T	depth-averaged excess temperature
T_o	maximum plume excess temperature
T_u	predicted excess temperature in an unbounded channel
u_x, u_y, u_z	local longitudinal, vertical and transverse velocities
u_a	river approach velocity
u_b	bypass velocity outside the slipstream
u_e	effluent velocity
u_n	slipstream velocity
u_t	r.m.s turbulent velocity fluctuation
u_x'	turbulent velocity fluctuation
u^*	local shear velocity
U^*	mean shear velocity
u_a'	effective approach velocity
$\langle u_x \rangle$	ensemble mean velocity
v_x, v_z	depth-averaged longitudinal and
v_{xi}	depth-averaged velocity at point i across the channel
v_α	longitudinal velocity in curvilinear coordinates
v_b^n, v_b^{n+1}	non-dimensional bypass velocity at iterations n and $n+1$
V_x	cross-sectional averaged velocity
x, y, z	longitudinal, vertical and transverse directions
Δx_j	length of sub-reach j
x^*, z^*	non-dimensional longitudinal and transverse distance
x_o, z_o	source location
$\Delta x, \Delta y$	finite difference mesh sizes
α, β	longitudinal and transverse distances in curvilinear coordinates
β	non-dimensional diffuser length
$\varkappa$	von Karman's constant
λ	non-dimensional diffuser length $= L/b$
λ_c	non-dimensional limiting diffuser length $= L_c/b$
ξ	non-dimensional approach flow $= q_a/q_e$
ν_a	non-dimensional approach velocity $= u_a/u_e$
ν_u	non-dimensional slipstream velocity $= u_n/u_e$
ν_b	non-dimensional bypass velocity $= u_b/u_e$
$\Delta\rho$	density difference
ρ	density
σ	plume contraction factor
σ_z^2	transverse variance
ψ	dimensionless shape factor

Chapter 4

Longitudinal Dispersion

4.1 INTRODUCTION

This chapter considers mixing in the far-field, which begins once the tracer has had the opportunity to mix across the entire channel. As discussed in Section 3.5.3 this occurs about 100–300 channel widths downstream from a point source located in mid-channel. If the tracer source and river flow rate are steady and the tracer is conservative, then concentrations in the far-field are constant

$$C_{\mathrm{m}} = \frac{m}{Q} \tag{4.1}$$

where C_{m} = fully mixed concentration; m = mass inflow rate and Q = river flow rate. If, however, the tracer source is rapidly varying then at a given time there are significant longitudinal concentration gradients and at a given site concentrations vary with time. Figure 4.1 shows the cross-sectional averaged dye concentrations measured at six sites downstream from an instantaneous release of dye to the Waikato River, New Zealand. The current carries the entire dye cloud downstream and there is a tendency for the tracer to disperse along the channel. The latter process is termed longitudinal dispersion and results in the peak concentration decreasing and the time of passage past each site increasing with distance below the source.

A common use of longitudinal dispersion models is to analyse pollutant concentrations in rivers downstream from a sudden discharge such as a road tanker accident, a storage tank collapse or the start up of a polluting discharge. In many practical problems the tracer input varies only gradually and longitudinal dispersion is of secondary importance compared with vertical and transverse dispersion (Thomann 1973). Surprisingly, there is much more published work on longitudinal dispersion than on the arguably more important problems of transverse and vertical dispersion.

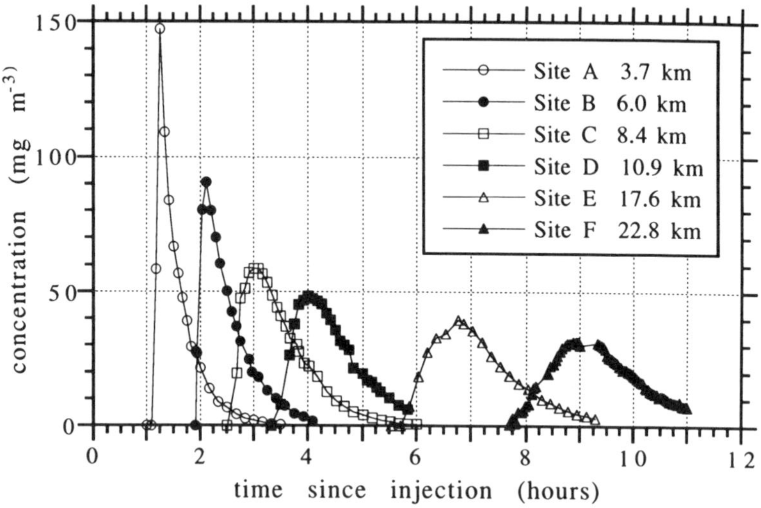

Figure 4.1. Cross-sectional averaged dye concentrations measured at six sites on the Waikato River below an instantaneous transverse line source

4.2 SUMMARY

A summary of the main conclusions reached in this chapter are given at the outset to help the reader appreciate where the ensuing discussions are leading and to act as a checklist subsequently.

1. Longitudinal dispersion arises because vertical and transverse velocity shear carry the tracer downstream more slowly near the bed and the banks than in mid-channel. Fischer (1967) showed that in river channels transverse velocity shear makes a greater contribution to longitudinal dispersion than vertical velocity shear. Turbulent diffusion counteracts the effects of velocity shear.
2. Taylor (1953, 1954) showed that at a point some distance downstream from a tracer injection in stationary, homogencous turbulence an equilibrium becomes established between the effects of velocity shear and turbulent diffusion. Thereafter the variance of the tracer cloud grows linearly with time. It is inferred from Taylor's analysis that some distance downstream from the source (in the so-called equilibrium zone) longitudinal dispersion can be modelled using Fick's law. The equilibrium zone commences once tracer has had an opportunity to sample the entire flow. The time taken for this to happen is closely related to the time taken for tracer from a steady source to become well mixed across the channel. There is

a common misconception that in the equilibrium zone tracer concentrations are uniform across the channel. Sayre (1968a) showed that this is not the case for an instantaneous tracer input because velocity shear continually creates concentration gradients which are counteracted but never entirely removed by turbulent mixing.

3. Considerable skewness (asymmetry) is imparted to the tracer profile close to the source (in the so-called advective zone). The Fickian model predicts that in the equilibrium zone asymmetry decays slowly until a very long way below the source (typically five to 10 times the length of the advective zone) the profiles become approximately Gaussian.
4. Few, if any, experimental data are Gaussian. There are three reasons for this. Firstly, most experimental data are concentration versus time data at a fixed site and the Fickian model predicts that only concentration versus distance profiles are Gaussian. We can, however, make allowance for this using Chatwin's transformation. Secondly, asymmetry imparted to the profile in the advective zone takes a long time to decay away. Thirdly, the presence of a laminar diffusive boundary layer in smooth laboratory channels and of dead zones in natural river channels extends considerably the length of the advective zone.
5. Although the Fickian model can model accurately the behaviour of the centroid and variance of a tracer profile it tends to underestimate the skewness. Nevertheless, the Fickian model is a very useful, if approximate, tool for analysing longitudinal dispersion especially at sites a long way below the source.
6. Longitudinal dispersion coefficients have been collated from a number of publications. It is found that most of the data fall within the range

$$2 < \frac{K_x}{bU^*} < 50 \tag{4.21}$$

where K_x = longitudinal dispersion coefficient; b = channel width; and U^* = average shear velocity. Channel curvature can markedly reduce the longitudinal dispersion coefficient by accelerating the rate of transverse mixing. In some channels K_x increases with increasing flow but in others there is little or no change with flow. The flow response depends on how the velocity distribution changes with flow. The value of K_x can be determined from the velocity distribution provided the transverse dispersion coefficient is known.

7. An analytical solution to the Fickian dispersion model for an instantaneous injection which neglects the advective zone is

$$S(x, t) = \frac{M}{A\sqrt{4\pi K_x t}} \exp\left[-\frac{(x - V_x t)^2}{4K_x t}\right] \tag{4.4}$$

where $S(x, t)$ = cross-sectional averaged concentration; M = mass of tracer injected and V_x = cross-sectional averaged velocity. An appreciation of the behaviour of this equation is useful in understanding longitudinal dispersion. It can also be used to predict tracer concentrations a long way below an instantaneous source. One of the advantages of the Fickian disperion model is that it is linear, which means that solutions [such as equation (4.4)] can be added together to solve problems with complex initial or boundary conditions. Superposition is used to predict concentrations below a time varying or a spatially distributed source and also to route the tracer profile observed at one site downstream to another site.

8. Field tests usually give concentration versus time profiles at fixed sites. We can use the frozen cloud approximation to convert these into concentration versus distance profiles at the mean time of passage and vice versa. This is particularly useful when routing tracer profiles. The Hayami solution to the Fickian dispersion equation can be used to route concentration versus time profiles without needing to use the frozen cloud approximation.
9. Within the advective zone concentrations can be predicted using an unsteady two-dimensional (depth-averaged) numerical model or an unsteady streamtube model.
10. The effects of dead zones can be included in the Fickian model to give the so-called transient storage model. The model equations must be solved numerically and further work needs to be done before model coefficients can be predicted from bulk channel parameters.
11. The aggregated dead zone model is a powerful technique for analysing longitudinal dispersion. The multi-order version of the model represents the channel by a number of well-mixed cells joined in series and/or in parallel. This gives flexibility but parameter identification can only be done using time series methods. In many channels a simplified first- or second-order model can be fitted to tracer data fairly easily. Further work is needed before model coefficients can be predicted from bulk channel parameters.

4.3 MECHANISMS CAUSING LONGITUDINAL DISPERSION

4.3.1 Effects of Vertical and Transverse Velocity Shear

Figure 4.2 shows the plan view of a channel in which the longitudinal velocity varies parabolically across the channel. We shall assume that tracer is released from an instantaneous transverse line source because this illustrates most clearly the processes important in longitudinal dispersion. Tracer travels downstream more slowly near the banks than in the main stream and as a

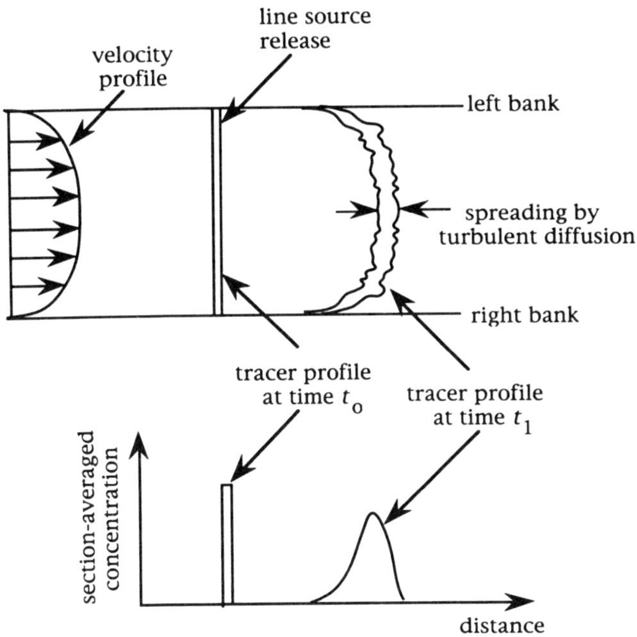

Figure 4.2. Schematic diagram showing the combined effects of transverse velocity shear and transverse turbulent diffusion on longitudinal spreading of tracer

result, what was originally a straight line of tracer across the channel, assumes a parabolic shape. Turbulent diffusion causes some localised spreading both along and across the channel. If we now calculate the cross-sectional averaged concentration, we see that transverse velocity shear and longitudinal turbulent diffusion cause the mean concentration to spread along the channel. In Figure 4.2 longitudinal diffusion causes very little mixing along the channel and the major contribution comes from the transverse velocity shear. Figure 4.2 shows only the early stages after tracer release, but the same processes continue to operate in the far-field. Transverse velocity shear increases the transverse concentration gradient so that in the leading edge of the tracer cloud concentrations are higher in mid-stream than near the banks, whereas in the trailing edge of the cloud the converse applies. These transverse concentration gradients promote transverse mixing. The rate of longitudinal dispersion reflects the balance between velocity shear (which acts to spread tracer along the channel) and transverse mixing (which promotes uniform concentrations and hence counteracts the effects of velocity shear). Thus if the rate of transverse mixing is very high (e.g. in a narrow sinuous channel) or the velocity is uniform, then the rate of longitudinal dispersion is low. Conversely, if the rate of transverse mixing is low or the velocity is highly non-uniform (e.g. in a wide channel with extensive shallows), then the rate of longitudinal dispersion is high. This point is important in explaining the effects of channel curvature on longitudinal dispersion and the differences in longitudinal dispersion coefficients between channels.

Much of the theoretical and laboratory work on longitudinal dispersion in open channels has been done in plane (two-dimensional) shear flow in which the velocity varies with depth but not with distance across the channel. One reason is that experimental work in plane shear flow can be done in relatively short laboratory flumes. A body of theory has been developed around this experimental work (Elder 1959, Sayre 1968a, Valentine and Wood 1977, Denton 1990) which quantifies the effects of vertical velocity shear and vertical diffusion on longitudinal dispersion. These studies shed light on the longitudinal dispersion process but care must be exercised using their results in natural rivers for reasons that are discussed more fully in Section 4.9. Nevertheless the theory developed from these studies is sound and has been used to quantify the effects of transverse velocity shear in river channels. Subsequent discussion in this chapter concentrates on the effects of transverse velocity shear, but the reader should bear in mind that vertical velocity shear operates in an analogous manner although it makes a relatively small contribution to longitudinal dispersion in most rivers.

4.3.2 Transverse Concentration Gradients in the Far-field

In the far-field attention is focused on the behaviour of the cross-sectional averaged concentration. This has led to the misconception that in the far-field tracer is fully mixed across the channel. While this is true for a steady source of conservative traces, it is not true for an unsteady source.

In Chapter 1 we derived an equation which describes the depth-averaged tracer concentration

$$\frac{\partial s}{\partial t} + v_x \frac{\partial s}{\partial x} + v_z \frac{\partial s}{\partial z} = \frac{1}{h}\frac{\partial}{\partial x}\left(hk_x \frac{\partial s}{\partial x}\right) + \frac{1}{h}\frac{\partial}{\partial z}\left(hk_z \frac{\partial s}{\partial z}\right) \qquad (1.106)$$

where x and z = longitudinal and transverse distances; s = depth-averaged tracer concentration; h = local depth; v_x and v_z = depth-averaged velocities in the x and z directions; and k_x and k_z = dispersion coefficients in the x and z directions. If the channel is uniform then h is constant and we can align the x axis with the main flow so that $v_z = 0$. If k_x and k_z are also assumed constant then equation (1.106) becomes

$$\frac{\partial s}{\partial s} + v_x \frac{\partial s}{\partial x} = k_x \frac{\partial^2 s}{\partial x^2} + k_z \frac{\partial^2 s}{\partial z^2} \qquad (4.2)$$

There are several published papers which use equation (4.2) and assume that the velocity v_x is constant (i.e. does not vary across the channel). This leads to very misleading results. If v_x and k_z are constant then equation (4.2) predicts that for $x > 0.536 v_x b^2 / k_z$ (i.e. at distances greater than the transverse mixing distance described in Chapter 3) transverse concentration gradients vanish. In practice we find that for a non-steady input, the concentration gradients do not vanish, even well within the far-field. Sayre (1968b) discussed the landmark paper of Fischer (1967) and showed that concentration gradients were apparent in Fischer's laboratory data even well into the far-field. Transverse concentration gradients are also a feature of experimental data from rivers as illustrated by Figure 4.3 (Rutherford *et al* 1980). The reason that transverse concentration gradients persist is that the velocity v_x is not constant across the channel. Although there is a tendency for tracer to mix across the channel as a result of transverse diffusion and dispersion, transverse velocity shear always results in tracer being carried downstream faster in mid-channel than near the banks. Consequently transverse concentration gradients never completely vanish, although they do become quite small in the far-field. By neglecting transverse velocity shear, equation (4.2) is only able to match observed tracer concentrations by using very large values of k_x and values fitted in this way can be very misleading.

In the far-field concentration gradients are fairly small (of the order of 5–10%) and in most situations are of no ecological significance. Consequently in the far-field we can usually afford to ignore any transverse and vertical

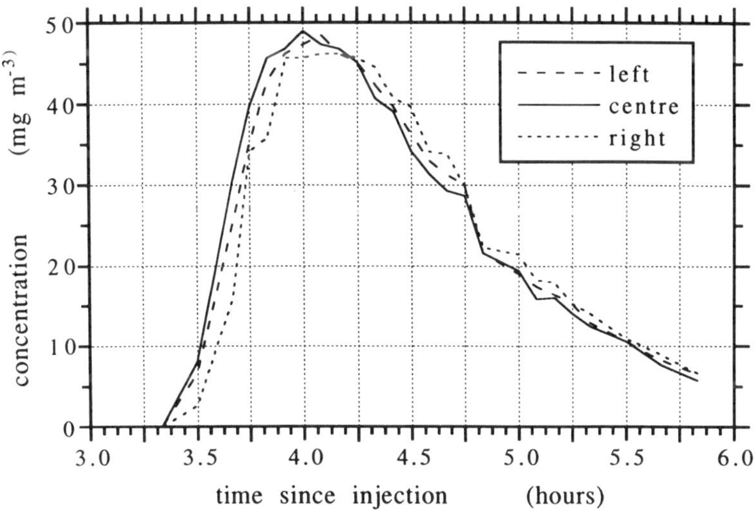

Figure 4.3. Dye concentrations measured at three different points across the channel showing persistent transverse non-homogeneity. Data are from the Waikato River, Site D, 10.9 km below the injection point (see also Figure 4.1)

concentration gradients and to examine just the cross-sectional averaged concentration. The question then arises whether a model exists for predicting the behaviour of the cross-sectional averaged concentration in the far-field.

4.4 FICKIAN MODEL OF LONGITUDINAL DISPERSION

4.4.1 Taylor's Analysis

When a tracer cloud is first released, its shape is largely dictated by the velocity differences across the channel, as is illustrated in Figure 4.2. The region close to the source is termed the advective zone to emphasise the importance of the velocity distribution. Within the advective zone the tracer profile becomes negatively skewed (i.e. the rising limb is shorter than the falling limb) as is apparent in Figure 4.1. Figure 4.4 shows how the skewness varies below a slug injection, increasing markedly in the advective zone and subsequently decreasing slowly. We will describe methods for estimating the length of the advective zone shortly.

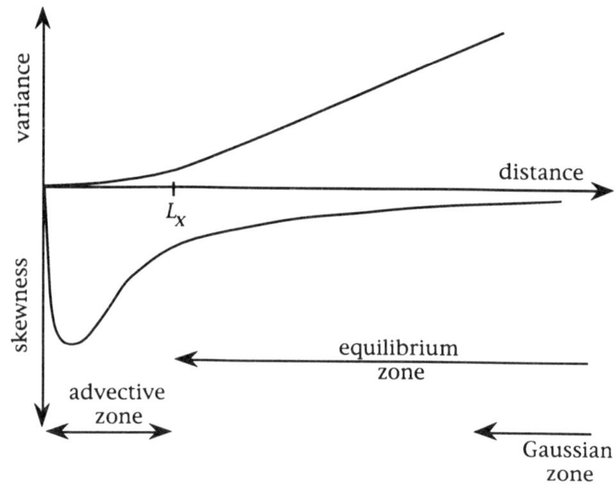

Figure 4.4. Fickian model predictions of how the variance and skewness of a concentration profile change with time. In the advective zone the variance increases non-linearly with time and the skewness increases rapidly. In the equilibrium zone the variance grows linearly with time and the skewness starts to decay slowly. In the Gaussian zone the skewness has decayed to a small value and tracer profiles comply with equation (4.4)

Theoretical and experimental work by Taylor in pipe flow (Taylor 1953, 1954) indicates that at some point downstream from the source an equilibrium becomes established between transverse velocity shear (which promotes longitudinal dispersion) and transverse diffusion (which counteracts longitudinal dispersion). Beyond this point two things happen. Firstly, the longitudinal variance of the cross-sectional averaged tracer concentration increases linearly with time. Secondly, any skewness introduced by velocity shear in the advective zone or by the initial distribution of tracer begins to decay slowly and eventually the spatial tracer distribution becomes Gaussian. The zone in which the variance increases linearly is known as the equilibrium zone. Figure 4.4 shows the behaviour of the variance and the skewness in the advective, equilibrium and Gaussian zones.

It is well known that Fick's law predicts that the variance of the tracer cloud increases linearly with time. The fact that Taylor's analysis shows that the variance increases linearly with time in the equilibrium zone suggests that the cross-sectional averaged concentration can be modelled using Fick's law. For a uniform channel

$$\frac{\partial S}{\partial t} + V_x \frac{\partial S}{\partial x} = K_x \frac{\partial^2 S}{\partial x^2} \tag{4.3}$$

where S = cross-sectional averaged tracer concentration; V_x = cross-sectional averaged velocity; and K_x = longitudinal dispersion coefficient (a constant). There is a more detailed derivation of equation (4.3) in Section 1.8.4.

Equation 4.3 makes no attempt to model the variation of tracer concentration within the cross-section. The effects of velocity shear and turbulent mixing (both transverse and vertical) are included in the longitudinal dispersion coefficient. Thus the value of K_x depends on the hydraulic properties of the channel, which determine the velocity shear and turbulent mixing rates. Equation (4.3) can be called the Fickian model of longitudinal dispersion.

If V_x and K_x are assumed to be constant then the solution to equation (4.3) for an instantaneous point source (slug infection) is

$$S(x, t) = \frac{M}{A\sqrt{4\pi K_x t}} \exp\left[-\frac{(x - V_x t)^2}{4K_x t}\right] \tag{4.4}$$

where M = mass of tracer injected at $x = 0$ and $t = 0$; and A = cross-sectional area of the channel. Historically equation (4.4) predates equation (4.3) (Chatwin 1980) and hereafter is called the Taylor Solution.

4.4.2 Gaussian Tracer Profiles

A feature of equation (4.4) is that the spatial concentration pattern (i.e. the distribution of concentration with distance at a given time) follows the

Gaussian bell-shaped profile. Few observed concentration profiles are Gaussian and almost invariably they are skewed with steep fronts and elongated tails (i.e. they exhibit negative skewness). Day and Wood (1976) state that they are:

> '... unaware of any Gaussian concentration distribution ever being recorded from flow in an open channel ...'

There are two reasons why experimental data could be non-Gaussian.

Firstly, the practicalities of sampling usually dictate that temporal tracer profiles (concentration versus time at a fixed site) are measured rather than spatial profiles (concentration versus distance at a fixed time). Equation (4.4) gives Gaussian spatial profiles but skewed temporal profiles. This skewness arises from the fact that longitudinal dispersion occurs during the time taken for the tracer cloud to pass a fixed sampling site. Thus asymmetry in temporal tracer profiles is not in itself evidence of inadequacies in equation (4.4). Temporal profiles can, however, be analysed using the transformation developed by Chatwin (1980). Equation (4.4) is rearranged in the form

$$\sqrt{t \log_e\left(\frac{a}{S\sqrt{t}}\right)} = \frac{x}{2\sqrt{K_x}} - \frac{V_x t}{2\sqrt{K_x}} \tag{4.5}$$

where

$$a = \frac{M}{A\sqrt{4\pi K_x}} \tag{4.6}$$

For concentration versus true data at a fixed site which conform with equation (4.4), a plot of

$$\sqrt{t \log_e\left(\frac{a}{S\sqrt{t}}\right)} \qquad \text{versus } t \tag{4.7}$$

yields a straight line with slope

$$-\frac{V_x}{2\sqrt{K_x}} \tag{4.8}$$

and intercept

$$\frac{x}{2\sqrt{K_x}} \tag{4.8}$$

In general a is not known in advance but to a close approximation

$$a = S_{\max}\sqrt{t_{\max}} \tag{4.10}$$

where $S_{\max}$ = the peak concentration which occurs at time $t_{\max}$. Both $S_{\max}$ and $t_{\max}$ can be estimated from experimental data. The practicalities of using this transformation are discussed later in Section 5.6.3.

The second reason why observed tracer profiles are non-Gaussian is that equation (4.3) is not valid in the advective zone which extends a distance L_x below the source (see Figure 4.4). Equation (4.4), however, is derived assuming that the Fickian model applies throughout the advective zone. Considerable skewness is introduced into tracer profiles in the advective zone and this is not reproduced by equation (4.4). The Fickian dispersion model predicts that in the equilibrium zone any asymmetry imparted to the tracer profile in the advective zone decays slowly (see Figure 4.4) and eventually spatial tracer profiles become Gaussian. Fischer *et al* (1979) suggest that equation (4.4) is valid at $x > 2.5L_x$, Denton (1990) suggests $x > 4–5L_x$, Sayre's results (Sayre 1968a) suggest $x > 10L_x$, while Liu and Cheng (1980) suggest $x > 50L_x$. Thus many experimental results may well be non-Gaussian simply because skewness introduced during the advective zone has not had enough time to decay.

If concentration versus time profiles are measured at several points along a channel in the equilibrium zone, then they can each be plotted using Chatwin's transformation [equation (4.5)]. The initial profiles might not be linear because of residual skewness inherited in the advective zone, but if equation (4.3) is valid then the skewness should decay and profiles should eventually plot as straight lines. This may only occur at distances below the source several times the length of the advective zone. It has been found, however, that when experimental data are plotted in this manner curvature persists for longer than expected (Day and Wood 1976, Rutherford *et al* 1980). Figure 5.12 illustrates the curvature which is typically found in field data. This persistent curvature in field data suggests that equation (4.4) is seldom valid in natural rivers and also raises doubts about the ability of the underlying equation (4.3) to model accurately the skewness of observed tracer clouds. The observed curvature is usually most severe at low concentrations in the rising and falling limbs of tracer profiles. For concentrations above about half the peak, equation (4.3) usually provides a tolerably good description of longitudinal dispersion.

4.4.3 Longitudinal Dispersion in a Non-uniform Channel

Equation (4.3) describes longitudinal dispersion in a uniform channel. In a non-uniform channel it seems logical to write

$$\frac{\partial S}{\partial t} + V_x \frac{\partial S}{\partial x} = \frac{1}{A} \frac{\partial}{\partial x} \left(A K_x \frac{\partial S}{\partial x} \right) \tag{4.11}$$

where A = cross-section area. There is, however, some conjecture about the validity of equation (4.11). Taylor's analysis suggests that longitudinal dispersion obeys Fick's law provided the turbulence is homogeneous and stationary and a sufficiently long time has elapsed since the release of the tracer. In an irregular channel it is unlikely that turbulence is homogeneous. This means that there is no guarantee that an equilibrium will become established between velocity shear and turbulent diffusion before the tracer cloud is carried downstream into a different regime. It is not clear, however, what effect violation of this condition has on the accuracy of equations (4.3) and (4.11). There is empirical evidence that these equations describe reasonably well the important features of observed tracer profiles in both laboratory and river channels. Overall equations (4.3) and (4.11) provide useful, if approximate, models for longitudinal dispersion in the equilibrium zone.

4.4.4 Spatial and Temporal Moments

Considerable insights into longitudinal dispersion can be gained by examining how the spatial and temporal moments of tracer profiles change with time. The longitudinal (spatial) moments of the cross-sectional averaged tracer concentration are estimated from

$$\mu_p(t) = \int_{x=-\infty}^{\infty} x^p S(x, t)\, \mathrm{d}x \tag{4.12}$$

where $\mu_p(t) = p$th spatial moment after time t; $x =$ longitudinal displacement; and $p = 0, 1, 2 \ldots$ etc. The first four moments describe, respectively, the tracer mass, the location of the centroid, the variance and the skewness of the tracer profile. Normally four moments provide an adequate description of a tracer profile. Of particular interest is the spatial variance

$$\sigma_x^2(t) = \frac{\int_{x=-\infty}^{\infty} (x - \bar{x})^2 S(x, t)\, \mathrm{d}x}{\int_{x=-\infty}^{\infty} S(x, t)\, \mathrm{d}x} = \frac{\mu_2(t)}{\mu_0(t)} \tag{4.13a}$$

where the location of the centroid is

$$\bar{x} = \frac{\int_{x=-\infty}^{\infty} xS(x, t)\, \mathrm{d}x}{\int_{x=-\infty}^{\infty} S(x, t)\, \mathrm{d}x} = \frac{\mu_1(t)}{\mu_0(t)} \tag{4.13b}$$

Sayre (1968a) conducted a numerical modelling study of solute dispersion in two-dimensional shear flow where there was a vertical velocity gradient but no transverse velocity gradient. He calculated the first four spatial moments numerically.

Initially the rate of growth of the variance was non-linear but after some time it was found to grow linearly with time. This behaviour is illustrated in Figure 4.4. The non-dimensional time at which the growth of variance became linear was

$$\tau = \frac{e_y t}{h^2} \approx 0.5 \tag{4.14}$$

where h = depth and e_y = vertical dispersion coefficient. Several other theoretical and experimental studies have given similar estimates of the point at which the variance starts to grow linearly (Valentine and Wood 1977, Tsai and Holley 1978, Denton 1990).

A unique feature of Sayre's study was that he also calculated the vertical distribution of the two spatial moments which describe the centroid and variance of the tracer profile. For non-dimensional time $\tau > 0.5$ the vertical distribution of these moments did not change. This finding provides support for Taylor's intuitive hypothesis that at the end of the advective zone a balance becomes established between velocity shear and diffusion. Sayre's analysis showed that the second moment is not uniform over the depth, which confirms that tracer is not vertically well mixed even well into the equilibrium zone. Sayre also found that although the variance grows linearly after $\tau > 0.5$, the skewness of the tracer cloud decreases only very slowly. Sayre's calculations only extend to $\tau = 3$, at which point there is still significant skewness present, but extrapolating his results one might expect tracer concentrations to become Gaussian at about $\tau = 10$. This emphasises the point made earlier that tracer concentrations only become Gaussian a long time after the end of the advective zone.

Evaluation of the spatial moments from field data requires measurements of spatial tracer profiles at fixed times. Data of this kind are occasionally available from aerial photographs but more commonly in rivers, tracer is measured over time at a fixed site (temporal profiles). In an analogous manner to equation (4.12) temporal moments can be estimated from such data

$$\nu_q(x) = \int_{t=-\infty}^{\infty} t^q S(x, t)\, \mathrm{d}t \tag{4.15}$$

where $\nu_q(x)$ = qth temporal moment at the fixed location x; t = time and $q = 0, 1, 2, \ldots$ etc. The temporal variance is

$$\sigma_t^2(t) = \frac{\int_{t=-\infty}^{\infty} (t - \bar{t})^2 S(x, t)\, \mathrm{d}t}{\int_{t=-\infty}^{\infty} S(x, t)\, \mathrm{d}t} = \frac{\nu_2(x)}{\nu_0(x)} \tag{4.16a}$$

where the time of passage of the centroid is

$$\bar{t} = \frac{\int_{t=-\infty}^{\infty} tS(x, t)\, \mathrm{d}t}{\int_{t=-\infty}^{\infty} S(x, t)\, \mathrm{d}t} = \frac{\nu_1(x)}{\nu_0(x)} \tag{4.16b}$$

Tsai and Holley (1978) undertook a modelling study in two-dimensional flow (in which tracer was assumed vertically well mixed) to investigate the effects of transverse velocity shear. The velocity was assumed to vary parabolically across the channel. Whereas Sayre (1968a) only examined the behaviour of the spatial moments, Tsai and Holley (1978) examined both spatial and temporal moments. When the temporal variance σ_t^2 is plotted against the time of passage of the centroid a familiar pattern emerges. Initially the variance grows rapidly and in a non-linear fashion, but after a certain time the rate of growth of temporal variance becomes linear. The spatial variance behaves in exactly the same way (see Figure 4.4). As might be expected, the time at which the temporal and spatial moments begin to grow linearly are the same. The value varies a little depending on the source location but from Tsai and Holley's study is typically

$$\tau = \frac{k_z t}{b^2} \approx 0.4\text{–}0.5 \tag{4.17}$$

Equations (4.14) and (4.17) have similar coefficients.

The temporal variance has the units (s^2) while the spatial variance has the units (m^2) and to make comparisons the temporal variance is scaled as $V_x^2\sigma_t^2$. If the temporal variance is determined at location x and the spatial variance is determined at the time of passage of the centroid then it is found that

$$V_x^2\sigma_t^2(x) - \sigma_x^2(\bar{t}) > 0 \tag{4.18}$$

Once the tracer cloud enters the equilibrium zone this difference becomes a constant so that it is valid to determine the longitudinal dispersion coefficient from either

$$K_x = \frac{1}{2}\frac{\mathrm{d}\sigma_x^2(t)}{\mathrm{d}t} = \frac{1}{2}\frac{\sigma_x^2(t_2) - \sigma_x^2(t_1)}{t_2 - t_1} \tag{4.19}$$

where t_1 and t_2 = times at which the spatial variances are determined, or

$$K_x = \frac{1}{2}\frac{\mathrm{d}}{\mathrm{d}t}\,[V_x^2\sigma_t^2(x)] = \frac{1}{2}\,V_x^2\,\frac{\sigma_t^2(x_2) - \sigma_t^2(x_1)}{\bar{t}_2 - \bar{t}_1}$$

$$= \frac{1}{2}\,V_x^3\,\frac{\sigma_t^2(x_2) - \sigma_t^2(x_1)}{x_2 - x_1} \qquad (4.20)$$

where $\bar{t}_1$ and $\bar{t}_2$ = times of passage of the centroid at sites 1 and 2; and x_1 and x_2 = locations of sites 1 and 2. It is worth noting, however, that the approximation

$$\sigma_x^2 = V_x^2\sigma_t^2 \qquad (4.21)$$

is not valid at small times although at large times the values of the spatial and temporal variance become large and their difference (a constant) becomes progressively less significant.

4.4.5 Non-Fickian Models

Taylor's analysis shows that provided sufficient time has elapsed since injection, the variance grows linearly with time and one feature of solutions to equation (4.3) is that the variance grows linearly with time. Does this mean that equation (4.3) is an exact model for longitudinal dispersion outside the advective zone?

Chatwin argues that it is not and based on a theoretical analysis (Chatwin 1970) suggests that a more exact equation describing tracer concentrations is

$$\frac{\partial S}{\partial t} + V_x\frac{\partial S}{\partial x} = K_2\frac{\partial^2 S}{\partial x^2} + K_3\frac{\partial^3 S}{\partial x^3} + K_4\frac{\partial^4 S}{\partial x^4} + \cdots \qquad (4.22)$$

where $K_2, K_3, K_4 \ldots$ etc are mixing coefficients. Only if $K_3 = K_4 \cdots = 0$ does equation (4.22) match equation (4.3) exactly. By taking spatial moments of equation (4.22) it can be shown that $K_3, K_4 \ldots$ do not affect the centroid or the variance of the tracer distribution but they strongly influence the skewness and higher order moments.

Chatwin (1980) re-examined Fischer's laboratory data and showed that although equation (4.3) can describe satisfactorily the behaviour of the centroid and the variance, it does not describe accurately the observed behaviour of the skewness and tends to predict that tracer profiles become Gaussian more quickly than is observed in practice. The most likely reason for this is the presence of a viscous sub-layer near the bed and walls. Tracer which diffuses into this sub-layer is temporarily trapped near the bed and only when the main body of the tracer cloud has passed does it diffuse slowly back into the main flow. As a consequence a long tail is induced in the tracer profile. Elder (1959) and Dewey and Sullivan (1977) provide experimental evidence to support the conclusion that the viscous sub-layer affects the behaviour of the tracer cloud in laboratory channels.

4.4.6 Dead Zones

In natural river channels there are regions of recirculating flow or stagnant water associated with irregularities in the bed and banks. These are collectively known as dead zones and they have the potential to trap tracer near the bed in much the same way as the viscous sub-layer. Dead zones are thought to explain the long tails observed in river tracer data by Thackston and Schnelle (1970) and Sabol and Nordin (1978).

If we reconsider Taylor's analysis then dead zones have two effects. Firstly, they invalidate the assumption of homogeneous turbulence. This probably does not present a major problem provided the characteristics of the dead zones are fairly uniform along the channel. In this situation turbulence may not be homogeneous vertically, but it will be similar at all points along the channel. Secondly, because the rate of diffusion into and out of the dead zones is small, the time taken for tracer particles to sample the entire flow is greatly increased. Thus the length of the advective zone will be longer in a natural channel where dead zones are present than in a smooth channel (Valentine and Wood 1977).

The presence of dead zones does not necessarily invalidate equation (4.3). Valentine and Wood (1977) conducted a numerical modelling study of a two-dimensional flow with regular dead zones along the bed. They determined the spatial moments of the tracer distribution and examined how these changed with time. Even when dead zones are present, a point is still reached where an equilibrium becomes established between advection in the overlying water and exchange with the dead zones. At this point the mean velocity of the tracer cloud becomes a constant, the ratio of the tracer mass in the dead zones to the mass in the overlying flow become constant and the growth of variance becomes linear. Valentine's analysis shows that within the equilibrium zone the skewness of the tracer profile decreases with time and eventually tracer profiles become Gaussian. The time required for this to happen is much longer in a rough channel than in a smooth one. The mean velocity of the tracer cloud in the equilibrium period is smaller than the mean velocity of the overlying water

$$V = V_x(1 - \eta) \tag{4.23}$$

where V = mean velocity of the centroid of the tracer cloud; V_x = mean velocity of the water (excluding the water in the dead zones) and η = ratio of the volume of the dead zones to the volume of water in the reach. [In Figure 4.13 the bed contains uniform dead zones whose depth is d, dead zones occupy 50% of the bed surface area and the water depth (including dead zones) is h. In this situation $\eta = d/2h$.] The length of the advective zone and the value of the dispersion coefficient in the equilibrium zone both depend on the value of η.

4.5 ESTIMATING THE LENGTH OF THE ADVECTIVE ZONE

Equation (4.3) only applies once an equilibrium becomes established between the effects of velocity shear and turbulent diffusion. Taylor argued that the time taken for this equilibrium to be established (the Eulerian time scale) is similar to the time taken for tracer to '... sample the entire flow...'. In Chapter 3 the constant-coefficient model predicts that tracer becomes fully mixed across the channel within a distance of

$$L_z = 0.536\,\frac{v_x b^2}{k_z} \tag{3.28}$$

for an outfall at either bank and

$$L_z = 0.134\,\frac{v_x b^2}{k_z} \tag{3.29}$$

for a source in mid-channel, where b = channel width; v_x = depth-averaged longitudinal velocity (assumed constant across the channel); and k_z = transverse dispersion coefficient. By analogy with equations (3.28) and (3.29) we would expect the length of the advective zone to be

$$L_x = \alpha\,\frac{V_x L_t^2}{k_z} \tag{4.24}$$

where L_x = length of the advective zone; V_x = cross-sectional averaged velocity; L_t = a transverse length scale; and α = a constant. Fischer suggested that L_t should be the transverse distance from the point of maximum velocity to the farthest bank. Thus in a symmetrical channel $L_t = b/2$, where b = channel width, while in natural channels a reasonable approximation is $L_t = 0.7b$. We have already discussed the fact that below an unsteady source tracer never becomes fully mixed across the channel and equation (4.24) is best viewed as quantifying the distance required for a tracer particle to sample the entire flow.

Fischer (1967) conducted laboratory experiments in uniform straight channels to determine the distance below an instantaneous transverse line tracer source at which a linear growth of variance became well established and estimated $\alpha = 0.3$. This is a lower bound estimate because we would expect tracer to take longer to sample the entire flow if it originates from a point source at one bank than from a transverse line source. On the basis of evidence from numerical experiments, Fischer (1968) suggests that for a bankside source $\alpha = 0.6$. Other numerical experiments with two-dimensional shear flows suggest that $\alpha = 0.4$–0.5 for a range of source locations (Sayre 1968, Tsai and Holley 1978). These values lie in the range $0.3 < \alpha < 0.6$ and are comparable with the coefficients in equations (3.28) and (3.29) of 0.13–0.54.

Fischer's experiments were conducted in smooth laboratory channels where dead-zone effects were small, although some tracer is likely to have been retained in the viscous sub-layer. The presence of dead zones in a natural river channel significantly increases the Eulerian time-scale (i.e. the time taken by a tracer particle to sample the entire flow field). Although dead zones also reduce the mean velocity of the tracer cloud [see equation (4.23)] this is more than offset by the increase in the Eulerian time-scale so that dead zones increase the length of the advective zone. Fischer *et al* (1979) discuss the laboratory data of Valentine (1978) and conclude that dead zones can increase the length of the advective zone by a factor of about two, giving $\alpha = 0.6$–1.2. My reading of Valentine's results is that in a rough laboratory channel where dead zones occupied 4 and 25% of the channel volume, the length of the advective zone was increased by factors of four and seven, respectively. This suggests that $\alpha = 1.6$–2.8. Valentine and Wood (1979a) also present data collected from an irrigation canal with encroaching vegetation and extensive dead zones.

Table 4.1. Estimates of the length of the advective zone

Reference	$\alpha = L_x k_z / L_t^2 V_x$	Comments
Smooth channels		
Fischer (1973)	0.2	Review of numerical experiments
Fischer (1967)	0.3	Laboratory channel, transverse line source
Tsai and Holley (1978)	0.4–0.5	Numerical experiments
Sayre (1968)	0.5	Numerical experiments
Fischer (1968)	0.6	Bankside injection
Chatwin (1972)	1	Theoretical analysis
Rough channels		
Denton (1990)	1.4	Numerical experiments, 5% dead zones
Valentine (1978)	1.6	Laboratory channel, 4% dead zones
Valentine (1978)	2.8	Laboratory channel, 25% dead zones
Valentine and Wood (1979b)	>3	Irrigation canal, 8–12% dead zones
Valentine and Wood (1979b)	>10	Irrigation canal, 27–38% dead zones

L_x = length of the advective zone; k_z = transverse dispersion coefficient; L_t = transverse length scale (normally 0.5–$0.7b$); b = channel width; and V_x = cross-sectional averaged velocity.

Assuming a fairly low value for the transverse dispersion coefficient ($k_z/HU^* = 0.23$) and a large transverse length scale ($L_t = b$) their data suggest lower bound estimates of $\alpha = 3$ and 10 for dead zones occupying 10 and 30% of the channel volume, respectively. Values of $\alpha > 5$ are probably appropriate for small and very rough streams, including the boulder-bed mountain streams studied by Day and Wood (1976) and Dobran (1982). Denton (1990) calculated a value of $\alpha = 1.4$ in numerical experiments with a dead zone volume of 5%. Table 4.1 summarises a number of published estimates of the length of the advective zone.

4.6 ESTIMATING THE LONGITUDINAL DISPERSION COEFFICIENT

4.6.1 Experimental Data

Table 4.2 contains a selection of longitudinal dispersion coefficients measured in natural channels. In collating these data the original references have been consulted wherever possible and in some cases figures have been checked with the original authors. Nevertheless some of the dispersion coefficients may be inaccurate either because of poor estimation procedures, tracer loss or because measurements were made within the advective zone. Another source of uncertainty lies with the reported values of water depth, channel width and shear velocity; these parameters are difficult to estimate accurately without painstaking survey work.

Data are usually reported as the non-dimensional coefficient K_x/HU^* where H = mean depth and U^* = average shear velocity [see equation (3.10)]. There is a very wide variation in the reported longitudinal dispersion coefficients and published values range over three orders of magnitude. Much of the data falls within the range (see Figure 4.5)

$$30 < \frac{K_x}{HU^*} < 3000 \qquad (4.25)$$

4.6.1.1 *Effects of flow*

Figure 4.5 suggests that the non-dimensional dispersion coefficient is largely independent of flow although the high scatter in the published data may disguise a tendency for K_x/HU^* to increase with river flow. In Figure 4.6 values of K_x collated in Table 4.2 from several different rivers are plotted against river flow. This figure shows fairly clearly that large rivers tend to have high values of K_x. This reflects the fact that large rivers tend to be deep and if both K_x/HU^* and U^* are independent of the size of the river then a large deep

Table 4.2. Reported longitudinal dispersion coefficients

Reference	Canal/River	H (m)	b (m)	V_x ($m\,s^{-1}$)	U^* ($m\,s^{-1}$)	Q ($m^3 s^{-1}$)	K_x ($m^2 s^{-1}$)	K_x/HU^*	K_x/bU^*
Canals									
1	Albert	4.40	100	0.029	0.0016	16.2	0.2	34	1.5
	Dessel-Herentals	2.50	35	0.037	0.0022	3.4	0.2	35	2.5
2	Yuma Mesa A	3.45	7.6	0.68	0.047	18	0.5	3	1.5
1	Bocholt-Dessel	2.50	35	0.107	0.0063	14.7	1.4	89	6.4
	Villemsvaart	2.50	34	0.13	0.0079	10.4	1.7	86	6.3
3	Chicago	8.07	49	0.27	0.019	106	3.0	19	3.2
4	Coachella	1.56	24	0.71	0.043	26.6	9.6	143	9.3
5	irrigation	0.19	1.4	0.38	0.11	—	9.6	469	64
		0.14	1.5	0.33	0.10	—	1.9	136	13
European rivers									
6	Derwent	0.25	—	0.38	0.140	—	4.6	130	—
7	Wolf	—	—	0.03	—	0.2	0.7	—	—
		—	—	0.09	—	0.9	1.3	—	—
		—	—	0.35	—	5.7	4.3	—	—
		—	—	0.68	—	19.8	9.0	—	—
	Lyd	—	—	0.16	—	0.6	4.0	—	—
		—	—	0.24	—	1.2	4.4	—	—
		—	—	0.48	—	3.9	7.4	—	—
		—	—	0.93	—	15.7	15.5	—	—
8	Miljacka	0.29	11	0.35	0.058	1.0	2.7	161	4.4
7	Nidd	—	—	0.13	—	2.0	4.1	—	—
	Tamar	—	—	0.13	—	2.8	8.3	—	—
		—	—	0.25	—	10.7	12.4	—	—
		—	—	0.60	—	37.5	15	—	—
		—	—	1.15	—	165	30	—	—
	Upper Tame	0.83	9.9	0.46	0.090	3.8	5.5	74	6.2
		0.92	9.9	0.52	0.10	4.8	5.1	55	5.2

	Tame	—	—	0.46	—	8.5	14.6	—	—
		—	—	0.53	—	10.6	10.0	—	—
	Wye	—	—	0.37	—	8.6	54	—	—
		—	—	0.38	—	27	51	—	—
		—	—	0.72	—	72	70	—	—
	Ure	—	—	0.10	—	15	3.5	—	—
USA rivers									
9	Comite	0.26	13	0.31	0.044	1.0	7.0	612	13
		0.43	16	0.37	0.056	2.6	13.9	577	16
10	Copper, VA	0.40	19	0.16	0.116	1.2	9.9	213	4.5
		0.49	16	0.26	0.080	2.0	9.5	242	7.4
		0.49	16	0.27	0.080	2.1	20	510	16
		0.85	18	0.60	0.100	9.2	21	247	12
9	Antietam	0.39	16	0.32	0.062	2.0	9.3	385	9.4
		0.52	20	0.43	0.071	4.4	16.3	441	12
		0.71	24	0.52	0.083	8.9	25.6	434	13
9	Monocacy	0.32	35	0.21	0.044	2.4	4.7	334	3.0
		0.45	37	0.32	0.051	5.2	13.9	606	7.4
		0.88	48	0.44	0.072	18.4	37.2	587	11
9	Anacoco	0.42	20	0.29	0.045	2.4	13.9	735	16
		0.94	26	0.34	0.068	8.2	32.5	508	19
		0.92	37	0.40	0.067	13.5	39.5	641	16
9	Muddy	0.81	13	0.37	0.081	4.0	13.9	212	13
		1.20	20	0.45	0.099	10.6	32.5	274	17
10	Powell	0.85	34	0.15	0.055	4.0	9.5	203	5.1
9	Elkhorn	0.30	33	0.43	0.046	4.3	9.3	674	6.1
		0.42	51	0.46	0.047	9.9	21	1060	8.8
10	Clinch, VA	0.58	36	0.21	0.049	6.8	8.1	285	4.6
11	Duwamish	1.10	20	—	0.049	—	7.5	139	7.7
12	South Platte	0.46	—	0.66	0.069	—	16.2	510	—
9	Sabine, TX	0.98	35	0.21	0.042	7.4	39.4	957	27

(continued)

Table 4.2. (*Continued*)

Reference	Canal/River	H (m)	b (m)	V_x ($m\,s^{-1}$)	U^* ($m\,s^{-1}$)	Q ($m^3 s^{-1}$)	K_x ($m^2 s^{-1}$)	K_x/HU^*	K_x/bU^*
13	Minnesota	2.74	80	0.034	0.0024	7.9	22.3	3390	120
		2.74	80	0.14	0.0097	27	34.9	1310	45
9	Amite	0.81	37	0.29	0.070	8.6	23.2	409	9.1
		0.80	42	0.42	0.069	14.2	30.2	547	10
9	White	0.55	67	0.35	0.044	12.7	30.2	1250	10
4	Clinch, TE	0.85	47	0.32	0.067	12.8	14	246	4.4
		2.10	53	0.83	0.107	92.4	47	209	8.3
		2.10	60	0.94	0.104	118	54	247	8.7
9	John Day	0.58	25	1.01	0.140	14.2	13.9	171	4.0
		2.46	34	0.82	0.181	69	65	146	11
9	Chattahoochee	1.13	66	0.39	0.076	29.0	32.5	378	6.59
9	Nooksack	0.76	64	0.67	0.27	32.6	34.8	170	2.0
		2.93	86	1.20	0.53	303	153	98	3.4
9	Wind-Bighorn	1.10	59	0.88	0.120	58	41.8	317	5.9
		2.16	69	1.55	0.166	231	163	455	14
9	Yadkin	2.33	70	0.43	0.100	71	111	476	16
		3.85	72	0.76	0.129	213	260	524	28
13	Mississippi	3.05	530	0.08	0.0056	74	13.0	761	4.4
		3.05	530	0.07	0.0050	92	19.5	1280	7.4
		3.05	530	0.12	0.0082	132	16.6	664	3.9
		3.05	530	0.14	0.0097	268	35.9	1210	7.0
14	Mississippi	—	—	—	—	10 310	232	—	—
		—	—	—	—	22 600	700	—	—
9	Susquehanna	1.35	203	0.39	0.065	106	92.9	1060	7.0
9	Sabine, LA	2.04	104	0.58	0.054	119	316	2870	56
		4.75	128	0.64	0.084	389	670	1680	62

15	Missouri	2.33	183	0.89	0.066	380	465	3020	39
		2.70	200	1.55	0.074	837	1500	7510	100
		3.56	201	1.28	0.084	913	837	2800	50
		3.11	197	1.53	0.078	935	892	3700	58
New Zealand rivers									
16	Punehu	0.28	5	0.26	0.21	0.4	7.2	122	6.9
	Kapuni	0.30	9	0.37	0.15	1.2	8.4	187	6.2
	Kapuni	0.35	10	0.53	0.17	1.7	12.4	208	7.3
	Manganui	0.40	20	0.19	0.18	2.3	6.5	90	1.8
	Waiongana	0.60	13	0.48	0.24	2.5	6.8	47	2.2
	Stony	0.63	10	0.55	0.30	3.4	13.5	71	4.5
17	Waiotapu	0.75	11.4	0.41	0.061	3.5	8.0	175	12
	Manawatu	0.72	59	0.37	0.070	15.6	32	635	7.7
	Manawatu	1.00	63	0.32	0.094	20.0	22	234	3.7
18	Manawatu	0.95	60	0.46	0.092	26.3	47	538	8.5
17	Tarawera	1.21	25	0.73	0.084	22.1	27	266	13
		1.92	20	0.62	0.123	23.7	11.5	49	4.7
		1.38	25	0.77	0.091	26.6	20.5	163	9.0
		1.40	25	0.78	0.091	27.3	15.5	122	6.8
		1.57	25	0.83	0.096	32.4	18	119	7.5
19	Waikato	2.60	85	0.69	0.060	153	52	333	10
		2.00	120	0.64	0.050	153	67	670	11

1 = Marvioet and Craenenbroeck (1986); 2 = Schuster (1965); 3 = Thomas (1958); 4 = Godfrey and Frederick (1970); 5 = Valentine (1978); 6 = Owens *et al* (1964); 7 = Cole, J. A., Water Research Centre, England, unpublished; 8 = Dobran (1982); 9 = McQuivey and Keefer (1974); 10 = Fischer *et al* (1979); 11 = Fischer (1968); 12 = Glover (1964); 13 = Stefan and Demetracopoulos (1981); 14 = McQuivey and Keefer (1976); 15 = Yotsukura *et al* (1970); 16 = Kitto and Rutherford (1982); 17 = Wilcock, R. J., NIWA Ecosystems, New Zealand, unpublished; 18 = Rutherford and Williams (1992); and 19 = Rutherford *et al* (1980).

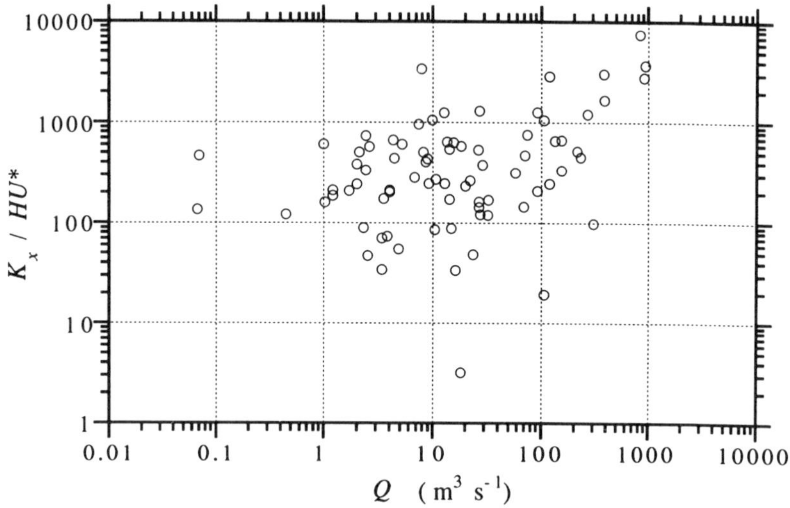

Figure 4.5. Variation of the non-dimensional longitudinal dispersion coefficient between rivers with different flows

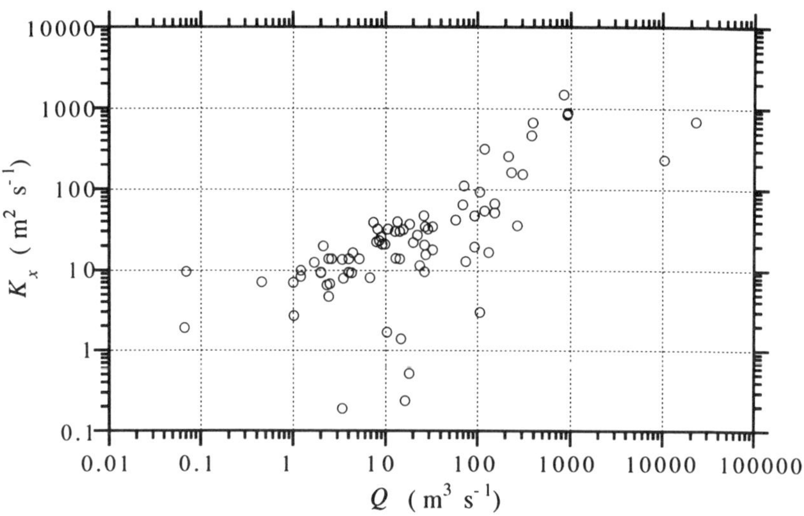

Figure 4.6. Variation of the longitudinal dispersion coefficient between rivers with different flows

river will have a high K_x. This is a useful piece of information when wishing to estimate the value of K_x in a river where only flow data are available.

It is also of interest to know how the value of K_x varies with flow in a particular river. There are two possibilities. Firstly, the non-dimensional dispersion coefficient K_x/HU^* may remain constant as the flow increases. If so then as depth and shear velocity both increase with flow, we can expect K_x to increase with flow. There are comparatively few studies in which the dispersion coefficient has been measured over the same river reach at a number of different flows. Table 4.2 does, however, include the results of four such experiments conducted in English rivers by the Water Research Centre. Data from the Rivers Wye, Tamar, Wolf and Lyd (the last two being tributaries of the Tamar) are plotted against flow in Figure 4.7. In the Rivers Tamar, Wolf and Lyd there is a strong positive correlation between K_x and Q. LeGrand-Marcq and Laudelout (1985) have also reported an increase in K_x with Q in a small Belgian river.

The second possibility is that as river flow increases the velocity becomes more uniform across the channel because near the banks the depth increases and bed friction has less effect. In addition the transverse dispersion coefficient often increases with increasing river flow. Both processes will cause a reduction of K_x with increasing flow rate. In Figure 4.7 K_x is markedly higher in the River Wye than in the Tamar, Wolf or Lyd. This is presumably because the transverse velocity shear is high and/or the rate of transverse mixing is

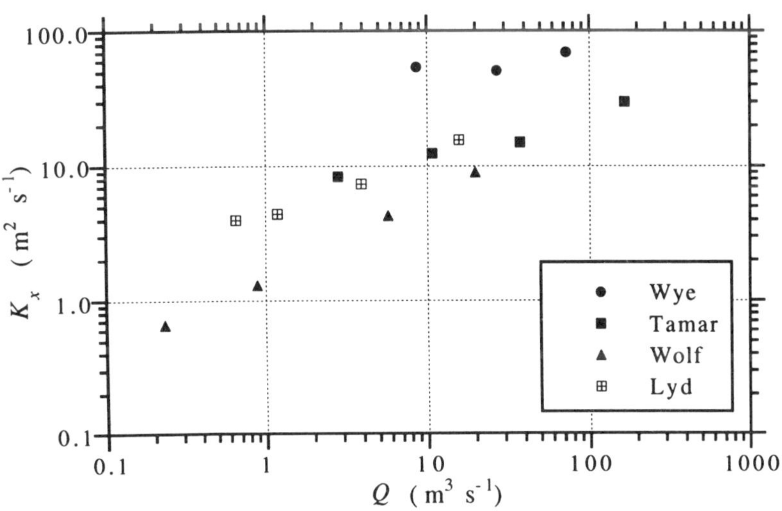

Figure 4.7. Variation of the longitudinal dispersion coefficient with flow over the same reach in four rivers

low. K_x shows only a slight increase with flow in the Wye, which is in marked contrast to the other three rivers. In Example 5.8 we examine velocity distributions in the Waikato River measured at a number of different flows. These show that in some parts of the river transverse velocity shear decreases significantly with increasing flow and in such circumstances we would expect the value of K_x to decrease.

Thus it is not possible to generalise about how K_x will change with flow in a given river and an assessment can only be made in a particular situation based on a knowledge of how the transverse velocity shear and the transverse dispersion coefficient change with flow.

4.6.1.2 Effects of channel width

Figure 4.8 shows that K_x/HU^* increases with the aspect ratio of the channel. There are two factors which contribute to this behaviour. Firstly, the rate of transverse mixing is high in narrow and deep rivers because secondary currents are strong. The longitudinal dispersion coefficient is inversely proportional to the transverse dispersion coefficient and so would be expected to be low in a narrow and deep river. Secondly, in wide and shallow channels the transverse variation of velocity is large, especially if there are extensive areas of shallows. The larger the velocity differences across the channel, the larger the longitudinal dispersion coefficient.

It is traditional to non-dimensionalise longitudinal dispersion coefficients

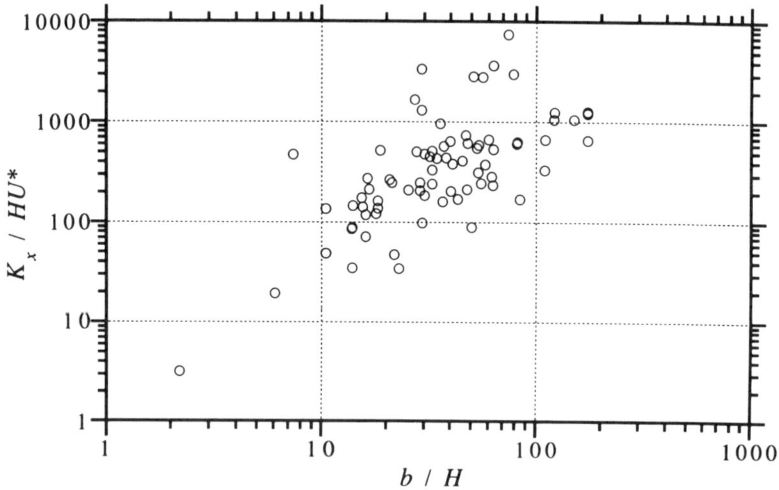

Figure 4.8. Variation of the non-dimensional longitudinal dispersion coefficient with channel aspect ratio

as K_x/HU^*. Figure 4.8 suggests an alternative non-dimensional dispersion coefficient, K_x/bU^*. The use of the channel width b as the length scale in place of the mean depth H seems reasonable in the light of Fischer's finding that transverse velocity shear has a greater effect on longitudinal dispersion in natural channels than vertical velocity shear.

Figure 4.9 shows that K_x/bU^* is largely independent of flow and ranges over two orders of magnitude. Most of the data lie in the range

$$2 < \frac{K_x}{bU^*} < 50 \tag{4.26}$$

Equation (4.26) represents a modest reduction in the scatter of published experimental data over equation (4.25).

4.6.1.3 *Effects of channel curvature*

Transverse mixing counteracts the effect of transverse velocity shear and the longitudinal dispersion coefficient is inversely proportional to the transverse dispersion coefficient. Channel curvature influences k_z strongly (as is discussed in Chapter 3) and hence can be expected to influence K_x. This is supported by Fukuoka and Sayre (1973) who found that in a sinuous laboratory channel K_x/HU^* was inversely proportional to the radius of curvature. It is also worth noting that the length of the advective zone is inversely

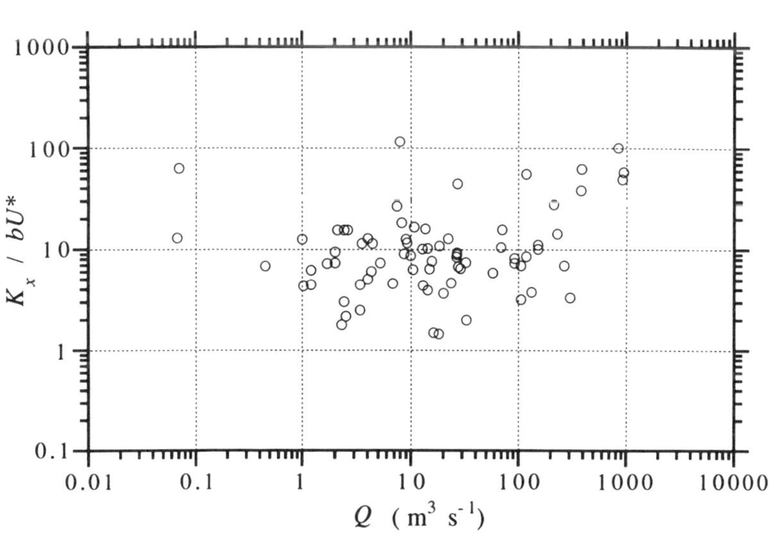

Figure 4.9. Variation of the non-dimensional longitudinal dispersion coefficient K_x/bU^* between rivers with different flows

proportional to the transverse dispersion coefficient [equation (4.19)] and would be expected to decrease with increasing channel curvature.

4.6.1.4 Effects of dead zones

Dead zones increase the length of the advective zone (see Table 4.1) and they also increase the value of the longitudinal dispersion coefficient. For example, data in Valentine (1978) indicate that dead zones occupying 4 and 25% of the channel volume increased the longitudinal dispersion coefficient by a factor of 2 and 10, respectively, above the value expected for a smooth channel. Similarly, reducing the dead zone fraction from 38 to 12% in a water race reduced the dispersion coefficient by a factor of five (Valentine 1978).

4.6.2 Empirical Formulae for the Dispersion Coefficient

Several workers have derived semi-empirical formulae for predicting the longitudinal dispersion coefficient. McQuivey and Keefer (1974) suggest that

$$K_x = 0.058 \frac{Q}{sb} \tag{4.27}$$

where Q = river flow; s = channel slope; and b = channel width. Fischer (1973) and Jain (1976) both give

$$K_x = \alpha \frac{V_x^2 b^2}{k_z} \tag{4.28a}$$

where V_x = mean velocity; k_z = transverse dispersion coefficient, and

$$0.001 < \alpha < 0.016 \tag{4.28b}$$

Liu (1977) suggests that

$$K_x = \beta \frac{Q^2}{U^* R^3} \tag{4.29a}$$

where R = hydraulic radius and

$$\beta = 0.18 \left(\frac{U^*}{V_x} \right)^{1.5} \tag{4.29b}$$

If these formulae are compared with published data it is found that at best they predict K_x to within a factor of about 10. Figures 4.5–4.9 also show that for a given flow or aspect ratio the reported range of K_x covers at least an order of magnitude. This is typical of the uncertainty inherent in estimating dispersion coefficients from published data (Fischer *et al* 1979).

When making predictions of tracer concentration using published dispersion coefficients, the practitioner should undertake a careful analysis of how sensitive the concentrations are to a wide variation in the choice of dispersion coefficient. This will indicate whether the conclusions based on predicted concentrations are robust to uncertainties in the estimated values of K_x (in which case published values should suffice) or are sensitive to uncertainties (in which case additional investigations using more mathematically sophisticated models, physical models and/or field tests are required).

Example 4.1

The Waikato River near Hamilton, New Zealand, has the channel characteristics summarised in Table 4.3. When planning a dye test in 1975 preliminary estimates were made of the longitudinal dispersion coefficient and the length of the advective zone.

From equation (4.25), $30 < K_x/HU^* < 3000$, which gives $4 < K_x < 428\ \mathrm{m^2\,s^{-1}}$. Figure 4.6 suggests a value of $20 < K_x < 200\ \mathrm{m^2\,s^{-1}}$ for a flow of $150\ \mathrm{m^3\,s^{-1}}$. From equation (4.26), $2 < K_x/bU^* < 50$, which gives $10 < K_x < 240\ \mathrm{m^2\,s^{-1}}$. There is a high uncertainty in the estimated value of K_x but this is typical of the scatter in published data. (The dye test eventually yielded estimates of $52 < K_x < 67\ \mathrm{m^2\,s^{-1}}$, which lie within the range predicted prior to the tests.)

The river channel is gently meandering in the study reach. From equation (3.12) we would expect the non-dimensional transverse dispersion coefficient to lie in the range $0.3 < k_z/HU^* < 0.9$. Thus $0.043 < k_z < 0.128\ \mathrm{m^2\,s^{-1}}$. For the dye test, tracer was introduced at six points equidistant across the channel, which approximates a transverse line source. Thus from Table 4.1 it is appropriate to assume that $\alpha = 0.3$. The main flow is near mid-channel and so $L_t = 0.5b = 42.5$ m. Equation (4.24) predicts the length of the advective zone to be $3.0 < L_x < 8.8$ km.

To help plan the dye test tracer, concentrations were simulated for a range of values ($20 < K_x < 150\ \mathrm{m^2\,s^{-1}}$, $0.6 < V_x < 0.8\ \mathrm{m\,s^{-1}}$). Sampling sites were then selected so that experimental work could be completed within daylight hours and most of the sites lay outside the advective zone. The mass of dye injected was calculated to ensure that the peak dye concentration would be an order of magnitude higher than the detection limit at the downstream site. We used Lissamine Red 4B dye, which was found to have a

Table 4.3. Hydraulic parameters of the Waikato River near Hamilton

River flow (Q)	$150\ \mathrm{m^3\,s^{-1}}$
Mean depth (H)	2.5 m
Average width (b)	85 m
Mean velocity (V_x)	$0.70\ \mathrm{m\,s^{-1}}$
Channel slope (s)	0.00013
Shear velocity (U^*)	$0.057\ \mathrm{m\,s^{-1}}$

detection limit of 0.1 mg m^{-3}. We decided to release 36 kg of dye made up in advance into 100 kg m^{-3} stock solution. Initial calculations suggested for the largest expected K_x peak concentrations at Site D (+10.9 km) would be higher than 25 mg m^{-3} and hence well above the detection limit. Figure 4.1 shows the observed dye concentrations. Peak concentrations at Site D were approximately 50 mg m^{-3}: twice the planned value and sufficiently high to get good resolution of the tracer profiles. It is likely that sites A and B lay within the advective zone but we can be fairly confident that sites D, E and F were in the equilibrium zone. From Section 4.4.2 we would not expect tracer profiles to comply with equation (4.4) within 5–10L_x, i.e. within 15–90 km of the source.

4.7 APPROXIMATE MODEL FOR LONGITUDINAL DISPERSION

4.7.1 Introduction

An analysis of longitudinal dispersion can be made using an approximate model which neglects the advective zone and assumes that the mean velocity and longitudinal dispersion coefficient are constant. The Taylor solution to equation (4.3) for an instantaneous point source (slug injection) is

$$S(x, t) = \frac{M}{A\sqrt{4\pi K_x t}} \exp\left[-\frac{(x - V_x t)^2}{4K_x t}\right] \tag{4.4}$$

Strictly equation (4.4) is only valid a long way below the source after the asymmetry introduced into the tracer profile in the advective zone has had a chance to decay away. As discussed in Section 4.4.2 this is unlikely to occur before $x = 5–10L_x$ where L_x is the length of the advective zone [given by equation (4.24)]. It is worth noting that equation (4.4) satisfies the initial condition that at $t = 0$ the concentration is a delta function of x (i.e. concentration everywhere is zero except at $x = 0$ where the mass M is concentrated).

4.7.2 Predictions for an Instantaneous Point Source

Some pollution discharges to rivers can be treated as slug injections. The most common example is a road or industrial accident which releases a large quantity of tracer into the river over a short period of time. When conducting a field test to measure the velocity and dispersion coefficient it is usual to make a slug injection of tracer. Equation (4.4) is a very useful approximate model which predicts fairly well important features of the tracer profile such as the rate of attenuation of the peak concentration and the time of passage of the tracer profile past a given site. Thus a good understanding of equation (4.4) is important.

4.7.2.1 *Behaviour of the peak concentration*

At a specified time after release t_n equation (4.4) predicts that the concentration versus distance profile is Gaussian and hence symmetrical about the peak concentration. The peak concentration occurs at

$$x_{max} = V_x t_n \tag{4.30}$$

and is

$$S_{max} = \frac{M}{A\sqrt{4\pi K_x t_n}} = \frac{M}{A\sqrt{\dfrac{4\pi K_x x_{max}}{V_x}}} \tag{4.31}$$

Thus the peak concentration decreases inversely with the square root of time. Equation (4.30) is used to replace t_n by x_{max}/V_x in equation (4.31) and it can be seen that the peak concentration decreases inversely with the square root of x_{max}.

The centroid of the spatial distribution is defined by equation (4.13b) and for the Gaussian profile of equation (4.4) is

$$\bar{x}(t_n) = x_{max} = V_x t_n \tag{4.32}$$

It is often of interest to know the length of channel in which concentrations exceed some critical value S_c. By setting $S = S_c$ in equation (4.4) and substituting from equation (4.31), two values of x_c are found whose difference is

$$\Delta x_c = 4\sqrt{K_x t_n \log_e\left(\frac{S_{max}}{S_c}\right)} \tag{4.33}$$

where $S_c < S_{max}$. Even though observed tracer concentration profiles are usually more highly skewed than is predicted by equation (4.4), equation (4.33) gives fairly accurate estimates outside the advective zone of the length of channel in which concentration exceeds S_c.

Equations (4.31)–(4.33) are useful when analysing concentration versus distance profiles measured at a fixed time. It is, however, more common to measure concentration versus time profiles at a fixed location x_n. Equation (4.4) predicts that the concentration versus time profile is not symmetrical. The peak concentration occurs at

$$t_{max} = \sqrt{\frac{K_x^2}{V_x^4} + \frac{x_n^2}{V_x^2}} - \frac{K_x}{V_x} \tag{4.34}$$

The peak concentration can be estimated by substituting t_{max} into equation

(4.4). The centroid of the concentration versus time profile can be calculated using equation (4.16b) and is found to occur at (Thackston *et al* 1967)

$$\bar{t} = \frac{x_n}{V_x} + \frac{2K_x}{V_x^2} \tag{4.35}$$

Note that the mean travel time x_n/V_x lies somewhere between the times of passage of the peak and the centroid of a concentration versus time profile.

Example 4.2

During the 1970s investigations were undertaken for a scheme to abstract drinking water from the lower Waikato River to supply the city of Auckland. One important design consideration was the time it would take for a pollutant accidentally discharged into the river (e.g. following a road tanker accident) to reach the intake site and the length of time concentrations would be above some critical value. To help design off-river storage, calculations were made using equation (4.4) of the arrival time and the departure time of tracer at the proposed abstraction site (Mercer) for instantaneous injections at several upstream bridge sites. The channel parameters in this reach are summarised in Table 4.4.

Consider the case of a bridge 30 km above the abstraction site and the instantaneous injection of 500 kg of a tracer which is likely to cause water treatment problems at concentrations exceeding 5 mg m^{-3}. From Table 4.2 the measured non-dimensional dispersion coefficient in the Waikato River near Hamilton is $K_x/bU^* = 10$. We shall assume that in the study reach $K_x = 75$ m^2 s^{-1}.

The problem is to solve equation (4.4) for the times when S first reaches $S = 5$ mg m^{-3} (the arrival time) and eventually drops below $S = 5$ mg m^{-3} (the departure time) given the values of x, M, A, K_x and V_x. Time t appears three times in equation (4.4) and we cannot solve for t directly. Table 4.5 shows the output from a spreadsheet program which solves equation (4.4) iteratively for the arrival and departure times.

The user inserts the input data in lines 1–7. The program then makes a

Table 4.4. Hydraulic parameters of the lower Waikato River

River flow (Q)	250 m^3 s^{-1}
Mean depth (H)	2.0 m
Average width (b)	200 m
Mean velocity (V_x)	0.625 m s^{-1}
Channel slope (s)	0.00007
Shear velocity (U^*)	0.037 m s^{-1}

Table 4.5. Results from a spreadsheet program which solves equation (4.4) iteratively to predict the arrival and departure times (i.e. the times when tracer concentration reaches and drops below some low but critical value) at a specified distance downstream from an instantaneous slug injection of tracer

		Line
Input data		1
Critical concentration (S_c)	5 mg m^{-3}	2
Tracer mass (M)	500 kg	3
Distance below source (x)	30 km	4
Channel area (A)	400 m^2	5
Velocity (V_x)	0.625 m s^{-1}	6
Dispersion (K_x)	75 m^2 s^{-1}	7
		8
Preliminary calculations		9
Mean travel time (t_{mean})	13.33 hrs	10
Maximum concentration (S_{max})	185.85 mg m^{-3}	11
Concentration ratio (S_{max}/S_c)	37.17 mg m^{-3}	12
Approx. arrival time (t_a)	10.13 hrs	13
Approx. departure time (t_b)	16.54 hrs	14
		15
Iterations		16
Arrival time (t_a)	10.5	17
Departure time (t_b)	16.8	18
Time step (Δt)	0.01	19
		20

		Rising limb		Falling limb		Line
						21
Eqn (4.4) predictions		Time	Conc.	Time	Conc	22
	1	10.40	4.35	16.70	6.90	23
	2	10.41	4.48	16.71	6.78	24
	3	10.42	4.62	16.72	6.66	25
	4	10.43	4.76	16.73	6.55	26
	5	10.44	4.90	16.74	6.43	27
	6	10.45	5.04	16.75	6.32	28
	7	10.46	5.19	16.76	6.21	29
	8	10.47	5.34	16.77	6.10	30
	9	10.48	5.49	16.78	6.00	31
	10	10.49	5.65	16.79	5.89	32
	11	10.50	5.82	16.80	5.79	33
	12	10.51	5.98	16.81	5.69	34
	13	10.52	6.15	16.82	5.59	35
	14	10.53	6.33	16.83	5.49	36
	15	10.54	6.50	16.84	5.39	37
	16	10.55	6.69	16.85	5.30	38
	17	10.56	6.87	16.86	5.21	39
	18	10.57	7.06	16.87	5.11	40
	19	10.58	7.26	16.88	5.02	41
	20	10.59	7.46	16.89	4.93	42
	21	10.60	7.66	16.90	4.85	43

first guess at the arrival and departure times. The mean travel time is

$$t_{\text{mean}} = \frac{x}{V_x}$$

Setting $t = t_{\text{mean}}$ in equation (4.4)

$$S(x, t_{\text{mean}}) = S_{\max} \exp\left[-\frac{(x - V_x t_{\text{mean}})^2}{4K_x t_{\text{mean}}} \right]$$

where

$$S_{\max} = \frac{M}{A\sqrt{4\pi K_x t_{\text{mean}}}}$$

Rearranging this equation and solving for the values of x at which concentration equals S_c we get

$$x_c = V_x t_{\text{mean}} \pm \sqrt{4K_x t_{\text{mean}} \log_e\left(\frac{S_{\max}}{S_c}\right)}$$

Now let us assume that the tracer cloud undergoes no dispersion during the time that it takes to pass the site (this is the frozen cloud assumption which is discussed more fully in the next section). We divide through by the velocity V_x, set $t_{\text{mean}} = x/V_x$ and $t_c = x_c/V_x$ to estimate the times at which concentration equals S_c

$$t_c = \frac{x \pm \sqrt{\dfrac{4K_x x \log_e\left(\dfrac{S_{\max}}{S_c}\right)}{V_x}}}{V_x}$$

Lines 13 and 14 of Table 4.5 show the resulting initial estimates of the arrival and departure times. These estimates are only approximate but they provide a starting point for an iterative solution procedure. A straightforward grid search produces satisfactory results. The user specifies the arrival and departure times in lines 17 and 18 and a time step in line 19. In lines 23–43 the program solves equation (4.4) at values of t either side of the specified arrival and departure times. The user inspects the predicted concentrations and if necessary changes the values in lines 17, 18 and/or 19 manually. This process continues until the user can identify satisfactorily the times at which the concentration first exceeds and subsequently drops below the critical concentration. In this example the arrival and departure times can be identified from Table 4.5 as being $t_a = 10.45$ hours and $t_b = 16.88$ hours giving a time of passage of 6.43 hours during which concentrations exceed 5 mg m^{-3}.

In my experience, provided the velocity and dispersion coefficient are known then this approach gives estimates of the time of passage which match closely those obtained from the tracer data itself. We can then use this method with some confidence to predict the times of passage for other flows, velocities and tracer masses. Equation (4.4) has a tendency, however, to predict arrival and departure times which are slightly earlier than tracer observations. This point is discussed in more detail by Rutherford *et al* (1980) where an empirical method is suggested to correct the arrival and departure times based on the results of tracer tests.

4.8 FROZEN CLOUD APPROXIMATION

4.8.1 Introduction

In many situations advection dominates dispersion

$$\frac{x}{V_x} \gg \frac{K_x}{V_x^2} \tag{4.36}$$

and to a close approximation

$$t_{\max} \approx \bar{t} \approx \frac{x_n}{V_x} \tag{4.37}$$

The frozen cloud approximation exploits equation (4.37) to estimate concentration versus distance profiles at a fixed time from concentration versus time profiles at a fixed location and vice versa. As its name suggests, the frozen cloud approximation assumes that no longitudinal dispersion occurs during the time taken for tracer to pass a sampling site. This is not strictly valid in most natural channels because although advection is more rapid than dispersion, the latter does have a discernible effect on tracer concentrations. Nevertheless the frozen cloud approximation is commonly made when analysing longitudinal dispersion data.

Figure 4.10 shows hypothetical concentration versus time data measured at a fixed point, Site 1, located a distance x_1 downstream from the source. The centroid passes Site 1 at time t_1. Below it is shown the concentration versus distance data measured at a fixed time t_1. The centroid of this profile occurs at a distance x_1 downstream from the source. Under the frozen cloud assumption

$$x_1 = V_x t_1 \tag{4.38}$$

$$S(x_1, t) = S[x_1 + V_x(t_1 - t), t_1] \tag{4.39a}$$

$$S(x, t_1) = S\left(x_1, t_1 + \frac{x_1 - x}{V_x}\right) \tag{4.39b}$$

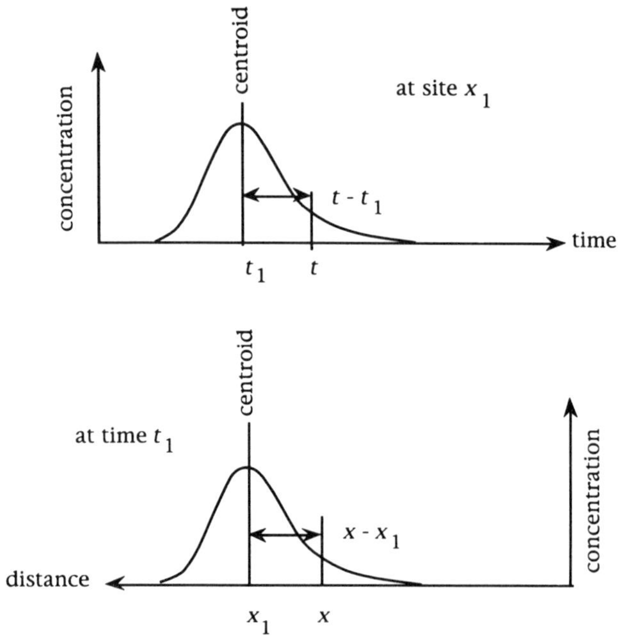

Figure 4.10. Schematic representation of the frozen cloud approximation

where V_x = mean velocity; x_1 = location of Site 1 = location of the centroid of the concentration versus distance profile measured at time t_1; and t_1 = time of passage of the centroid at Site 1 = time that the concentration versus distance profile is measured at Site 1. Equation (4.39a) is used to estimate concentration versus time data from concentration versus distance data and equation (4.39b) to go the other way. Note the inversion of the axes in Figure 4.10: concentrations measured at negative values of $t - t_1$ (i.e. in the leading edge) have positive values of $x - x_1$.

4.8.2 Predictions for a Time-varying Point Source

Equation (4.3) is linear, which means that solutions such as equation (4.4) can be added together to address complex problems. This is the superposition principle. Consider the problem where the source is time-varying. We can approximate the time-varying source $m(t)$ by a series of instantaneous releases each of mass $m(t_i)\Delta t$ occurring at times t_i where $i = 1, 2 \ldots N$. The tracer profile is then predicted for each release separately using equation (4.4) and the results added together. This can easily be programmed on a small computer.

Example 4.3

Consider the hypothetical channel in Table 4.6. The tracer inflow is zero except between 1.0 and 2.0 hours when it is 4 kg hr^{-1} (a top hat inflow). Determine the period during which concentrations exceed 1 mg m^{-3} and the maximum concentration at a site 20 km below the source.

We can approximate the top hat discharge pattern by four instantaneous point discharges of 1 kg at times 1.125, 1.375, 1.625 and 1.875 hours respectively. Equation (4.4) is used to compute the concentration profile at $x = 20$ km for each instantaneous point discharge separately as shown in Table 4.7 and Figure 4.11. These separate profiles are then summed.

From Table 4.7 and Figure 4.11 the maximum concentration at 20 km is 107 mg m^{-3} and the concentration exceeds 1 mg m^{-3} for just over two hours. Note the slight asymmetry in the predicted concentrations versus time profiles. Also note that slightly more accurate estimates of the concentration could be obtained by considering eight slug inputs at 7.5 minute intervals.

Table 4.6. Hydraulic parameters in a hypothetical channel

Depth (H)	1 m
Width (b)	10 m
Mean velocity (V_x)	1 m s^{-1}
Dispersion coefficient (K_x)	20 m^2 s^{-1}

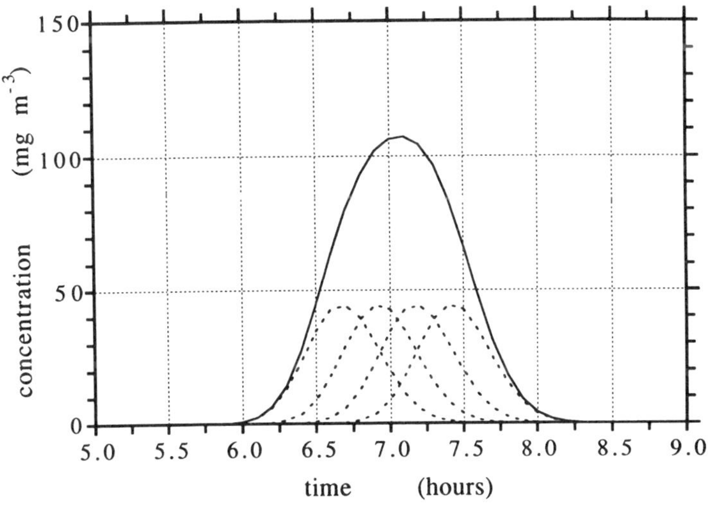

Figure 4.11. Predicted tracer profiles 20 km downstream from a top hat source. The broken lines are profiles for slug injections of 1 kg at 1.125, 1.375, 1.625 and 1.875 hours predicted using equation (4.4) and the solid line is the sum of the four profiles

Table 4.7. Predicted tracer concentrations in a hypothetical channel

Time (hour)	1st slug ($mg\,m^{-3}$)	2nd slug ($mg\,m^{-3}$)	3rd slug ($mg\,m^{-3}$)	4th slug ($mg\,m^{-3}$)	Sum ($mg\,m^{-3}$)
5.80	0.03	0.00	0.00	0.00	0.03
5.90	0.15	0.00	0.00	0.00	0.16
6.00	0.66	0.01	0.00	0.00	0.67
6.10	2.24	0.07	0.00	0.00	2.30
6.20	6.02	0.33	0.00	0.00	6.36
6.30	13.12	1.25	0.03	0.00	14.40
6.40	23.39	3.77	0.15	0.00	27.31
6.50	34.51	9.12	0.66	0.01	44.31
6.60	42.60	17.95	2.24	0.07	62.85
6.70	44.39	29.07	6.02	0.33	79.81
6.80	39.41	39.19	13.12	1.25	92.97
6.90	30.06	44.39	23.39	3.77	101.61
7.00	19.85	42.65	34.51	9.12	106.13
7.10	11.43	35.06	42.60	17.95	107.04
7.20	5.78	24.86	44.39	29.07	104.10
7.30	2.58	15.32	39.41	39.19	96.50
7.40	1.03	8.26	30.06	44.39	83.73
7.50	0.36	3.92	19.85	42.65	66.79
7.60	0.12	1.65	11.43	35.06	48.26
7.70	0.03	0.62	5.78	24.86	31.29
7.80	0.00	0.21	2.58	15.32	18.12
7.90	0.00	0.06	1.03	8.26	9.35
8.00	0.00	0.02	0.36	3.92	4.30
8.10	0.00	0.00	0.12	1.65	1.77
8.20	0.00	0.00	0.03	0.62	0.65
8.30	0.00	0.00	0.00	0.21	0.22
8.40	0.00	0.00	0.00	0.06	0.07
8.50	0.00	0.00	0.00	0.02	0.02

4.8.3 Routing a Spatial Concentration Profile

A valuable use of the superposition principle is in routing downstream an observed tracer concentration profile, often so that it can be compared with an observed profile at another site further downstream to verify velocity and dispersion coefficient estimates. Equation (4.4) is the Taylor solution to equation (4.3) which satisfies the initial condition of the concentration being everywhere zero except at $x = 0$ where the mass M is concentrated (see page 204). Given the variation with distance along the channel of the cross-sectional averaged tracer concentration at a specified time $t = t_1$, then a routing procedure can be developed as follows. The channel is sub-divided into a number of short segments each Δx long and centred on x_i where $i = 1, 2, 3 \ldots N$. The mass of tracer in segment i is $S(x_i, t_1)\Delta x$ and is assumed to behave like an instantaneous source at $x = x_i$ released at $t = t_1$. This source results in concentrations at a later time $t = t_2$ which are predicted using equation (4.4). The process is repeated for each value of i and the separate concentration profiles are summed. Allowing Δx to become very small we obtain

$$S(x, t_2) = \int_{\xi=-\infty}^{\infty} \frac{S(\xi, t_1)}{\sqrt{4\pi K_x(t_2 - t_1)}} \exp\left\{ -\frac{[x - \xi - V_x(t_2 - t_1)]^2}{4K_x(t_2 - t_1)} \right\} \mathrm{d}\xi \quad (4.40)$$

where $S(x, t_1)$ = observed concentration as a function of distance at time t_1; $S(x, t_2)$ = predicted concentration as a function of distance at time t_2; and ξ = dummy distance variable of integration. Equation (4.40) is valid provided equation (4.4) accurately predicts the concentration profile arising from each of the instantaneous point sources. There is a slight difficulty here because, as discussed above, equation (4.4) is an asymptotic solution which is accurate only a long distance below an instantaneous point source. Thus it may not be strictly accurate to use this equation to route downstream each element of the upstream concentration profile if the time interval $t_2 - t_1$ is small. As this time interval increases, however, the accuracy of equation (4.40) increases.

4.8.4 Routing a Temporal Concentration Profile

Equation (4.40) requires knowledge of the instantaneous tracer concentration versus distance profile, such as might be determined by aerial photography. It is more common, however, to know the tracer concentration as a function of time at a fixed sampling site downstream. This constitutes a boundary condition and strictly equation (4.4) can no longer be used. There are two alternative methods for routing an observed concentration versus time profile downstream.

4.8.4.1 Frozen cloud method

One approach is to make the frozen cloud assumption and to transform the concentration versus time profile at Site 1 (Profile 1) into a concentration versus distance profile (Profile 1A) at the mean time of passage using equation (4.39b). This equation is then routed downstream using equation (4.40) to give a concentration versus distance profile (Profile 2A) at the mean time of passage past Site 2. Profile 2A is then transformed into a concentration versus time profile (Profile 2) at Site 2 using equation (4.39a). This entire process can be summarised in the equation

$$S(x_2, t) = \int_{\tau=-\infty}^{\infty} \frac{S(x_1, \tau) V_x}{\sqrt{4\pi K_x(\bar{t}_2 - \bar{t}_1)}} \exp\left[-\frac{V_x^2(\bar{t}_2 - \bar{t}_1 - t + \tau)^2}{4K_x(\bar{t}_2 - \bar{t}_1)} \right] \mathrm{d}\tau \qquad (4.41)$$

where $S(x_1, t)$ = observed concentration as a function of time at Site 1 (located at x_1); $S(x_2, t)$ = predicted concentration as a function of time at Site 2 (located at x_2); $\bar{t}_1$ and $\bar{t}_2$ = mean times of passage at Sites 1 and 2, respectively, where

$$\bar{t}_2 = \bar{t}_1 + \frac{x_2 - x_1}{V_x} \tag{4.42}$$

where V_x = mean velocity and τ = dummy time variable of integration.

If the concentration versus time profile has also been measured at Site 2, then the observed and predicted profiles can be compared and if necessary the values of V_x and K_x adjusted. Example (5.5) illustrates this method of estimating V_x and K_x from field results.

It is important to remember four points about the routing procedure. Firstly, Site 1 must be outside the advective zone. Secondly, the entire concentration versus time profile must be measured to Site 1 (i.e. sampling must cover both the leading edge and the long tail). Thirdly, tracer loss must be negligible or a correction must be made *a priori* to the tracer data. Finally, the routing procedure gives reach-averaged values of V_x and K_x.

4.8.4.2 Hayami solution

Barnett (1983) suggests an alternative method for routing a concentration versus time profile. In addition to equation (4.4) there exists another solution to equation (4.3), the so-called Hayami solution

$$S(x, t) = \frac{Mx}{AV_x t\sqrt{4\pi K_x t}} \exp\left[-\frac{(x - V_x t)^2}{4K_x t}\right] \tag{4.43}$$

Equation (4.43) is simply equation (4.4) multiplied by $x/V_x t$. It satisfies the boundary condition that at $x = 0$, $S = 0$ for all values of t except $t = 0$ when a tracer slug of mass M is introduced over an infinitesimally small time interval. The advantage of equation (4.43) is that it can be used to route downstream an observed concentration versus time profile without needing to invoke the frozen cloud approximation. Thus given a concentration versus time profile at a fixed site then applying the superposition principle and making use of equation (4.43)

$$S(x_2, t) = \int_{\tau=-\infty}^{\infty} \frac{S(x_1, \tau)(x_2 - x_1)}{(t - \tau)\sqrt{4\pi K_x(t - \tau)}} \exp\left[-\frac{[x_2 - x_1 - V_x(t - \tau)]^2}{4K_x(t - \tau)}\right] d\tau \tag{4.44}$$

For the values of V_x and K_x typical of rivers the Hayami (4.43) and Taylor (4.4) solutions give similar solutions at moderately large x and t. They differ significantly in their behaviour at very small t where the Taylor solution predicts positive concentrations upstream from the source (which are not observed in field experiments in rivers) and the Hayami solution predicts small

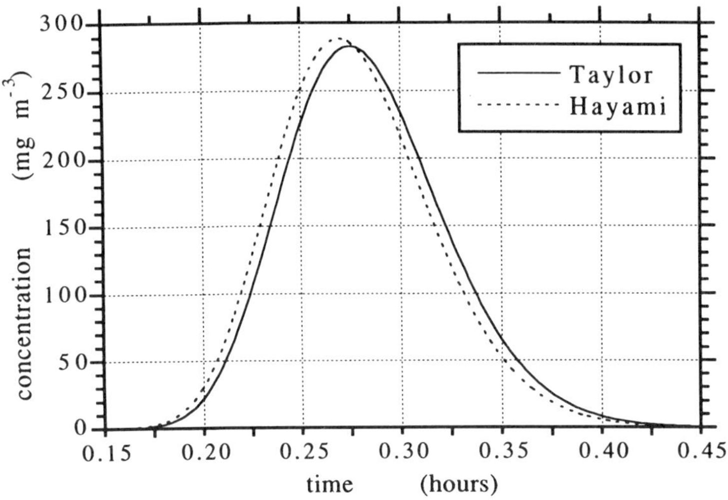

Figure 4.12. Comparison of concentration profiles 1 km below an instantaneous tracer injection of 10 kg predicted using the Taylor and Hayami equations with $V_x = 1\ \mathrm{m\,s^{-1}}$, $K_x = 10\ \mathrm{m^2\,s^{-1}}$ and $A = 10\ \mathrm{m^2}$

negative concentrations upstream from the source (which are physically unrealistic). These defects are of little practical significance. Figure 4.12 compares concentrations profiles predicted by the Taylor and Hayami equations 1 km downstream from an instantaneous tracer injection in a hypothetical channel with $M = 10$ kg, $V_x = 1\ \mathrm{m\,s^{-1}}$, $K_x = 10\ \mathrm{m^2\,s^{-1}}$ and $A = 10\ \mathrm{m^2}$. The Hayami solution is slightly more skewed than the Taylor solution but the difference between profiles is small.

4.9 ESTIMATING K_x FROM VELOCITY MEASUREMENTS

Longitudinal dispersion arises principally as a result of transverse velocity shear and the longitudinal dispersion coefficient can be estimated knowing the velocity distribution. This approach was first developed by Taylor for pipe flow and is discussed in Section 1.8.4. Taylor (1953, 1954) applied equation (1.99) to pipe flow and derived the well-known formula

$$K_x = 10.1 r u^* \tag{4.45}$$

where r = pipe radius. This equation is not readily transferred to rivers. Elder (1959) applied Taylor's analysis to two-dimensional shear flow and found that in an infinitely wide channel

$$K_x = 5.93HU^* \tag{4.46}$$

where H = mean depth. Equation (4.46) quantifies the effects of vertical velocity shear and the counteracting vertical turbulent diffusion. Consequently it underestimates K_x in most natural river channels where transverse velocity shear dominates longitudinal dispersion. Fischer (1967) extended the Taylor and Elder method to natural channels and showed that

$$K_x = -\frac{1}{A}\int_{z=0}^{b} q'(z)\,\mathrm{d}z \int_{z'=0}^{z} \frac{\mathrm{d}z'}{k_z(z')h(z')} \int_{z''=0}^{z'} q'(z'')\,\mathrm{d}z'' \tag{4.47a}$$

where

$$q'(z) = \int_{y=0}^{h(z)} u_x'(y, z)\,\mathrm{d}y \tag{4.47b}$$

and

$$u_x'(y, z) = u_x(y, z) - v_x(z) \tag{4.47c}$$

In equations (4.46) A = cross-sectional area; b = channel width; $u_x(y, z)$ = longitudinal velocity; $v_x(z)$ = local depth-averaged velocity; $u_x'(y, z)$ = velocity deviation from the local depth-averaged velocity; $q'(z)$ = the integral over the depth of the velocity deviation; $h(z)$ = local depth and $k_z(z)$ = transverse dispersion coefficient. Equations (4.47) can be approximated by

$$K_x = \frac{Ib^2\overline{u_x'^2}}{k_z} \tag{4.48}$$

where I = dimensionless integral which quantifies the velocity variation over the cross-section. In practice it is found that I is close to $I = 0.07$ (Fischer *et al* 1979). The overbar denotes the mean square over the cross-section of the velocity deviation defined in equation (4.47c). Equation (4.48) illustrates two points which have already been discussed: the value of K_x is inversely proportional to the transverse dispersion coefficient and increases with increasing transverse velocity shear.

In a pipe and in two-dimensional shear flow the analytical form of the velocity distribution is known which enables the integrals in equations (4.47) to be evaluated analytically, giving the results shown in equations (4.45) and (4.46). In natural rivers, however, the transverse velocity distribution cannot normally be written in a simple analytical form and equations (4.47) and (4.48) cannot be evaluated analytically. It is relatively straightforward to evaluate the integrals numerically provided the transverse distributions of velocity and transverse dispersion coefficient are known. In the absence of any other information, k_z can be estimated using the data summarised in Chapter 3. As written, equation (4.47a) allows k_z to vary across the channel although in most practical problems it may only be feasible to estimate the cross-sectional averaged value of k_z.

Measurements of the depth and velocity distributions in natural channels are commonly made at water level recording stations to derive rating curves between water level and discharge. Such data yield estimates of K_x at the gauging sites. Given estimates at several sites on a river the reach-averaged dispersion coefficient is

$$\bar{K}_x = \frac{1}{L} \sum_{i=1}^{n} K_i \Delta x_i \tag{4.49}$$

where K_i = dispersion coefficient at cross-section i; Δx_i = length of the sub-reach containing cross-section i and L = length of the study reach.

From this discussion it is clear that there are two drawbacks to estimating the longitudinal dispersion coefficient in irregular natural channels from equations (4.47) or (4.48). Firstly, the transverse dispersion coefficient must be known accurately. In Chapter 3 it was shown that k_z varies significantly between channels and along a given channel. The uncertainty in estimating k_z from published data is of the order of ±50% and this leads to a comparable uncertainty in estimates of K_x. Secondly, equations (4.47) and (4.48) yield point estimates of K_x and in order to obtain a reliable reach-averaged estimate, gauging data and k_z estimates are required at a number of points along the channel. Nevertheless the cost of conducting several gaugings may be lower than the cost of conducting a tracer test, especially in a large river where the advective zone is very long. Example (5.8) discusses the estimation of K_x in the Waikato River, New Zealand using velocity data.

4.10 MODELLING LONGITUDINAL DISPERSION IN THE ADVECTIVE ZONE

In many problems the river reach of interest extends well beyond the advective zone and we can use the Fickian model of longitudinal dispersion, equation (4.3), with some confidence. In other rivers (especially very rough and very wide ones) the advective zone is very long and we find ourselves wishing to model longitudinal dispersion in a region where equations (4.3) and (4.11) are not valid.

A pragmatic approach would be to conduct tracer tests and estimate values of the mean velocity and the longitudinal dispersion coefficient within the advective zone. It is possible to fit equation (4.3) to tracer data collected within the advective zone although observed tracer profiles will be more highly skewed than predicted profiles. The value of K_x within the advective zone is not constant but increases with distance downstream. This means that values estimated from tracer data collected within the advective zone underestimate the true value of K_x in the equilibrium zone. It also means that K_x cannot be assumed constant and simple analytical solutions like equation (4.4) cannot be used. Some of the published longitudinal dispersion coefficients were probably estimated from tracer data collected within the advective zone and hence could be misleading. Clearly alternative models to equation (4.3) are required for estimating longitudinal dispersion within the advective zone.

One approach to modelling longitudinal dispersion within the advective zone is to use the unsteady two-dimensional depth-averaged mixing equation (3.1). This equation may not be accurate in the near-field where tracer is poorly mixed vertically and source momentum is important, but it can be used to model the effects of transverse velocity shear and transverse dispersion in the mid-field. This type of model is discussed in Chapter 3 where three methods of solution were outlined: rectangular finite difference modelling, curvilinear numerical methods and streamtube models. The streamtube models developed by Yotsukura and Cobb (1972) and Yotsukura and Sayre (1976) assume steady tracer input but the method can be extended to non-steady inputs at the expense of requiring numerical methods of solution. Fischer (1966) developed a simple non-steady streamtube model to predict tracer dispersion in the advective zone which is also described in Fischer *et al* (1979). Harden and Shen (1979) and Luk *et al* (1990) have also developed non-steady streamtube models. As is discussed in Chapter 3, each of these two-dimensional models requires extensive field data on channel bathymetry and velocity distributions. The model equations must be solved numerically taking care to ensure that mass is conserved, the numerical scheme is stable and numerical dispersion is minimised (Abbott and Basco 1989). Thus a substantial effort is required to develop, calibrate and test a two-dimensional numerical model. This effort can only be justified on fairly major problems.

4.11 ALTERNATIVE MODELS FOR LONGITUDINAL DISPERSION

4.11.1 Introduction

It is clear from the preceding discussion that even in the equilibrium zone, equations (4.3) and (4.11) provide only an approximate model which is capable of predicting the behaviour of the tracer centroid and variance but not its skewness. This has led some researchers to abandon the Fickian model and search for better alternatives. We shall discuss three alternative types of model: the transient storage, cells in series and aggregated dead zone models.

4.11.2 Transient Storage Model

Several attempts have been made to include in equation (4.3) terms which model the movement of tracer into and out of dead zones. Such models have been termed transient storage models by Bencala and Walters (1983). Consider the problem of plane shear flow in a uniform channel with dead zones at the bed (Figure 4.13). If the depth is assumed constant and neither flow nor concentration varies across the channel then the tracer concentration is given by

$$\frac{\partial c}{\partial t} + u_x \frac{\partial c}{\partial x} = \frac{\partial}{\partial x}\left(e_x \frac{\partial c}{\partial x}\right) + \frac{\partial}{\partial y}\left(e_y \frac{\partial c}{\partial y}\right) \tag{4.50}$$

where x and y = longitudinal and vertical distances; $c(x, y, t)$ = tracer concentration; $u_x(y)$ = longitudinal velocity; and $e_x(y)$ and $e_y(y)$ = turbulent eddy diffusivities. Note that in equation (4.50) we are allowing the concentration to vary vertically, longitudinally and with time. At the water surface there is zero flux and so at $y = h$

$$e_y \frac{\partial c}{\partial y} = 0 \tag{4.51}$$

At the bed ($y = 0$) there is exchange between the dead zones and the overlying water. We can postulate that the dead zones are locally well mixed. The tracer flux between the bed and the river water must depend on the difference in concentration between the dead zones and the overlying water. It will also depend on the intensity of turbulence at the bed and the proportion of the bed containing dead zones. The flux can be estimated from

$$J_{\mathrm{d}} = f\theta[c(x, 0, t) - c_{\mathrm{d}}(x, t)] \tag{4.52}$$

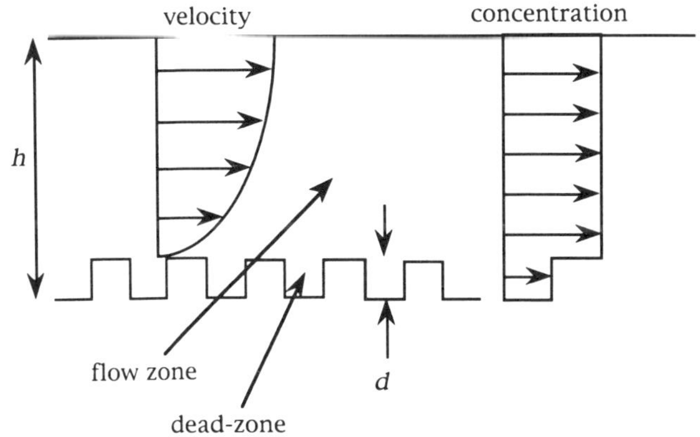

Figure 4.13. Diagram of two-dimensional shear flow with dead zones on the bed

where J_d = tracer flux from the overlying water to the dead zones (units $g\,m^{-2}\,s^{-1}$); f = proportion of the bed containing dead zones (i.e. the surface area of the bed containing dead zones divided by the total surface area of the bed); $c(x, 0, t)$ = tracer concentration in the water just above the bed; $c_d(x, t)$ = tracer concentration in the dead zones; and θ = the mass transfer coefficient at the bed (units $m\,s^{-1}$). In Figure 4.13 dead zones occupy 50% of the length of the bed and if each dead zone extends across the entire channel then $f = 0.50$. The rate of change of concentration in the dead zone is

$$\frac{\partial c_d}{\partial t} = \frac{J_d}{d} = \frac{f\theta}{d}\,[c(x, 0, t) - c_d(x, t)] \tag{4.53}$$

where d = mean depth of the dead zones. The bottom boundary condition for equation (4.50) is

$$-\left(e_y \frac{\partial c}{\partial y}\right)_{y=0} = f\theta\,[c(x, 0, t) - c_d(x, t)] \tag{4.54}$$

Valentine and Wood (1979b) found that the mass transfer coefficient was proportional to the velocity of the overlying flow and wrote

$$\theta = k' v_x \tag{4.55}$$

where k' = non-dimensional mass transfer coefficient; and v_x = mean velocity of the overlying water (i.e. flow divided by the area of the channel excluding the dead zones).

The model equations (4.50) and (4.53) together with boundary conditions (4.51) and (4.54) cannot be solved analytically for the general case where the velocity and diffusivity vary with depth. Valentine and Wood (1979a, 1979b) and Denton (1990) have, however, used numerical methods to evaluate the spatial moments of the concentration distributions predicted by these equations.

In a natural channel we can expect to find dead zones along the banks as well as the bed. The former gives rise to transverse, and the latter to vertical, concentration gradients. In this situation equation (4.50) would need to be extended to three dimensions. It is a daunting task to solve the full three-dimensional equations although Valentine (1978) discusses some aspects of the behaviour of the solutions.

The model equations can be simplified considerably if we neglect concentration variations in the overlying water. We can infer from Sayre's work (Sayre 1968a) that vertical and transverse concentration gradients will be present downstream from an instantaneous tracer injection in a natural channel containing dead zones. We can, however, still choose to develop an equation for the cross-sectional averaged concentration and to write

$$\frac{\partial S}{\partial t} + V_x \frac{\partial S}{\partial x} = \frac{1}{A} \frac{\partial}{\partial x} \left(A K_x \frac{\partial S}{\partial x} \right) - \frac{f\theta}{H} (S - c_d) \tag{4.56a}$$

$$\frac{\partial c_d}{\partial t} = \frac{\theta}{d} (S - c_d) \tag{4.56b}$$

where $S(x, t)$ = cross-sectional averaged tracer concentration in the river water; V_x = cross-sectional averaged velocity (Q/A); A = cross-sectional area of the channel excluding dead zones; H = mean depth excluding dead zones; K_x = longitudinal dispersion coefficient; f = proportion of the bed occupied by dead zones; c_d = dead zone concentration; d = depth of the dead zones; and θ = dead zone mass transfer coefficient.

Equations (4.56) were developed by Thackston and Schnelle (1970) to explain the long tails observed in tracer profiles and were used subsequently by Pedersen (1977) and Nordin and Troutman (1980). Bencala and Walters (1983) have solved equations (4.56) using both finite-difference and finite-element numerical methods and were able to find values of f, θ and d which gave a reasonably good fit between observed and predicted tracer concentrations.

Considerable insights have been gained into the effects of dead zones on longitudinal dispersion using transient storage models but they have not been adopted for routine predictions of longitudinal dispersion. There are two main reasons for this. Firstly, the model equations must be solved numerically and it is not easy to obtain satisfactory solutions which are substantially free from numerical dispersion. Secondly, the model coefficients are difficult to determine. Once the effects of dead zones are included in the mixing equations, two mechanisms contribute to longitudinal dispersion: velocity shear within the water column and trapping within the dead zone elements. It is normal practice to calibrate model coefficients using the results of tracer tests. It is not easy to obtain a unique calibration of K_x, f, θ and d from a set of tracer results. It requires painstaking survey work to measure f and d in a particular channel. There is as yet a rather poor understanding of the rates of exchange between dead zones and the overlying water and it is not feasible to estimate θ theoretically. Some progress has been made in deriving model coefficients from tracer data and relating these coefficients to bulk channel parameters (Pedersen 1977, Bencala and Walters 1983) but to date comparatively few data sets have been analysed and more work is needed before the practitioner has sufficient information to use the transient storage model routinely.

4.11.3 Cells in Series Model

The cells in series (CIS) model represents a river reach by a sequence of sub-reaches (hereafter called cells) which flow into each other. The model owes much to continuously stirred tank reaction models which are widely used in chemical engineering. Each cell is assumed to be well mixed and hence the outflow concentration equals the average cell concentration. The concentration of a conservative tracer is determined solely by the concentration of the inflow and the residence time of the cell. A river reach is modelled by a number of cells in series and the outflow from one cell becomes the inflow to the next cell downstream. The analogy has some clear physical basis in pool–riffle rivers (Beltaos 1980a, 1982) but has also been applied in uniform channels where it is more tenuous (Banks 1974, Stefan and Demetracopoulos 1981).

The mass balance for a non-conservative tracer which decays at a rate proportional to the remaining concentration (first-order decay) in each cell leads to the equation

$$V\frac{\mathrm{d}S(t)}{\mathrm{d}t} = QS_u(t) - (Q + kV)S(t) \tag{4.57}$$

where t = time; S = average cell tracer concentration; S_u = inflowing concentration (i.e. concentration of the cell next upstream); V = cell volume; Q = river flow rate; and k = first-order decay rate coefficient. If an instantaneous tracer injection is made to the inflow of a cell at time $t = 0$ then the concentration of that cell is

$$S(t) = \frac{M}{V}\exp(-\alpha t) \qquad t > 0 \tag{4.58}$$

where M = mass of tracer injected; $\alpha = k + 1/T$; and $T = V/Q$ = cell residence time. It is left as an exercise for the reader to show that for the case of a conservative tracer ($k = 0$) the total mass flow out of the cell is M.

If two identical cells are placed in series then equation (4.58) gives the inflow to the second cell and we can solve equation (4.57) to yield

$$S_2(t) = \alpha t S_1(t) = \alpha t\frac{M}{V}\exp(-\alpha t) \tag{4.59}$$

where $S_1(t)$ and $S_2(t)$ = concentrations in cells 1 and 2, respectively. If $n + 1$ identical cells are placed in series then

$$S_{n+1}(t) = \frac{\alpha^n t^n}{n!}\frac{M}{V}\exp(-\alpha t) \tag{4.60}$$

In Figure 4.14 the solid lines show concentrations calculated from equation (4.60) assuming $\alpha = 0.01$ (i.e. cell residence time $T = 100$ seconds) for $n = 5$, 10, 15 and 20. The number of cells n is directly related to the distance downstream. It can be seen in Figure 4.14 that as n increases the time of passage of the centroid increases and the peak concentration drops. This matches qualitatively the behaviour of tracer following a slug injection shown in Figure 4.1. In the first cell ($n=0$) equation (4.58) indicates that the concentration versus time profile decays exponentially with time and is highly skewed. As n increases (i.e. as the tracer moves downstream) the concentration profiles become progressively more symmetrical. The centroid, variance and skewness coefficient of the concentration versus time profile in a given cell are

$$\bar{t} = \frac{n}{\alpha} = nT \tag{4.61}$$

$$\sigma_t^2 = \frac{n}{\alpha^2} = nT^2 \tag{4.62}$$

$$\varepsilon = \frac{2}{\sqrt{n}} \tag{4.63}$$

where $\bar{t}$ = time of passage of the centroid, σ_t^2 = variance and ε = skewness coefficient. Equation (4.62) indicates that the variance grows linearly with distance downstream whereas equation (4.63) indicates that the skewness coefficient decreases inversely with the square root of distance. This is the same as the behaviour of solutions to the Fickian dispersion model equation (4.3).

There are two main drawbacks of the CIS model. Firstly, it does not reproduce the persistent skewness which is usually observed in tracer data from rivers (Stefan and Demetracopoulos 1981). This probably stems from the assumption that tracer is well mixed throughout each cell, which is unlikely when there are dead zones present. Secondly, the number of cells, n, determines both the travel time, the rate of dispersion and the skewness [see equations (4.61)–(4.63)]. These parameters cannot be varied independently which restricts the usefulness of the model. Although the CIS model is widely used in chemical engineering, it has not proved to be of great practical use in modelling longitudinal dispersion in natural channels.

4.11.4 Aggregated Dead Zone Model

Beer and Young (1984) introduced a variant on the CIS model called the aggregated dead zone (ADZ) model. It is postulated that a sub-reach (cell) length can be found in which the river tracer concentration is governed by a first-order model with the same form as equation (4.57). Within this cell it is assumed that the complex interaction of dead zones can be approximated by a single aggregated dead zone element. The main difference from the CIS model is that a pure time delay is introduced into the input concentration which allows advection and dispersion to be decoupled. The mass balance equation (4.57) becomes

$$V\frac{dS(t)}{dt}=QS_u(t-\tau)-(Q+kV)S(t) \tag{4.64}$$

where τ = cell time delay. The solution to equation (4.64) for an instantaneous injection of mass M at $t=0$ is

$$S(t)=\frac{M}{V}\exp[-\alpha(t-\tau)] \qquad t>\tau \tag{4.65}$$

It should be noted that the time delay τ is not the time taken for tracer to travel the length of the cell. When equation (4.16b) is used to calculate the centroid of the tracer concentration given by equation (4.60) we find that

$$\bar{t}=T+\tau \tag{4.66}$$

where $T=V/Q$ = residence time of the cell.

If $n+1$ identical cells are placed in series then by analogy with equation (4.60) the concentration is

$$S_{n+1}(t)=\frac{\alpha^n(t-\tau)^n}{n!}\frac{M}{V}\exp[-\alpha(t-\tau)] \tag{4.67}$$

Figure 4.14 compares concentrations predicted using equations (4.60) and (4.67) with $k=0$ (i.e. a conservative tracer), $\alpha=0.01\ \mathrm{s}^{-1}$ (i.e. $T=100$ s) and $\tau=10$ s. The profiles are identical except that each of the ADZ solutions is delayed by $n\tau$ where n = number of cells. The centroid occurs at

$$\bar{t}=n(T+\tau) \tag{4.68}$$

while the variance and skewness coefficient are given by equations (4.62) and (4.63).

Equation (4.64) can be rewritten in discrete form as

$$S^m=\exp(-\alpha\Delta t)S^{m-1}+(1-\exp(-\alpha\Delta t))S_u^{m-\delta-1} \tag{4.69}$$

where S^m and S^{m-1} = tracer concentrations at times $m\Delta t$ and $(m-1)\Delta t$; T = residence time of the aggegated dead zone = V/Q; S_u = inflow concentration = concentration at the upstream boundary of the dead-zone; and δ = integer $(\tau/\Delta t)$ = cell time delay. Figure 4.15 shows the results of routing a known upstream tracer profile through a single cell using equation (4.69) assuming a conservation tracer. Two different values of T are used and there is less dispersion with $T=100$ s than with $T=200$ s. This is in accord with equation (4.62) which states that the variance is proportional to T^2. The peak (and hence also the centroid) occurs earlier for $T=100$ s which is in accord with equation (4.68).

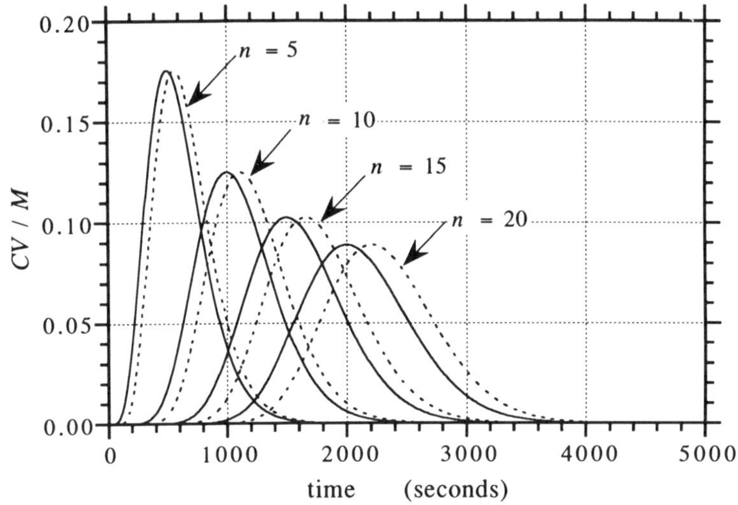

Figure 4.14. Non-dimensional concentration versus time profiles predicted using the CIS (solid) and ADZ (dashed) models for n identical cells in series. The coefficients are: $T = 1/\alpha = 100$ s, $\tau = 10$ s and $k = 0$. The ADZ profiles are delayed by $n\tau$ more than the CIS profiles where n = number of cells. Note that n is equivalent to the distance downstream

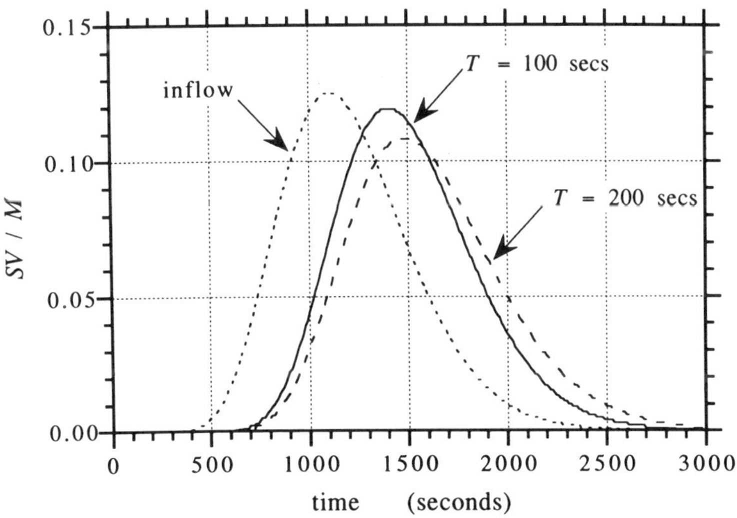

Figure 4.15. Non-dimensional concentration versus time profiles predicted using equation (4.69) for two values of cell residence time T with a prescribed inflow distribution. The other model coefficients are: $\tau = 200$ s, $\Delta t = 10$ s and $k = 0$

It is possible to generalise the method outlined above and to develop a multi-order form of the ADZ model in which several first-order ADZ elements with different time constants (α) and time delays (τ) are combined in series and/or parallel. The tracer can also be considered non-conservative. The generalised multi-order model when written in discrete form is

$$S^{m+1} = a_0 S^m + a_1 S^{m-1} + a_2 S^{m-2} + \cdots + a_i S^{m-i} + b_0 S_u^{m-\delta} + b_1 S_u^{m-\delta-1} + b_2 S_u^{m-\delta-2} + \cdots + b_j S_u^{m-\delta-j} \tag{4.70}$$

i and j = number of past values of S and S_u needed to explain the observed tracer profile. The coefficient b_0 determines whether or not the solute is conservative.

Equation (4.70) is a powerful and flexible model but the practical difficulty is to determine the number of coefficients (i and j) which need to be included and to estimate each of the coefficients $a_1 \ldots a_i$ and $b_0 \ldots b_j$. Young and co-workers advocate using time series methods to estimate these coefficients from observed tracer profiles such as those available in the computer package CAPTAIN (Wallis *et al* 1989b). Some experience is required to fit complex models to field data effectively and especially to interpret the physical significance of the coefficients. In the hands of an experienced person the CAPTAIN package can help gain insights into the behaviour of the river. For example, Wallis *et al* (1989b) found that tracer data collected in two Lancashire rivers could best be described using two first-order aggregated dead zones in parallel. They inferred that the dead zone with the shorter residence time represented flow in the main channel whereas the dead zone with the longer residence time represented flow near or in the river bed.

Equation (4.69) is a much simplified version of equation (4.70) in which $k = 0$ and only a_o and b_o are non-zero. It turns out, however, that in many natural channels tracer profiles can be routed from one site to another fairly satisfactorily using equation (4.69) (Wallis *et al* 1989b). At first sight it might seem surprising that the complexities of a natural channel can be modelled satisfactorily using a single aggregated dead zone. Taylor's analysis showed, however, that the complex interaction of shear and turbulent diffusion can be modelled asymptotically by two coefficients: the advective velocity and a Fickian dispersion coefficient. It appears that in many channels the effects of irregularities in the bed and banks can be lumped together and described by two reach-averaged coefficients: the cell residence time T and the cell time delay τ.

Before equation (4.69) can be used to make predictions in a given reach the values of T and τ must be estimated. At present this requires field tracer data. Given observed tracer versus time profiles at two sites then the coefficients can be estimated as follows. Firstly the reach time of travel is calculated

$$\bar{t} = \bar{t}_2 - \bar{t}_1 \tag{4.71}$$

where $\bar{t}$ = time of travel along the reach; and $\bar{t}_2$ and $\bar{t}_1$ = times of passage of the dye centroids past the downstream and upstream sites, respectively.

By definition the time delay is

$$\tau = t_\mathrm{b} - t_\mathrm{a} \tag{4.72}$$

where τ = reach time delay; t_b and t_a = first arrival times of tracer at the downstream and upstream sites, respectively. In practice it is sometimes difficult to determine exactly when tracer first arrives at a site because concentrations are low and experimental uncertainties are high. Assuming that we can estimate τ then equation (4.66) can be rearranged to give

$$T = \bar{t} - \tau \tag{4.73}$$

where T = reach residence time.

This estimation procedure is susceptible to errors in the data especially to errors in identifying the arrival time of tracer at each site. Consequently it is desirable to check the coefficients by using them in equation (4.69) to route the observed upstream concentration profile to the downstream site. If the fit between observed and predicted concentrations at the downstream site is adequate then the simple first-order model can be used for predictive purposes. If, however, the model fit is not satisfactory then there are two possible courses of action. A trial and error adjustment can be made of the coefficients to try and improve the fit of the first-order model. If this still does not achieve satisfactory results then a second-order model needs to be considered. Wallis *et al* (1989a) suggest trying two identical first-order sub-models in series in which the time delay τ and residence time T are half of the values estimated from equations (4.72) and (4.73). Again some trial and error adjustment of the model coefficients may be required. If a satisfactory fit still cannot be achieved then recourse will probably be required to the time series methods of identifying and fitting the best ADZ model.

Example 4.4

Dye profiles measured in the Manawatu River, New Zealand, were originally analysed using the frozen cloud routing method to estimate V_x and K_x (see Example 5.5). Data from Sites B (2.7 km below the source) and D (6.4 km) were subsequently re-examined using the ADZ model (equation 4.69). Figure 4.16 shows the observed concentrations profiles at each site.

An initial estimate was made of the coefficients of the ADZ model as follows. Dye first arrived at Site B at $t_\mathrm{a} = 0.85$ hours and at Site D at $t_\mathrm{b} = 2.75$ hours. Thus, from equation (4.72), the time delay of the reach is approximately $\tau = 1.9$ hours. The centroids of the tracer profiles occurred at $\bar{t}_2 = 4.2$ at Site D and at $\bar{t}_1 = 1.8$ hours at Site B. Thus the travel time of the reach is $t = 2.4$ hours. From equation (4.73) the residence time of the aggregated dead zone is $T = t - \tau = 0.5$ hours.

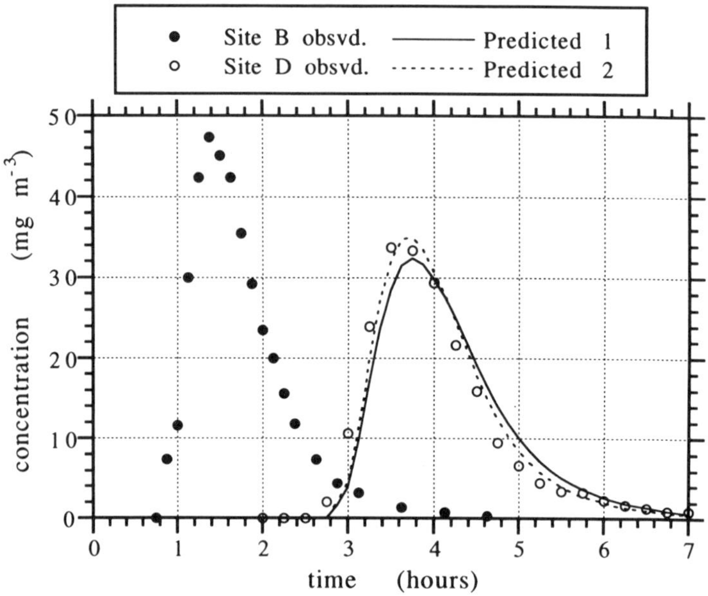

Figure 4.16. Comparison of observed concentrations in the Manawatu River at Site D with predictions made using a single-cell first-order ADZ model. The observed profile at Site B is input to the model. The model coefficients are $\tau = 2$ hours, $T = 0.5$ hours and $k = 0$ (conservative tracer) for Predicted 1 and $\tau = 1.875$, $T = 0.4$ hours for Predicted 2

The coefficients $T = 0.5$ hours, $\tau = 2$ hours and $\Delta t = 0.125$ hours were then used in equation (4.69) to make the Predictions 1 shown in Figure 4.16. The observed and predicted concentrations at Site D have the same general shape but the coefficient estimates need to be refined slightly. It is clear from Figure 4.15 that decreasing T will simultaneously increase the peak concentration and make it occur earlier. By comparison, reducing the value of τ will make the peak occur earlier but will not affect the peak concentration. Predictions 2 were made using $T = 0.4$ hours and $\tau = 1.875$ hours: values arrived at by trial and error. The match with the observed data is very good indeed. Thus a single-cell first-order ADZ model provides a good description of longitudinal dispersion in this reach of the Manawatu River. It is worth noting that Site B lies within the advective zone so that it would be unwise to use the Fickian dispersion model starting at Site B. There is no similar restriction on the use of the ADZ model.

The ADZ model has shown itself to be a powerful tool for analysing tracer data on longitudinal dispersion. Beer and Young (1984) re-examined the classical data set of Godfrey and Frederick (1970) from the Copper Creek,

Virginia, and found that they could route a tracer profile between two sites with a very high coefficient of determination. They detected that the tracer was non-conservative and that previous benchmark estimates of mean travel time and the longitudinal dispersion coefficient made by Fischer (1968) using the frozen cloud routing method (described in Section 4.8.3) had serious defects.

One disadvantage of the ADZ model is that it is not derived from a detailed physical understanding of the mechanisms which cause longitudinal dispersion but rather is an intuitive description of the combined effects of dead zones and advective transport. Consequently it is not easy to interpret the physical significance of the model coefficients (viz., cell residence time, cell time delay and the number and arrangement of cells). This reduces the possibility of relating the model coefficients to simple hydraulic parameters. Some progress has been made, however, analysing field data collected in several Lancashire streams (Young and Wallis 1986, Wallis *et al* 1989b) and data from laboratory flumes (Wallis *et al* 1989a). Two important findings come from this analysis.

The first conclusion relates to the variation of model coefficients with flow. It is well known that as the river flow increases the velocity increases and hence the travel time along a fixed reach decreases. Commonly it is found that

$$\bar{t} = c_1 Q^{c_2} \tag{4.74}$$

where c_1 and c_2 are constants and $c_2 < 1$. Experimental results indicate that the reach residence time T and the reach time delay τ also decrease with flow as might be expected. It turns out, however, that the ratios

$$\frac{T}{\bar{t}}, \frac{\tau}{\bar{t}} \quad \text{and} \quad \frac{\tau}{T} \tag{4.75}$$

are approximately constant with flow rate (Sabol and Nordin 1978, Smith 1987, Wallis *et al* 1989b).

The second finding relates to the ratio

$$\frac{T}{\bar{t}} \tag{4.76}$$

which is termed the dispersive fraction. T is the residence time of tracer in the aggregated dead zone and $\bar{t}$ the residence time of tracer in the reach as a whole. Thus the dispersive factor is related to the ratio of dead zone volume to total reach volume. The dispersive fraction varies from channel to channel but is consistently lower in smooth, man-made channels than in irregular natural rivers. As yet no satisfactory relationships have been found for predicting its value in a particular river although as more experience is gained using the ADZ model on a larger number of rivers this may change. Faced with the need for an accurate ADZ model there seems no alternative to conducting one or more tracer tests and analysing the results as outlined above.

NOTATION

a	coefficient
$a_0, , a_1, a_2$	dead-zone model coefficients
A	channel cross-sectional area
b	channel width
b_0, b_1, b_2	dead-zone model coefficients
c	tracer concentration
c_d	dead-zone tracer concentration
c_1, c_2	flow coefficients
C_m	fully mixed tracer concentration
d	mean depth of dead zones
e_x, e_y, e_z	eddy diffusivities in the x, y, z directions
f	proportion of the bed covered by dead-zones
h	local water depth
H	mean water depth
I	dimensionless integral
J_d	flux to/from the dead zones
k	first-order decay coefficient
k_x, k_z	longitudinal and transverse dispersion coefficients in depth-averaged mixing equations
k'	non-dimensional dead-zone mass transfer coefficient
K_x	longitudinal dispersion coefficient
$\overline{K_x}$	reach averaged longitudinal dispersion coefficient
$K_2, K_3, K_4 \ldots$	mixing coefficients
L	reach length
L_x	length of advective zone
L_t	transverse length scale
L_z	distances for complete transverse mixing
m	tracer mass inflow rate
M	total tracer mass injected
Q	total river flow
q'	flow deviation per unit width
r	pipe radius
R	hydraulic radius
s	channel slope
s	depth-averaged concentration
S	cross-sectional averaged concentration
S_c	critical concentration
S^m	concentration at time step m
S_{max}	maximum concentration at a site
S_u	cell inflow concentration
S_1, S_2, S_n	concentrations in cells $1, 2 \ldots n$

t	time
t_a, t_b	arrival and departure times of tracer at a fixed site
t_c	times at which concentration equals S_c
t_{max}	time of maximum concentration at a site
t_{mean}	mean travel time x/V_x
t_n	time of sampling spatial distribution n
t_1, t_2	times at which spatial variances are estimated
$\bar{t}$	temporal centroid of the concentration versus time distribution
$\bar{t}_1, \bar{t}_1$	centroid at sites 1 and 2
Δt	time increment
T	cell residence time
u_x	local longitudinal velocity
u'_x	velocity deviation
$\overline{u'^2_x}$	mean square velocity deviation
U^*	mean shear velocity
v_x, v_z	depth-averaged longitudinal velocity
V	cell volume
V	mean velocity of the centroid of the tracer cloud
V_x	cross-sectional averaged velocity
x	longitudinal distance
x_c	distances at which concentration equals S_c
Δx_c	length of channel in which concentrations exceed S_c
x_n	location of site n
x_{max}	location of concentration maximum
x_1, x_2	location of sites 1 and 2
$\bar{x}$	spatial centroid of the concentration versus distance distribution
α, β	coefficients
α	non-dimensional advective zone length
α	cell time constant $k + 1/T$
δ	cell delay in time steps
ε	skewness coefficient
μ_p	pth spatial moment
η	dead-zone volume fraction
θ	dead-zone mass transfer coefficient
ξ	dummy variable of integration
σ_t^2	temporal variance
σ_x^2	spatial variance in the x direction
τ	dummy time variable of integration
τ	non-dimensional time
τ	cell time delay
υ_q	qth temporal moment

Chapter 5

Field Measurements of Mixing

5.1 INTRODUCTION

When faced with a new mixing problem, it is sensible to start by making preliminary calculations aimed at 'getting a feel for the problem'. Such calculations are usually made with rather sparse data on channel bathymetry, slope and velocity together with dispersion coefficients estimated from previously published work and should be ascribed a high uncertainty. It is advisable to make predictions for various combinations of coefficients within the expected range to assess the sensitivity of predictions to uncertainties in the input data. Clearly for preliminary calculations of this kind, a sophisticated model is not cost effective and the simplified models outlined in Chapters 2–4 are optimal. Sometimes a desk exercise of this nature is sufficient to identify the required course of action. For example, preliminary calculations may indicate that pollutant concentrations are likely to be well below dangerous levels even under worst case mixing conditions, in which instance no further mixing investigations are required. In many cases, however, it will be necessary to reduce the uncertainty in the preliminary mixing calculations which usually requires field work to measure channel bathymetry, velocities and dispersion coefficients. The initial calculations are still valuable because they indicate which mixing mechanisms are important, what essential data are missing and hence they aid design of the fieldwork. This chapter provides a brief outline of field survey methods with particular emphasis on ways to estimate the dispersion coefficient from field results.

5.2 SUMMARY

1. It is worth spending time planning fieldwork and the simplified mixing models described in Chapters 2–4 are useful for predicting likely tracer concentrations.

2. Good bathymetric data are needed to make reliable mixing predictions especially when using two-dimensional numerical models.
3. Position fixing in rivers can rarely be done with sufficient accuracy using satellites or remote shore-based systems because river meanders and riparian vegetation interfere with lines of sight. Traditional land-based surveying methods are time consuming but reliable.
4. Vertical and transverse dispersion coefficients are most easily measured downstream from a steady source of tracer such as Rhodamine WT dye.
5. Velocity cannot be estimated using a steady source and requires either a slug injection, a large step change in mass inflow or drogues. Flow gaugings may give reliable estimates of velocity in a uniform channel but tracer determinations are preferable in irregular natural channels.
6. Tracer mass flow needs to be estimated to quantify tracer loss, experimental error or tributary dilution. Ideally flow and concentration should be measured simultaneously, although this can be labour intensive.
7. The vertical and transverse diffusion coefficients can be estimated from measurements of the mixing distance (i.e. the distance downstream from the source where tracer becomes well mixed) or the crossing distance (i.e. the distance required for tracer to travel from one boundary to another).
8. The dispersion coefficient is traditionally estimated from the rate of change of variance of the tracer distribution. The variance is strongly affected by experimental errors in low concentrations at the edges of the tracer cloud and the resulting estimates of the dispersion coefficient can be unreliable. Elder (1959) and Nokes (1986) suggest graphical methods for estimating the variance which avoid this problem.
9. Other graphical methods for estimating the dispersion coefficient are described based on the approximate constant-coefficient models for vertical and transverse mixing.
10. Methods based on the Cartesian mixing equations work well in uniform flow but when estimating transverse dispersion coefficients in non-uniform flow the streamtube model is preferable.
11. Holley's generalised method of moments allows the transverse dispersion coefficient to be estimated when tracer has impinged on the banks, but there are serious difficulties in obtaining reliable estimates from sparse field data.
12. Drogues are commonly used to measure velocity and can be used to measure transverse and longitudinal dispersion coefficients in rivers.
13. Slug injections of tracer are required to measure the longitudinal dispersion coefficient although care must be taken to ensure that measurements are made in the equilibrium zone and not the advective zone where K_x is not constant.
14. The most reliable method for estimating the longitudinal dispersion coefficient is the routing method using either the frozen cloud approximation

or the Hayami solution. There are, however, several other rapid procedures which can be used to make preliminary estimates of K_x.

15. The longitudinal dispersion coefficient can be calculated from measured velocity distributions provided the transverse dispersion coefficient is known. In some rivers it is more cost effective to measure the velocity distribution at a number of cross-sections than to conduct a tracer test, especially if the advective zone is very long and/or the river flow is large.

5.3 FIELD SURVEY METHODS

5.3.1 Tracers

The most common method of measuring velocities and dispersion rates is to inject a soluble tracer, such as fluorescent dye, which can be measured quickly to very low concentrations. Different dyes have different limits of detection but typically they can be measured in the field to better than 0.1 $mg\,m^{-3}$. This enables a transverse dispersion experiment to be conducted in a moderately large river (flow = 200 $m^3\,s^{-1}$) with about 10 l of stock solution. There are several dyes available commercially and it is important to choose one which is not unduly affected by photochemical decay or adsorption (to sediment or vegetation) and which is non-toxic. Smart and Laidlaw (1977) give a detailed comparison of commonly used dyes. I have found Rhodamine WT to be conservative in most river situations, moderately easy to handle and readily measurable to very low levels. Salt is sometimes used but its detection limit is high so that very large amounts are required in a river of any size. Radioactive tracers can be used but these require specialist injection and detection equipment and their use often arouses public disquiet.

One advantage of injecting an artificial tracer is that the tracer input rate is controlled. Occasionally, however, there is a conservative natural tracer present in a tributary (e.g. Stallard 1987) which can be used to quantify mixing rates, and effluents sometimes contain a suitable tracer. Temperature can be used to quantify mixing rates in rivers downstream from large cooling water discharges (e.g. power stations) but it is not an ideal tracer because of buoyancy effects in the near-field and non-conservative behaviour in the far-field.

5.3.2 Tracer Injection

When injecting an artificial tracer to measure vertical and transverse mixing rates it is advantageous for the source to be steady. Longitudinal dispersion can then be neglected and there is plenty of time to make measurements across the plume at each transect. One disadvantage of steady tracer injections is that they do not give estimates of velocity, which must be measured separately.

The three most commonly used devices for making a steady injection of tracer are the Marriott vessel, the floating siphon and the peristaltic pump.

The Marriott vessel (see Figure 5.1) is airtight except for the vent tube. Regardless of the level of tracer solution, there is a constant pressure head at the outlet and hence a constant flow rate. Concentrated dye solution is very

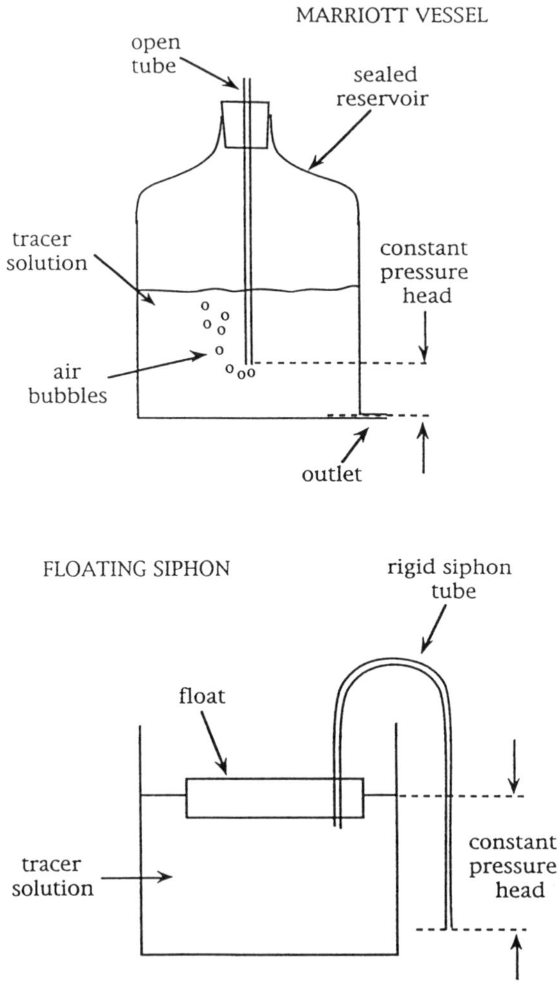

Figure 5.1. Sketch of two devices for injecting tracer at a steady rate. The Marriott vessel is airtight except for the open tube and the outlet. Pressure at the bottom of the open tube is always atmospheric and so the pressure head driving the outflow is constant, which ensures a constant outflow rate. The rigid siphon tube is supported by a float and the flow rate is proportional to the pressure head which is constant regardless of the water level

messy to handle and the Marriott vessel is troublesome to fill and seal. It is often more straightforward to siphon dye into an open container and use a pump to inject the dye into the river. A good quality peristaltic pump which is well maintained is required to achieve a reliable, steady rate of injection. The floating siphon (Figure 5.1) is a low cost device which works well, although it can be difficult to prime. It is desirable to measure the delivery rate of any system carefully before each deployment, both when the container is full and nearly empty.

5.3.3 Surveying River Bathymetry

Accurate estimates of river width, depth and slope are needed to make reliable mixing calculations. One of the major practical difficulties with developing a two-dimensional numerical flow model is obtaining an adequate representation of channel bathymetry.

The traditional method for measuring river bathymetry in a large river is to establish benchmarks at strategic points along the bank whose position (x, z) and height above datum (y) are known. Cross-sections need to be surveyed wherever the channel changes shape and in a river such as the Waikato this requires surveys every few hundred metres. At each cross-section the water level is measured and an echo sounding made across the channel by boat. Water level and boat position can be measured accurately using a tracking bearing/range survey instrument set up on the bank. This process gives excellent results but is very labour intensive.

In some situations automatic position fixing systems can be installed in the boat. One advantage is that the boat can then follow a random path thereby obtaining saturation coverage of the channel. Automatic range/range and range/bearing systems require line of sight between the shore beacons and the boat. This does not present a problem in coastal waters and open estuaries, but where the river channel meanders and there is dense vegetation on the bank, line of sight to all parts of a river is difficult to achieve. Satellite position fixing potentially overcomes this difficulty. The accuracy of many systems (even when using a reference sensor set up at a benchmark on the bank) is only of the order of metres, which is not sufficient for most bathymetry and dispersion work. Dense bankside vegetation obscures the satellites when working close to the banks and further reduces accuracy. Many automatic position fixing systems establish the (x, z) location of the boat but not the water level (y), which must be known to deduce bed level from echo sounding.

5.4.3 Sampling the Tracer

In a moderate sized river, the tracer plume must be sampled from a boat. A good strategy is to establish in advance a number of transects at strategic distances downstream from the source by clearing bankside vegetation and surveying benchmarks. The location of these transects should be selected on the basis of preliminary calculations to ensure that: (1) the maximum tracer concentration is well above the detection limit; (2) there is adequate time to measure the concentration distribution at each transect; and (3) there is a large enough change in concentration between sites to calculate the velocity and/or the dispersion coefficient.

In turbulent flow, tracer concentrations fluctuate considerably and it is desirable to collect enough field data to be able to define the average concentration. Mixing theory deals in terms of ensemble averages which are discussed in Section 1.6.2. Strictly, to measure an ensemble average we would need to conduct a number of identical field tests and average the results. This is not practical, but for a steady tracer source it is possible to collect replicate samples over a reasonably long time period and calculate time averaged concentrations. Thus when studying transverse dispersion below a steady source it is desirable to traverse the plume several times at the same transect and average the results. The replicate samples are also used to estimate extreme concentrations. For example, consider an experiment in which results indicate that extreme concentrations lie within ±25% of the average concentration. Subsequently mixing calculations are done using dispersion coefficients estimated using the average concentrations. It would be wise to consider the ecological implications of concentrations 25% higher than the predicted average concentrations.

When studying longitudinal dispersion it is necessary to sample for long enough to measure the entire concentration versus time profile at each site and check for tracer loss. This can be very time consuming because there are often long tails associated with tracer becoming trapped in dead zones (Chapter 4). There are usually small but detectable concentration differences across the channel even in the far-field (see Figure 4.3) and it is desirable to collect a number of replicate samples across the channel at each sampling time.

During the early stages of a test, dye is visible to the naked eye which facilitates sample collection. As the dye spreads out it becomes more difficult to assess the location of the plume by eye and a field fluorimeter mounted in the boat can help to optimise sampling. It is desirable to both record and display the fluorimeter output. A number of discrete water samples should be collected from time to time, analysed later in the laboratory and used to check for fluorimeter drift and/or possible interference by compounds in the river water.

When sampling the tracer plume, accurate position fixing is essential. In small rivers it is convenient to use a tag line stretched across the channel to measure transverse distances, a tape for longitudinal distance, and a calibrated rod for vertical distance. I have used tag lines for measuring transverse distance in moderately large rivers (such as the Waikato, width 200–300 m) but they are cumbersome and can impede navigation. For the reasons discussed in the previous section, automatic position fixing systems are often not suitable in rivers and a shore-based manual tracking range/bearing instrument is preferable.

5.3.5 Discharge Measurements

Accurate measurements of river discharge are essential for reliable mixing and mass balance calculations and it is advisable during tracer tests to have one or more hydrological field parties undertake river gaugings. These give data not only on river flow (needed to calculate tracer dilution) but also on channel bathymetry and velocity at the gauging sites.

In an irregular channel gauging data from a small number of cross-sections can give very misleading estimates of the reach-averaged velocity and depth. It is preferable in such channels to: (1) measure the discharge Q; (2) estimate the reach-averaged mean velocity $\bar{V}_x$ from tracer time-of-travel; (3) measure the mean channel width $\bar{b}$ from measurements made at a large number of cross-sections (e.g. from aerial photographs); and (4) and calculate the reach-averaged depth $\bar{H}$ from

$$\bar{H} = \frac{Q}{\bar{b}\bar{V}_x} \tag{5.1}$$

5.3.6 Velocity Determinations

Reliable estimates of velocity are very important for determining the time of travel of tracer. Gaugings yield reliable estimates of the velocity at the gauging cross-section. In an irregular channel, however, a large number of gaugings is needed to determine the reach-averaged velocity accurately and tracer methods are normally more cost effective. An instantaneous injection of tracer is made and concentrations measured at two (or more) sites at known locations downstream. The mean (reach-averaged) velocity is defined

$$\bar{V}_x = \frac{x_2 - x_1}{\bar{t}_2 - \bar{t}_1} \tag{5.2}$$

where x_2 and x_1 = distances of sites 1 and 2 from some reference point; $\bar{t}_1$ and $\bar{t}_2$ = times of passage of the centroid of the tracer profile past sites 1 and 2. At site 1

$$\bar{t}_1 = \frac{\int_{\tau=0}^{\infty} \tau S(x_1, \tau)\, \mathrm{d}\tau}{\int_{\tau=0}^{\infty} S(x_1, \tau)\, \mathrm{d}\tau} \tag{5.3}$$

where $S(x_1, \tau)$ = cross-sectional averaged concentration at site 1; and τ = a dummy time variable of integration. In the far-field where tracer is well mixed across the channel $S(x_1, t)$ can be estimated reasonably well using single grab samples taken near mid-channel although it is better to collect samples from several points across the channel and either bulk the sample or average the separate concentrations. In the mid-field (where tracer is not well mixed across the channel) it is essential to measure the transverse tracer distribution and evaluate

$$S(x, t) = \frac{1}{A} \int_{z=0}^{b} h(x, z) s(x, z, t) \, dz \tag{5.4}$$

where S = cross-sectional averaged concentration; s = local depth-averaged concentration; h = local depth and A = cross-sectional area. In practice the times of passage of the centroid can be approximated by the times of passage of the peak of the tracer profile. This approximation is useful if only part of the tracer profile is sampled and the integral in equation (5.3) cannot be evaluated.

The mean velocity can be determined at the beginning (or end) of a steady tracer injection from the time taken for tracer concentrations to reach a steady state at a measured distance below the source. Figure 5.2 shows the cross-sectional averaged tracer concentration versus time profile measured at a site downstream from a source which starts discharging at a steady rate beginning at time $t = 0$. If the velocity was uniform across the channel and there was no turbulent diffusion then we would have plug flow and the concentration would suddenly jump from zero to the steady-state concentration at time t_2. In

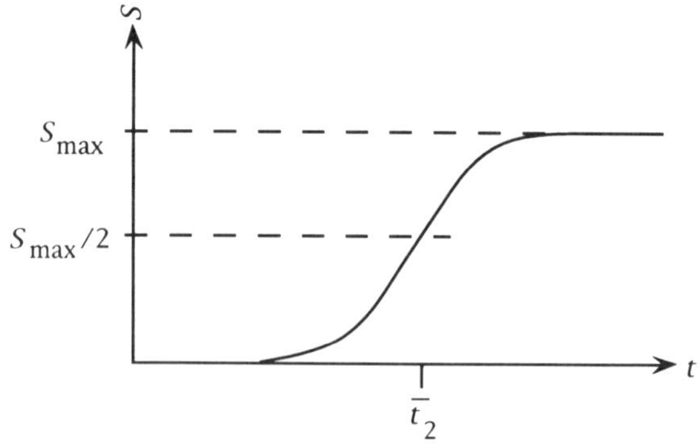

Figure 5.2. Idealised concentration versus time profile downstream from a source which starts discharging at a steady rate at time $t = 0$. S_{max} is the steady-state concentration and $\bar{t}_2$ is the time at which the concentration reaches $S_{max}/2$

practice the concentration increases gradually because of longitudinal dispersion. To calculate the mean velocity in this situation using equation (5.2), $\bar{t}_1 = 0$ and $\bar{t}_2$ is set equal to the time at which the concentration reaches $S_{max}/2$. The observed tracer versus time profile is often uneven because of experimental errors and turbulent eddies in the plume so that some smoothing may be required to estimate $\bar{t}_2$.

Sub-surface drogues and surface floats are commonly used in the oceans to study advection and dispersion. Surface floats were used in a number of early laboratory dispersion studies (e.g. Sayre and Chang 1968) and are occasionally used in rivers. Surface floats suffer from the obvious disadvantage of only sampling surface currents which tend to be higher than the depth-averaged current. An approximate value for the depth-averaged velocity is

$$V_x \approx \tfrac{2}{3}\,\bar{V}_s \tag{5.5}$$

where $\bar{V}_s$ = average velocity measured using a number of surface floats. Surface currents and hence surface floats are affected by the wind especially in wide river channels. Sub-surface drogues partly overcome these problems but they are difficult to use in natural rivers because of snags. Four drogues were released in the Waikato River simultaneously with dye in the test described in Example 5.3. It was anticipated they would travel faster than the dye and give advance warning to the sampling teams. Indeed the drogues were found to travel about 50% faster than the dye centroid and about 25% faster than the dye front (Rutherford *et al* 1980). The drogues were 20 l surface floats (50% submerged) with a sub-surface curtain approximately 1 m^2 in area centred at a depth of 0.5 m. Of the four drogues released, only one reached the downstream sampling site (20 km below the source); the others either became snagged on bankside vegetation or grounded on sandbanks. Drogues need not be very sophisticated and a box of windfall apples provides a cheap supply of expendable drogues suitable for small and moderate sized rivers.

5.3.7 Tracer Mass Flow

Several of the equations outlined in the following for estimating dispersion coefficients require knowledge of the tracer mass inflow rate. This is best determined by careful calibration of the flow rate delivered by the discharge apparatus together with careful measurement of the input dye concentration. Before analysing tracer data from the mid-field or far-field it is advisable to check for tracer loss due to either photochemical decay, adsorption or poor sampling. Tracer loss is rarely a problem in the near-field. In the mid- and far-field the mass flux is

$$m(x) = \int_{z=0}^{b} h(x, z) v_x(x, z) s(x, z) \, \mathrm{d}z \tag{5.6}$$

where v_x = local depth-averaged velocity, h = local depth, s = local depth-averaged concentration and b = channel width. Accurate evaluation of equation (5.6) requires careful measurements of the transverse distributions of not only depth-averaged tracer concentration but also depth and velocity. In an irregular channel this means that flow gauging needs to be carried out simultaneously with tracer measurements, which is labour intensive. Methods are also outlined below for estimating dispersion coefficients which avoid needing to determine the tracer mass flux.

5.3.8 Planning Field Work

Good planning is essential to get the most out of a field experiment. It is worth spending time carefully selecting sampling sites and optimising sampling procedures. The simple mixing models together with published mixing coefficients described in Chapters 2–4 are useful for doing this. Having decided where and when to collect samples, it is important to check what each field party is being asked to do. They will often have to work long hours in bad weather under trying physical conditions and it is worth spending time optimising their efforts. If several parties are working independently then each needs a clear set of instructions detailing the exact location of each sampling site (on more than one occasion field parties have sampled in the wrong place because my instructions were not clear enough!) and the sampling procedure to be followed. Good radio communication between field parties is desirable so that sampling procedures can be changed during the course of an experiment if it becomes clear that tracer is behaving differently from expected (e.g. if the mean velocity is higher or the dispersion coefficient is smaller). Possible changes in sampling procedure should have been thought out in advance (e.g. extra transects established in advance in case they are needed, extra sample bottles carried in each boat, etc). Careful book keeping is essential to label each gauging, echo sounding, dye sample and fluorimeter trace with site, date, time, location and (for transverse profiles) the direction of travel.

5.4 VERTICAL DIFFUSION COEFFICIENT

5.4.1 Introduction

Vertical mixing is one of the few problems where theoretical estimates can be made of the diffusivity. Thus for neutrally buoyant tracers discharged passively it is reasonable to estimate the depth-averaged vertical diffusivity from

$$\bar{e}_y \approx 0.067hu^* \tag{5.7}$$

There may, however, be situations where field measurements of the vertical mixing rate are desirable—for example, if the tracer is weakly buoyant or if the river flow is complex and secondary currents are likely to affect vertical mixing.

5.4.2 Using the Vertical Mixing Distance

Field tests can be designed to measure the distance downstream from the source at which vertical concentration gradients become negligibly small L_y. Then for a point source

$$\overline{e_y} = \beta \frac{v_x h^2}{L_y} \tag{5.8}$$

where $\overline{e_y}$ = depth-averaged vertical diffusivity; h = local depth; and v_x = local depth-averaged longitudinal velocity (assumed constant). The value of β depends on the location of the source and the degree of mixing (e.g. for a point source at mid-depth and for $P_m = c_{min}/c_{max} = 0.98$, $\beta = 0.134$; see equation 2.32). Values of β for other situations can be deduced from Figures 2.2 and 2.3.

5.4.3 Using Concentration Distributions

Field tests can be designed to measure concentration versus depth profiles at several sites downstream from the source. If tracer originates from a steady transverse line source then concentrations will be uniform across all or part of the channel and transverse variations can be neglected. If tracer originates from a point source then concentrations will vary both vertically and transversely and a considerable effort will be needed to quantify the spread of the plume accurately. Okoye (1970) has shown that the vertical diffusivity can be estimated satisfactorily by integrating the observed tracer concentrations across the flow. The variable

$$\bar{c}(x, y) = \frac{1}{b} \int_{z=0}^{b} c(x, y, z)\, dz \tag{5.9}$$

satisfies the equation

$$u_x \frac{\partial \bar{c}}{\partial x} = \frac{\partial}{\partial y}\left(e_y \frac{\partial \bar{c}}{\partial y}\right) \tag{5.10}$$

provided that both u_x and e_y are independent of transverse distance z. This requirement is met in the central region of many river channels although not near the banks. If the vertical diffusivity is assumed to be independent of depth then equation (5.10) becomes

$$u_x \frac{\partial \bar{c}}{\partial x} = e_y \frac{\partial^2 \bar{c}}{\partial y^2} \tag{5.11}$$

The solution for a steady source at $x = 0$, $y = y_o$ in an unbounded channel is

$$\bar{c}(x, y) = \frac{m}{b\sqrt{4\pi e_y x u_x}} \exp\left[-\frac{u_x (y - y_o)^2}{4 e_y x} \right] \tag{5.12}$$

It is important to note that equation (5.12) is only valid provided tracer has not impinged on either the bed or the surface.

Equation (5.12) indicates that

$$\bar{c}_{\max}(x) = \frac{m}{b\sqrt{4\pi e_y x u_x}} \tag{5.13}$$

where $\bar{c}_{\max}$ = maximum value of $\bar{c}$ over the depth at distance x. A plot of

$$\bar{c}_{\max}(x) \qquad \text{versus} \qquad \frac{1}{\sqrt{x}} \tag{5.14}$$

has the slope

$$\frac{m}{b\sqrt{4\pi e_y u_x}} \tag{5.15}$$

which can be estimated together with its confidence limits using linear models (e.g. regression). Then provided m, b and u_x are known, e_y can be estimated. This method assumes that tracer loss is negligibly small and that m is known.

Equation (5.12) can be rewritten

$$\log_e \left[\frac{\bar{c}_{\max}(x)}{\bar{c}(x, y)} \right] = \frac{u_x (y - y_o)^2}{4 e_y x} \tag{5.16}$$

At a given value of x, $\bar{c}_{\max}$ can be estimated from the measured depth distribution of $\bar{c}$ without needing to know the value of m. A plot of

$$\sqrt{\log_e \left[\frac{\bar{c}_{\max}(x)}{\bar{c}(x, y)} \right]} \qquad \text{versus } y \tag{5.17}$$

has the slope

$$\sqrt{\frac{u_x}{4 e_y x}} \tag{5.18}$$

which can be estimated together with its confidence interval using linear models (e.g. regression). Then only u_x need be known to estimate e_y. Clearly this method can be applied at several values of x provided tracer does not impinge on the bed or the water surface.

The work of McNulty (1983) and Nokes (1986), discussed in Section 2.6.2, indicates that the constant-coefficient model overestimates the vertical mixing distance (typically by 10–20%) for a source near the bed or the water surface. Thus estimates of e_y made using the methods outlined above (which use the constant-coefficient model) are likely to overestimate the true value of the vertical diffusivity. This does not present a problem if the derived diffusivities are then used in the constant-coefficient model to make further concentration predictions because any errors are likely to cancel each other out. If, however, the objective is to obtain the most reliable estimate of e_y then it is probably satisfactory in most cases to decrease the derived value of e_y by 10–20%. Note, however, that for a source near mid-depth the constant-coefficient model gives a reliable estimate of the vertical mixing distance (see Section 2.6.2).

5.5 TRANSVERSE DISPERSION COEFFICIENT

5.5.1 Use of the Transverse Mixing Distance

Field tests can be designed to measure the transverse mixing distance (i.e. the distance downstream from the source at which tracer becomes well mixed transversely) or the crossing distance (i.e. the distance downstream from a bankside source at which tracer first reaches the opposite bank). Such information may also come from aerial photographs of a river where there is a tributary inflow containing a visible tracer (e.g. humic compounds or suspended solids) or infra-red photographs of a cooling water plume. Knowing the transverse mixing distance, a rapid estimate can be made of the reach-averaged transverse dispersion coefficient. If tracer originates from a point source then

$$k_z = \beta \frac{v_x b^2}{L_z} \tag{5.19}$$

where L_z = transverse mixing distance; b = channel width; and v_x = longitudinal velocity (assumed constant). In an irregular channel we would use V_x in place of v_x. The value of β depends on the source location and the degree of mixing. If $P_m = s_{min}/s_{max} = 0.98$, then $\beta = 0.536$ and 0.134 for a bankside outfall and an outfall in mid-channel (see equations 3.28 and 3.29). Values of β for other source locations and other values of P_m can be estimated from Figures 3.7 and 3.8. If the crossing distance is known below a bankside source then from equation (3.31) $\beta = 0.054$ for $P_m = 0.02$ and values of β for other values of P_m can be estimated from Figure 3.7. Obviously equation (5.19) gives an average value of k_z between the source and the point of mixing and within that reach local values may be higher or lower by a substantial amount. Equation (5.19) requires that the width and velocity are known.

If we use the simplified streamtube model then we can estimate the factor of diffusion from the mixing distance using

$$D = \beta \frac{Q^2}{L_z} \tag{5.20}$$

Again if $P_m = 0.98$ then $\beta = 0.536$ and 0.134 for a bankside outfall and an outfall in mid-channel. The factor of diffusion is

$$D = \psi H^2 V_x k_z \tag{5.21}$$

where H = mean depth; V_x = reach-averaged velocity; ψ = shape factor and k_z = dispersion coefficient. Substituting equation (5.21) into (5.20), replacing Q by $HV_x b$, and rearranging gives

$$k_z = \beta \frac{V_x b^2}{\psi L_z} \tag{5.22}$$

As $\psi > 1$, equation (5.22) suggests that equation (5.19) overestimates k_z slightly in an irregular channel.

5.5.2 Uniform Flow

We now consider ways of estimating the transverse dispersion coefficient from measurements of tracer concentration made at one or more transects through the plume below a steady source. The classical method of determining the dispersion coefficient is from the rate of change of spatial variance of the tracer profile. For transverse dispersion

$$k_z = \frac{1}{2} \frac{\mathrm{d}}{\mathrm{d}t} (\sigma_z^2) \tag{5.23}$$

where t = time of travel. The transverse variance is given by

$$\sigma_z^2(x) = \frac{\int_{z=0}^{b} (z - \bar{z})^2 s(x, z)\, \mathrm{d}z}{\int_{z=0}^{b} s(x, z)\, \mathrm{d}z} \tag{5.24a}$$

and the centroid $\bar{z}$ is

$$\bar{z}(x) = \frac{\int_{z=0}^{b} z s(x, z)\, \mathrm{d}z}{\int_{z=0}^{b} s(x, z)\, \mathrm{d}z} \tag{5.24b}$$

There are four restrictions on the use of equations (5.23) and (5.24). Firstly, tracer dispersion must obey Fick's law with a constant dispersion coefficient. Secondly, tracer must not impinge on either bank which restricts the application to tracer plumes near mid-channel (Holley *et al* (1972) have generalised the change of moment method to include the case where tracer impinges on the bank). Thirdly, the tracer source must be steady. Fourthly, the channel must be uniform (the generalised change of moment method also handles non-uniform channels) so that the plume does not expand and contract as shown in Figure 3.1. Equation (5.24) does accommodate some slight meandering of the plume because $\bar{z}$ is evaluated from the tracer data and need not exactly match the source location.

In a uniform channel, provided tracer does not impinge on the banks, the transverse variance increases linearly with time and the plume velocity is constant. This means that equation (5.23) can be approximated by the finite difference form

$$k_z = \frac{1}{2} V_x \frac{\sigma_z^2(x_2) - \sigma_z^2(x_1)}{x_2 - x_1} \tag{5.25}$$

where V_x = velocity, which is assumed constant because the channel is uniform. A plot of

$$\sigma_z^2(x) \qquad \text{versus } x \tag{5.26}$$

has a constant slope of

$$\frac{2k_z}{V_x} \tag{5.27}$$

which can be estimated by regression. For equations (5.23)–(5.27) to be valid tracer need not originate from a point source. The upstream transect must, however, be placed far enough downstream from the source for tracer to be well mixed vertically.

It is apparent from the limits of integration in equation (5.24) that the entire transverse tracer profile must be measured. This leads to a serious practical difficulty because the variance estimates are greatly affected by experimental errors in concentration measurements at the edges of the tracer profile where s is small but $(z - \bar{z})^2$ is large. It is advisable to plot experimental results and either drop or smooth out any suspicious concentration measurements at the edges of the plume before using equation (5.24) to estimate the variance. Elder (1959) circumvented this problem by measuring the width of each profile at mid-height z^*. This width is linearly related to the standard deviation and so σ_z can be replaced by z^* in equation (5.25) to determine k_z. Elder's method depends on making accurate estimates of z^* which may not be easy if there is a lot of scatter in the experimental data. It also ignores a lot of potentially valuable information elsewhere in the tracer profile. Nokes (1986) suggests a graphical estimation method for analysing data below a steady vertical line source in a rectangular channel which makes use of all the concentration measurements. The normalised cumulative concentration is defined as

$$s_c(z) = \frac{\int_{z'=0}^{z} s(z')\, dz'}{\int_{z'=0}^{b} s(z')\, dz'} \tag{5.28}$$

When $s_c(z)$ is plotted against z on arithmetic probability paper, Gaussian distributions plot as a straight line. A straight line can be fitted (by eye or by regression) to a plot of experimental data and if necessary less weight can be given to inaccurate low concentrations at either end of the plot. A robust estimate of the variance of the tracer distribution is

$$\sigma_z^2 = \tfrac{1}{4}(z_{0.84} - z_{0.16})^2 \tag{5.29}$$

where $z_{0.84}$ and $z_{0.16}$ = transverse locations at which $s_c(z) = 0.84$ and 0.16, respectively. This variance estimate can then be substituted into equation (5.25) to estimate k_z. Nokes' method assumes that the tracer concentration distributions are Gaussian, which is only valid below a point source in an unbounded two-dimensional flow.

The transverse tracer distribution below a steady vertical line source given by the approximate constant-coefficient model (see Section 3.4.1) is

$$s(x, z) = \frac{m}{h\sqrt{4\pi k_z x v_x}} \exp\left[-\frac{v_x(z - z_o)^2}{4k_z x}\right] \tag{5.30}$$

Now

$$s_{max}(x) = \frac{m}{h\sqrt{4\pi k_z x v_x}} \tag{5.31}$$

is the maximum concentration anywhere across the channel at a given value of x. A plot of

$$\sqrt{\log_e\left[\frac{s_{max}(x)}{s(x, z)}\right]} \quad \text{versus } z \tag{5.32}$$

has a slope

$$\sqrt{\frac{v_x}{4k_z x}} \tag{5.33}$$

which can be estimated by regression. For $z < z_o$

$$\sqrt{\log_e\left[\frac{s_{\max}(x)}{s(x, z)}\right]}$$

is taken as negative. This method also makes use of all the measured tracer concentrations. It is advisable to inspect the plot visually to check for outliers especially among low concentrations at the edges of the tracer cloud.

Example 5.1
Table 5.1 and Figure 5.3 show dye concentrations below a vertical line source in a hypothetical rectangular channel. Concentrations were

Table 5.1. Tracer concentrations at two transects downstream from a steady vertical line source in a hypothetical rectangular channel. Concentrations were predicted using equation (5.30) and then random errors were added to simulate experimental uncertainties (see text for details)

Transverse distance (m)	Concentration at 500 m ($mg\,m^{-3}$)	Concentration at 1000 m ($mg\,m^{-3}$)
0.08	−0.02	0.05
2.09	−0.05	0.09
4.06	0.01	0.17
6.03	0.05	0.40
7.93	0.13	0.68
9.97	0.20	1.12
11.99	0.68	1.89
13.92	1.32	3.07
16.05	2.69	4.30
17.90	5.22	5.92
20.09	8.45	7.37
21.94	11.29	8.29
24.08	12.54	9.09
25.91	12.78	9.61
27.93	10.80	8.01
30.00	7.91	7.79
32.04	4.77	5.49
33.95	2.96	4.39
35.96	1.50	2.82
37.97	0.56	2.12
39.96	0.18	1.17
41.98	0.03	0.61
44.06	−0.01	0.36
45.97	−0.04	0.17
47.96	−0.02	0.07
49.90	−0.01	0.04

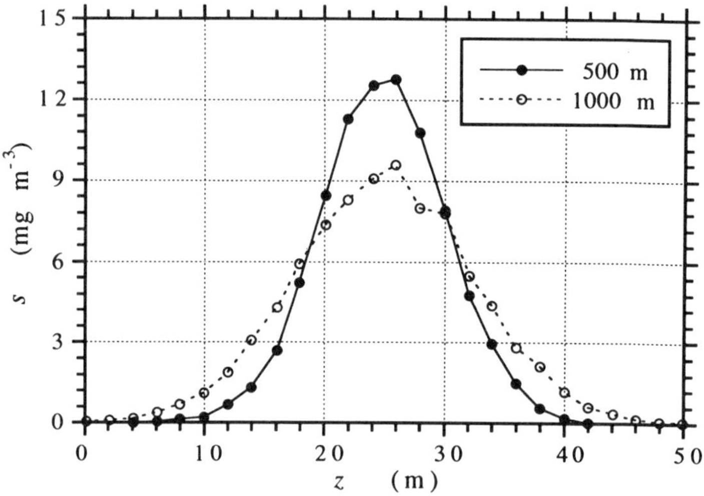

Figure 5.3. Tracer concentration versus transverse distance profiles at two transects downstream from a steady vertical line source in a hypothetical rectangular channel. Concentrations are those in Table 5.1

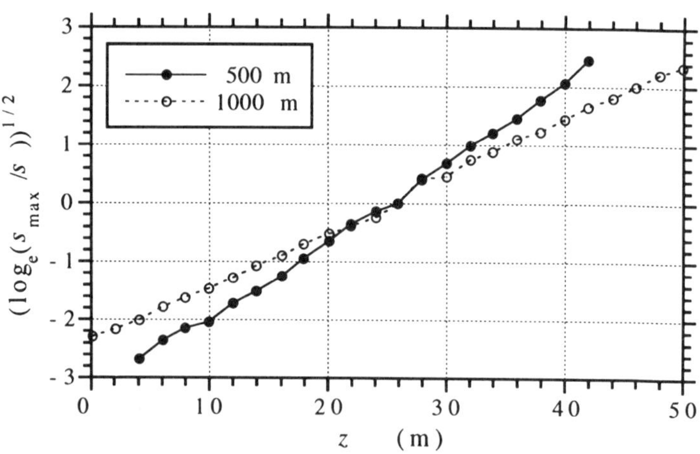

Figure 5.4. Plot of transformed concentration versus transverse distance. Data which comply with equation (5.30) plot as a straight line. The value of k_z is estimated from the slope (see text for details)

calculated from equation (5.30) using $k_z = 0.0167 \text{ m}^2\text{s}^{-1}$ and $v_x = 0.6 \text{ m s}^{-1}$. Random uncertainties of $\pm 10\%$ were added to the predicted concentration (to simulate sampling errors) together with a random error of $\pm 0.05 \text{ m}^2\text{s}^{-1}$ (to simulate the detection limit of the fluorimeter). The channel is 50 m wide and it is clear from Figure 5.3 that tracer has not impinged on either bank.

Figure 5.4 shows a plot of

$$\sqrt{\log_e\left(\frac{s_{max}}{s}\right)} \qquad \text{versus } z$$

For $z < z_o$

$$\sqrt{\log_e\left(\frac{s_{max}}{s}\right)}$$

is taken to be negative and concentrations below 0.05 mg m^{-3} are neglected. The slopes and 95% confidence intervals of the two lines in Figure 5.4 estimated by linear regression are: 0.133 ± 0.003 at $x = 500$ m and 0.096 ± 0.002 at $x = 1000$ m. Solving (5.33) gives $k_z = 0.017 \pm 0.004 \text{ m}^2\text{s}^{-1}$ at $x = 500$ m and $k_z = 0.016 \pm 0.004 \text{ m}^2\text{s}^{-1}$ at $x = 1000$ m.

The transverse variances calculated using the discrete version of equations (5.24) are

$$\sigma_z^2 = \frac{\sum_{i=1}^{n} (z_i - \bar{z})^2 s_i}{\sum_{i=1}^{n} s_i} \qquad \bar{z} = \frac{\sum_{i=1}^{n} z_i s_i}{\sum_{i=1}^{n} s_i}$$

where n = total number of concentration measurements. This gives $\sigma_z^2 = 27.38 \text{ m}^2$ at $x = 500$ m and $\sigma_z^2 = 55.70 \text{ m}^2$ at $x = 1000$ m. Thus from equation (5.25) $k_z = 0.017 \text{ m}^2\text{s}^{-1}$.

It is clear that these estimation methods are fairly successful at recovering the true value of k_z in this particular example where the channel is uniform and tracer does not impinge on either bank.

5.5.3 Generalised Method of Moments

When the channel bathymetry varies longitudinally then the plume undergoes cyclical expansions and contractions. This behaviour is discussed in Section 3.3.3 and illustrated in Figure 3.1. The methods outlined in the previous section should not be used to analyse data from irregular channels. They will underestimate k_z in regions of converging flow (negative k_z values are possible) and will overestimate k_z in regions of diverging flow.

Holley *et al* (1972) developed the generalised method of moments for a steady tracer input which takes into account transverse and longitudinal variations of depth and velocity, non-zero transverse velocities, and allows for tracer impinging on the banks. For steady-state, equation (3.1) becomes

$$\frac{\partial}{\partial x}(hv_x s) + \frac{\partial}{\partial z}(hv_z s) = \frac{\partial}{\partial z}\left(hk_z \frac{\partial s}{\partial z}\right) \tag{5.34}$$

where the longitudinal dispersion term has been neglected and e_z has been incorporated into k_z. We multiply each term in equation (5.34) by $(z - \bar{z})^2$ where $\bar{z}$ is the location of the centroid of the tracer flux given by

$$\bar{z} = \frac{\displaystyle\int_{z=0}^{b} zhv_x s \, \mathrm{d}z}{\displaystyle\int_{z=0}^{b} hv_x s \, \mathrm{d}z} \tag{5.35}$$

We then integrate each term across the channel. The resulting equation can be simplified in a few lines of algebra by interchanging the order of integration and differentiation, integrating the second and third terms by parts and noting that at the boundaries both the transverse velocity and the transverse tracer flux are zero. This is left as an exercise for the reader. The final result is

$$\frac{\mathrm{d}}{\mathrm{d}x}\int_{z=0}^{b} (z - \bar{z})^2 hv_x s \, \mathrm{d}z - 2\int_{z=0}^{b} (z - \bar{z}) hv_z s \, \mathrm{d}z + 2\int_{z=0}^{b} (z - \bar{z}) hk_z \frac{\partial s}{\partial z} \, \mathrm{d}z = 0 \tag{5.36}$$

Each term is then normalised by dividing by the total longitudinal tracer flux

$$\int_{z=0}^{b} hv_x s \, \mathrm{d}z \tag{5.37}$$

which is constant for a conservative tracer. It is a simple matter to show that in the case of constant h, v_x and k_z equations (5.36) and (5.23) are identical. Holley wrote equation (5.36) in the form

$$\frac{\mathrm{d}\sigma_3^2(x)}{\mathrm{d}x} - g(x) + 2Kf(x) = 0 \tag{5.38}$$

where

$$\sigma_3^2(x) = \frac{\displaystyle\int_{z=0}^{b} (z - \bar{z})^2 hv_x s \, \mathrm{d}z}{\displaystyle\int_{z=0}^{b} hv_x s \, \mathrm{d}z} \tag{5.39}$$

$$g(x) = \frac{2 \int_{z=0}^{b} (z - \bar{z}) h v_z s \, dz}{\int_{z=0}^{b} h v_x s \, dz} \tag{5.40}$$

and

$$f(x) = \frac{\int_{z=0}^{b} (z - \bar{z}) h \varphi \frac{\partial s}{\partial z} \, dz}{\int_{z=0}^{b} h v_x s \, dz} \tag{5.41}$$

The transverse dispersion coefficient is assumed to vary across the channel as

$$k_z = K\varphi(z) \tag{5.42}$$

where

$$\int_{z=0}^{b} \varphi(z) \, dz = 1 \tag{5.43}$$

Although the derivation of these equations is exact, there are several practical difficulties with applying them to experimental results. The first difficulty is that the transverse velocity is rarely measured in the field but v_z is required in equation (5.40). The value of v_z can be calculated if the longitudinal velocity distribution and the channel bathymetry are known (e.g. from a flow gauging). The continuity equation is

$$\frac{\partial}{\partial x}(h v_x) + \frac{\partial}{\partial z}(h v_z) = 0 \tag{5.44}$$

Integrating across the channel starting from one bank (where $v_z = 0$) and rearranging

$$v_z(z) = -\frac{1}{h}\frac{\partial}{\partial x}\left(\int_{z'=0}^{z} h v_x \, dz'\right) \tag{5.45}$$

The second difficulty is that in an irregular channel the variance of the tracer flux $\sigma_3^2(x)$ does not vary linearly with distance. Consequently the transverse dispersion coefficient cannot be estimated using the simple finite difference formula which is appropriate in a rectangular channel (see equation 5.25). In a non-uniform channel both $g(x)$ and $f(x)$ also vary non-linearly. Holley suggests that this problem can be overcome by integrating equation (5.38) along the channel to give

$$\sigma_3^2(x_2) - \sigma_3^2(x_1) - \int_{x=x_1}^{x_2} g(x)\,\mathrm{d}x + 2K \int_{x=x_1}^{x_2} f(x)\,\mathrm{d}x = 0 \qquad (5.46)$$

Holley asserts that numerical integration yields reliable estimates of the two integrals in equation (5.46) even from sparse field data. My experience is, however, that in irregular channels it is not easy to make reliable estimates of these integrals when, as is often the case in field studies, measurements are made at only a small number of widely separated transects. The function $f(x)$ is well behaved and decreases monotonically with distance. It is possible to fit an exponential function to known values of $f(x_1), f(x_2), \ldots f(x_n)$ using linear regression. The integral of this exponential function can then be evaluated analytically. The function $g(x)$ is more troublesome. It quantifies the first moment of the transverse tracer flux and its accurate estimation is critical in a channel where the flow is either diverging or converging. To integrate $g(x)$ accurately it is necessary to measure the depth, velocity and tracer concentration distributions at regular intervals along the channel. This requirement increases substantially the logistics of the fieldwork. The final difficulty is that equations (5.38) and (5.46) require accurate estimation of the variance of the tracer flux and measurement errors in concentrations at the edges of the tracer cloud affect the variance as discussed earlier in this chapter.

It has been my experience that the generalised method of moments does not produce robust estimates of the dispersion coefficient for field measurements typical of those collected in irregular natural channels. The streamtube methods (outlined below) are much to be preferred in natural channels. Equations (5.38)–(5.41) do, however, work well in two-dimensional flows (rectangular channels) where $v_z = 0$. Their major attraction is that they allow k_z to be estimated when tracer has impinged on the banks, which the simple methods outlined above do not.

Equations (5.38) and (5.46) also allow for k_z to vary across the channel (see equation 5.42) although the form of this variation must be specified *a priori* to within a constant. This led Holley to suggest a method for identifying the transverse variation of k_z. If tracer concentrations are measured below a steady source, then several plausible functions $\varphi(z)$ can be chosen and used to determine K at several points along the channel. The function $\varphi(z)$ for which K does not vary along the channel will describe the transverse variation of k_z. This is a potentially powerful technique but I am not aware of any systematic attempts to use the generalised method of moments in this way.

5.5.4 Streamtube model

The streamtube model of Yotsukura and Sayre (1976) is a more popular tool than the generalised method of moments for analysing transverse mixing below a steady tracer source in a non-uniform channel. As a first approximation the factor of diffusivity can be assumed constant and the solution for a steady vertical line source is

$$s(x, q) = \frac{m}{\sqrt{4\pi Dx}} \exp\left[-\frac{(q - q_o)^2}{4Dx}\right] \tag{5.47}$$

where m = tracer injection rate and q_o = cumulative discharge at the source. Equation (5.47) can be rewritten

$$s(x, q) = s_{\max}(x) \exp\left[-\frac{(q - q_o)^2}{4Dx}\right] \tag{5.48}$$

where

$$s_{\max}(x) = \frac{m}{\sqrt{4\pi Dx}} \tag{5.49}$$

is the peak tracer concentration at a given value of x. If tracer does not impinge on either bank then by analogy with (5.17) and (5.18) a plot of

$$\sqrt{\log_e\left[\frac{s_{\max}(x)}{s(x, q)}\right]} \quad \text{versus } q \tag{5.50}$$

has the slope

$$\frac{1}{\sqrt{4Dx}} \tag{5.51}$$

To estimate D in this manner the cumulative flow distribution must be known at each transect. Ideally a flow gauging should be carried out at each transect but gaugings require a lot more manpower than concentration profiling and this is often not feasible (with the advent of acoustic Doppler current profilers this situation may change). Although gaugings are time consuming, it is possible to measure the depth distribution quickly using an echo sounder. Yotsukura and Sayre (1976) suggest a semi-empirical method for estimating the transverse distribution of depth-averaged velocity from the measured depth profile and the mean (width-averaged) velocity. The well-known Manning's equation for steady, uniform flow is

$$V_x = \frac{R^{2/3} s^{1/2}}{n} \tag{5.52}$$

where V_x = width-averaged velocity; s = water slope; R = hydraulic radius and n = Manning friction coefficient (Henderson 1966). If equation (5.52) applies locally and both s and n are constant then

$$\frac{v_x}{V_x} = \left(\frac{h}{H}\right)^{2/3} \tag{5.53}$$

where v_x = local depth-averaged velocity; h = local depth; H = width-averaged depth and the hydraulic radius is approximated by the depth. Knowing the mean velocity v_x at a transect the transverse velocity distribution can then be deduced approximately from the measured depth distribution using equation (5.53).

The generalised method of moments was originally developed in Cartesian coordinates but it can equally well be applied to the streamtube equations and yields

$$\frac{\mathrm{d}\sigma_q^2}{\mathrm{d}x} = -2K_q f(x) \tag{5.54}$$

where

$$\sigma_9^2(x) = \frac{\int\limits_{q=0}^{Q} (q - \bar{q})^2 s(x, q)\, \mathrm{d}q}{\int\limits_{q=0}^{Q} s(x, q)\, \mathrm{d}q} \tag{5.55a}$$

$$\bar{q}(x) = \frac{\int\limits_{q=0}^{Q} q s(x, q)\, \mathrm{d}q}{\int\limits_{q=0}^{Q} s(x, q)\, \mathrm{d}q} \tag{5.55b}$$

and

$$f(x) = \frac{\int\limits_{q=0}^{Q} q\varphi(q) \frac{\partial s(x, q)}{\partial q}\, \mathrm{d}q}{\int\limits_{q=0}^{Q} s(x, q)\, \mathrm{d}q} \tag{5.56}$$

By definition the transverse flow between streamtubes is zero so that $g(x)$ in equation (5.40) vanishes. It is assumed that the factor of diffusion varies across the channel as

$$D(q) = K_q \varphi(q) \tag{5.57}$$

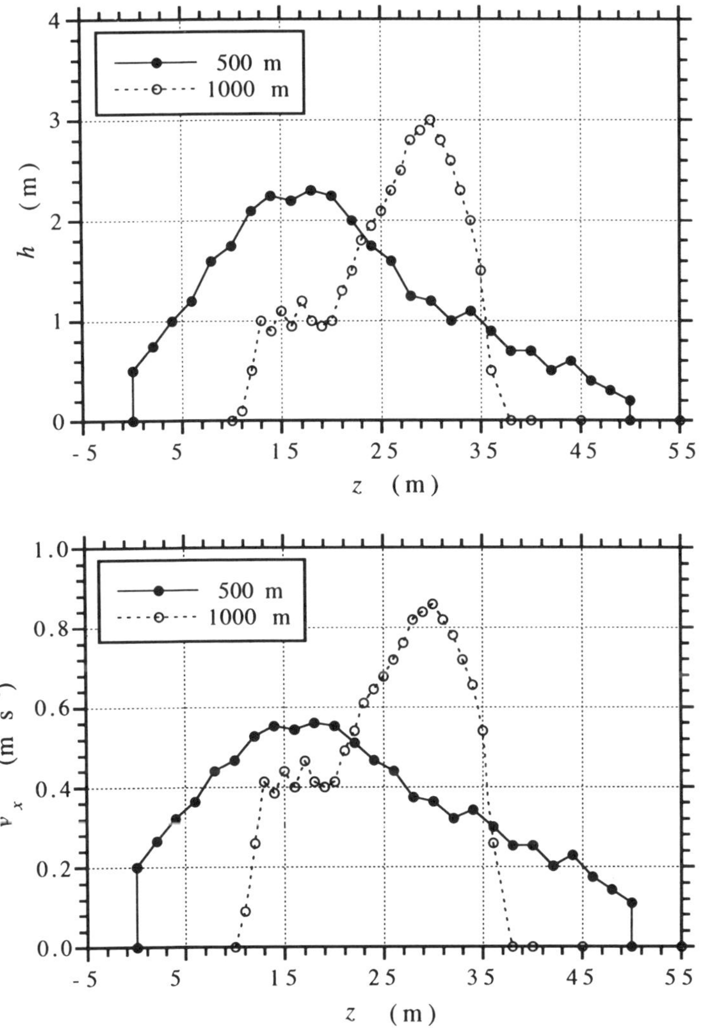

Figure 5.5. Depth and velocity distributions in a hypothetical channel. The channel converges between 500 and 1000 m

$\varphi(q)$ has the same meaning as in (5.42) and K_q = width-averaged value of D. If $\varphi(q) = 1$ and tracer does not impinge on either bank then integrating equation (5.56) by parts it can be seen that $f(x) = 1$. From equation (5.54) it then follows that

$$K_q = \frac{1}{2} \frac{\mathrm{d}}{\mathrm{d}x} (\sigma_q^2) \tag{5.58}$$

which is the ordinary method of moments solution. The main advantage of equation (5.54) is that it accommodates the situation where tracer impinges on the bank. When tracer concentrations are measured at widely separated sites it is advisable to integrate equation (5.54) along the channel which gives

$$K_q = -\frac{1}{2} \frac{\sigma_q^2(x_2) - \sigma_q^2(x_1)}{\int_{x=x_1}^{x_2} f(x)\, \mathrm{d}x} \tag{5.59}$$

Example 5.2

Figures 5.5 and 5.6 and Tables 5.2 and 5.3 show transverse depth, velocity and tracer concentration profiles in a hypothetical channel of irregular cross-section. Concentrations were predicted using equation (5.47) with

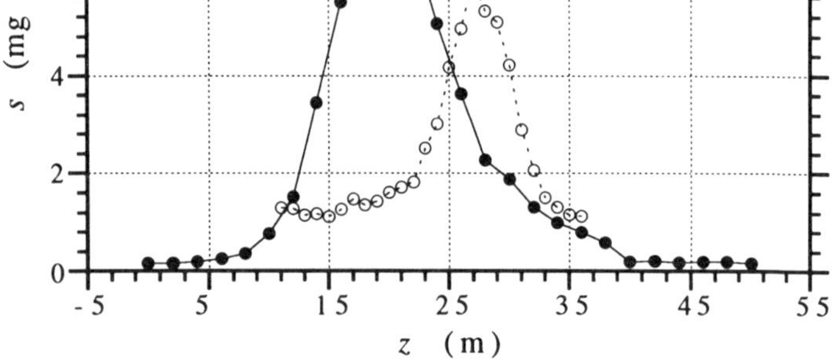

Figure 5.6. Tracer concentrations versus transverse distance downstream from a steady vertical line source in an irregular channel. Concentrations are those in Tables 5.2 and 5.3. As the channel is irregular, the thalweg and the tracer plume meander across the channel

$D = 0.025\ m^5 s^{-2}$, $Q = 30\ m^3 s^{-1}$, $m = 100\ mg\,s^{-1}$ and $q_o = 15\ m^3 s^{-1}$. A random error of $\pm 10\%$ was added to the predicted concentrations as was done in Example 5.1. Estimates were then made of the transverse dispersion coefficient. It is clear from Figure 5.5 that the flow converges markedly and it would be incorrect to apply the methods outlined in Example 5.1 in this channel.

Figure 5.7 shows concentrations plotted against cumulative flow which allows the effects of transverse mixing to be distinguished more easily than in Figure 5.6. Figure 5.8 shows the plot of

$$\sqrt{\log_e\left(\frac{s_{max}}{s}\right)} \qquad \text{versus } q$$

Table 5.2. Depth, velocity and dye concentrations 500 m below a steady vertical line source in an irregular hypothetical channel calculated using equation (5.47). A random error is added to the predicted concentrations to simulate experimental uncertainties. See text for details

Transverse distance (m)	Local depth (m)	Local velocity ($m\,s^{-1}$)	Cumulative flow ($m^3 s^{-1}$)	Dye concentration ($g\,m^{-3}$)
0	0.50	0.20	0	0.16
2	0.75	0.26	0.3	0.16
4	1.00	0.32	0.8	0.19
6	1.20	0.36	1.5	0.25
8	1.60	0.44	2.7	0.36
10	1.75	0.46	4.2	0.77
12	2.10	0.52	6.1	1.52
14	2.25	0.55	8.5	3.45
16	2.20	0.54	10.9	5.50
18	2.30	0.56	13.4	7.34
20	2.25	0.55	15.9	7.56
22	2.00	0.51	18.2	6.75
24	1.75	0.46	20.0	5.06
26	1.60	0.44	21.6	3.62
28	1.25	0.37	22.7	2.28
30	1.20	0.36	23.6	1.88
32	1.00	0.32	24.4	1.31
34	1.10	0.34	25.1	1.00
36	0.90	0.29	25.7	0.80
38	0.70	0.25	26.2	0.59
40	0.70	0.25	29.0	0.20
42	0.50	0.20	29.3	0.21
44	0.60	0.22	29.5	0.17
46	0.40	0.17	29.7	0.19
48	0.30	0.14	29.9	0.19
50	0.20	0.11	29.9	0.16

Table 5.3. Depth, velocity and dye concentrations 1000 m below a steady vertical line source in an irregular hypothetical channel calculated using equation (5.47). A random error is added to the predicted concentrations to simulate experimental uncertainties. See text for details

Transverse distance (m)	Local depth (m)	Local velocity ($m\,s^{-1}$)	Cumulative flow ($m^3\,s^{-1}$)	Dye concentration ($g\,m^{-3}$)
1	0.10	0.08	0	1.29
2	0.50	0.26	0.1	1.28
3	1.00	0.41	0.3	1.15
4	0.90	0.38	0.7	1.17
5	1.10	0.43	1.1	1.12
6	0.95	0.39	1.5	1.26
7	1.20	0.46	2.0	1.48
8	1.00	0.41	2.5	1.35
9	0.95	0.39	2.9	1.43
10	1.00	0.41	3.3	1.61
11	1.30	0.49	3.8	1.71
12	1.50	0.54	4.5	1.82
13	1.80	0.61	5.5	2.52
14	1.95	0.64	6.6	3.00
15	2.10	0.67	10.4	4.17
16	2.30	0.71	11.9	4.95
17	2.50	0.76	13.7	5.96
18	2.80	0.82	15.8	5.31
19	2.90	0.83	18.2	5.09
20	3.00	0.85	20.7	4.21
21	2.80	0.82	23.1	2.89
22	2.60	0.78	25.3	2.07
23	2.30	0.71	27.1	1.50
24	2.00	0.65	28.6	1.30
25	1.50	0.54	29.7	1.16
26	0.50	0.26	30.0	1.13

at the two transects. For $q < q_o$

$$\sqrt{\log_e\left(\frac{s_{max}}{s}\right)}$$

is taken as negative. The data plots very close to a straight line at 500 m. The slope of plot is 0.135 ± 0.003 which from equation (5.51) gives $D = 0.029 \pm 0.001\ m^5\,s^{-2}$ (mean $\pm 95\%$ confidence interval). This estimate is fairly close to the value $D = 0.025\ m^5\,s^{-2}$ used to generate the concentrations.

At the 1000 m transect data do not plot as a straight line in Figure 5.8. The reason is that tracer has impinged on the banks (as is clear in Figures 5.6 and 5.7) and it is no longer valid to estimate D using equations (5.50) and

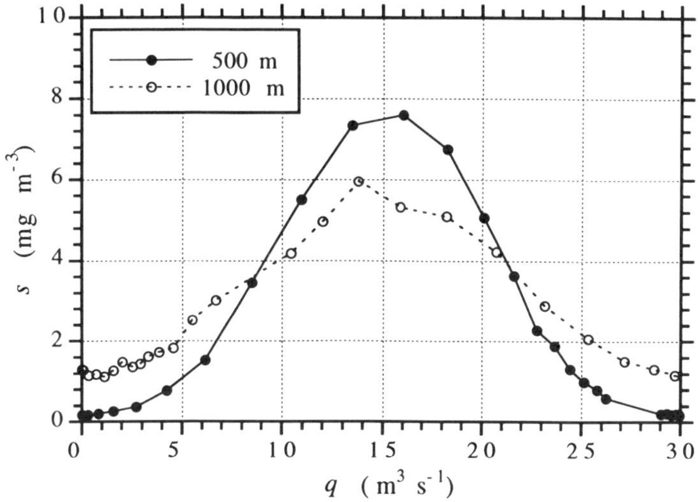

Figure 5.7. Tracer concentrations plotted versus cumulative flow. The use of cumulative flow as the x axis removes the meandering seen in Figure 5.6

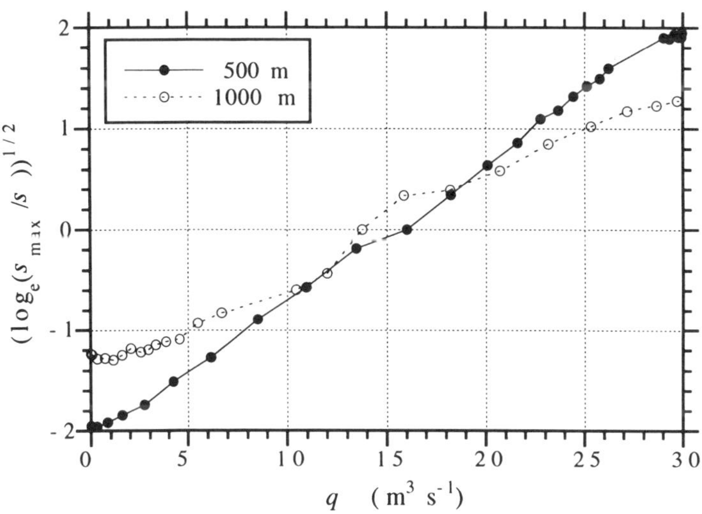

Figure 5.8. Transformed concentration versus cumulative flow. Data which comply with equation (5.47) plot as a straight line. The slope is used to estimate the value of the factor of diffusion D

(5.51). We can, however, use the generalised method of moments regardless of whether or not tracer impinges on the bank. If we assume that $\varphi(q) = 1$ (i.e. the factor of diffusion is constant) and make use of equations (5.55)–(5.56)

$$\text{at } x_1 = 500 \text{ m}, \ \sigma_9^2(x_1) = 24.92, \ f(x_1) = -0.979$$
$$\text{at } x_2 = 1000 \text{ m}, \ \sigma_9^2(x_2) = 45.21, \ f(x_2) = -0.647$$

If we assume that $f(x)$ decreases exponentially between sites then we can estimate that

$$\int_{x=x_1}^{x_2} f(x) \, dx = -400.8$$

Substituting these values into equation (5.59) gives $K_q = 0.0253 \text{ m}^5 \text{ s}^{-2}$. This is very close to the value used to generate the concentrations.

5.5.5 Non-steady Source

The methods for estimating the transverse dispersion coefficient outlined so far apply only to steady tracer injections. Steady injections are preferred for the reasons outlined in Section 5.3.2 but instantaneous tracer injections are commonly made to study longitudinal dispersion in the far-field and such experiments can yield valuable information about transverse mixing in the mid-field. If the depth, velocity and dispersion coefficients are constant then for an instantaneous source the solution to equation (3.1) in a laterally unbounded channel is

$$s(x, z, t) = \frac{M}{h4\pi t\sqrt{k_x k_z}} \exp\left[-\frac{(z - z_o)^2}{4k_z t}\right] \exp\left[-\frac{(x - v_x t)^2}{4k_x t}\right] \tag{5.60}$$

where s = depth-averaged tracer concentration; M = total mass of tracer injected at $x = 0$, $z = z_o$, $t = 0$; k_x and k_z = longitudinal and transverse dispersion coefficients; v_x = mean velocity; and t = time since release. Aerial photography is often used to take snapshots of surface concentration during the early phases of a tracer test and these data can be analysed using equation (5.60). An x–z grid is laid over the measured tracer cloud and concentrations estimated as a function of x and z. The tracer concentration data are integrated along each x grid line to yield the tracer mass per unit width at time t_n, $n = 1, 2 \ldots N$.

$$\bar{s}(z, t_n) = \int_{x=-\infty}^{\infty} s(x, z, t_n) \, dx \tag{5.61}$$

Integrating equation (5.60) in the x direction gives

$$\bar{s}(z, t_n) = \frac{M}{h\sqrt{4\pi k_z t_n}} \exp\left[-\frac{(z - z_o)^2}{4k_z t_n}\right]$$

$$= \bar{s}_{\max}(t_n)\exp\left[-\frac{(z - z_o)^2}{4k_z t_n}\right] \tag{5.62}$$

as

$$\int_{x=-\infty}^{\infty} \exp\left[-\frac{(x - v_x t_n)^2}{4k_x t_n}\right] \mathrm{d}x = \sqrt{4\pi k_x t_n} \tag{5.63}$$

Equation (5.62) is similar in form to equation (5.12) and a plot of

$$\sqrt{\log_e\left[\frac{\bar{s}_{\max}(t_n)}{\bar{s}(t_n)}\right]} \qquad \text{versus } z \tag{5.64}$$

has the slope

$$\frac{1}{\sqrt{4k_z t_n}} \tag{5.65}$$

In a practical problem the integral (5.61) can be evaluated with reasonable precision from field measurements using numerical methods (e.g. trapezoidal integration).

The method of moments can also be used to estimate k_z from two (or more) snapshots of the tracer concentration. If the tracer mass per unit width is determined from equation (5.61) then by analogy with equations (5.24)

$$\sigma_z^2(t) = \frac{\int_{z=0}^{b} (z - \bar{z})^2 \bar{s}(z, t)\, \mathrm{d}z}{\int_{z=0}^{b} \bar{s}(z, t)\, \mathrm{d}z} \tag{5.66a}$$

where

$$\bar{z} = \frac{\int_{z=0}^{b} z\bar{s}(z, t)\, \mathrm{d}z}{\int_{z=0}^{b} \bar{s}(z, t)\, \mathrm{d}z} \tag{5.66b}$$

and from equation (5.23)

$$k_z = \frac{1}{2}\frac{\mathrm{d}\sigma_z^2}{\mathrm{d}t} \tag{5.67}$$

It is important to remember that a single instantaneous tracer injection gives information about only one of the ensemble of possible outcomes. In practice results from a single instantaneous injection are usually assumed to describe the typical plume behaviour. A steady source is preferred when studying transverse dispersion because the plume can be traversed several times and the results time averaged.

Example 5.3

A dye test was conducted in Waikato River, near Hamilton, New Zealand. A 2 kg mass of Lissamine Red 4B was released simultaneously at each of six points equidistant across a road bridge. Channel characteristics are summarised in Table 5.4. In aerial photographs taken from an elevation of 600 m during the first five minutes after discharge, six dye clouds could be distinguished. Clouds 5 and 6 near the right bank were partly obscured by bankside vegetation and were not analysed. At each time t_n a line was drawn across the channel through the middle of each dye profile and the maximum width of each cloud was measured using the bridge as a scale. Setting $x = v_x t_n$ and $z_o = 0$ in equation (5.60) the concentration variation along this line is

$$s(z, t_n) = s_{\max} \exp\left(-\frac{z^2}{4k_z t_n}\right)$$

where $s_{\max}$ = peak concentration measured at time t_n. If the tracer cloud extends from $z = -L$ to $z = +L$, then rearranging the previous equation

$$\frac{L^2}{4k_z t_n} = \log_e\left[\frac{s_{\max}(t_n)}{s(L, t_n)}\right]$$

Estimates of k_z were made assuming that the ratio $r = s_{\max}(t_n)/s(L, t_n)$ was either 10 or 20 (i.e. the concentration at the edge of the cloud was either 10 or 5% of the peak). Results are summarised in Table 5.5.

Table 5.4. Channel characteristics of the Waikato River near Hamilton

Slope (s)	0.000013
Mean depth (H)	2.45 m
Mean velocity (V_x)	0.75 m s^{-1}
Shear velocity (U^*)	0.056 m s^{-1}
Channel width (b)	85 m
Dispersion coefficient (k_z ($r = 20$))	0.058 m^2 s^{-1}
Dispersion coefficient (k_z ($r = 10$))	0.076 m^2 s^{-1}

Table 5.5. Transverse dispersion coefficients estimated from aerial photographs of initial phases of a dye test. r = ratio of maximum concentration to concentration at the edge of the tracer cloud on a transverse line drawn through the middle of the cloud

Time (s)	Dye cloud	Transverse dispersion coefficient (k_z; $m^2 s^{-1}$)	
		$r = 20$	$r = 10$
120	1	0.076	0.098
120	2	0.041	0.053
120	3	0.057	0.074
120	4	0.048	0.063
150	1	0.078	0.100
150	2	0.045	0.059
150	3	0.061	0.079
150	4	0.061	0.079
150	4	0.061	0.079
300	4	0.058	0.075

The dispersion coefficient for cloud 1 was higher than for the other clouds. It was noted that the current was fairly strong near the left bank (where cloud 1 was injected), turbulence was high as a result of velocity shear near the river bank and there were strong eddies behind the left-hand bridge pier. The estimates of k_z showed no dependence on t but were slightly higher for $r = 10$ than $r = 20$. The average values of the non-dimensional transverse dispersion coefficient were $k_z/hu^* = 0.42$ ($r = 20$) and 0.55 ($r = 10$) which lie within the range of values expected.

Dye was mixed with the top 1 m of water at the point of discharge. For a point source at three-quarter depth results in Chapter 2 indicate that the time to achieve 90% vertical mixing for a surface discharge is

$$t = 0.3\,\frac{h^2}{e_y}$$

Taking $e_y = 0.067hu^*$, $u^* = 0.056\ \mathrm{m\,s^{-1}}$ and $h = 2.5$ m then $t = 200$ seconds. This indicates that the dye was not completely mixed vertically throughout the experiment and the inferred transverse dispersion coefficients relate only to the surface waters. Nokes and Wood (1988) suggest that k_z and v_x have a similar depth distribution. Velocity is distributed logarithmically with depth with a surface maximum and so it is likely that the depth-averaged value of k_z is lower than the values derived here.

If the longitudinal velocity varies across the channel and the source is non-steady then the estimation of the transverse dispersion coefficient becomes rather difficult. Consider what happens when an instantaneous tracer injection is made into a natural river. For a short time after injection, the tracer cloud is small and it is advected downstream at the local velocity which can be assumed constant. The simple methods outlined here can be used to estimate the transverse dispersion coefficient. Once the tracer cloud extends over more than about 30% of the channel width, however, the effects of transverse velocity shear become apparent. The tracer cloud assumes an ox-bow shape which enhances the transverse concentration gradients (and hence accelerates transverse mixing) at the head and tail of the tracer cloud. In this situation data are best analysed using a two-dimensional numerical model of the sort described in Chapter 3. Model calibration involves trial and error adjustments of the friction coefficient (which affects the velocity and depth) and the dispersion coefficients k_x and k_z until there is a satisfactory match between observed and predicted tracer concentrations.

5.5.6 Drogues for Measuring Transverse Dispersion

Surface floats were used in some of the early laboratory studies on transverse mixing and can be used to measure k_z in rivers. The transverse dispersion coefficient is higher near the surface than near the bed (Okoye 1970, Nokes 1986) and so surface floats tend to over-estimate the depth-averaged transverse dispersion coefficient. Nevertheless, drogues offer a convenient tool for making low cost estimates of V_x and k_z.

One convenient way to study transverse mixing is to release drogues just upstream from a bridge, mark off distance along the handrail of the bridge and record the transverse location and time of travel of each drogue as it passes under the bridge. A tag line across the river will do the same job. The velocity can be estimated from the time taken for drogues to reach the bridge using only a small number of drogues (five to 10). To obtain a good estimate of the transverse dispersion coefficient, however, a larger number of drogues is required (30–50). It is best to release drogues sequentially from the same point so that they do not interfere with each other.

At the measuring transect the channel is divided transversely into m segments of equal width where the centre of each segment is a distance z_i from the bank. The total number of drogues passing through each segment is then counted, $n(z_i)$, $i = 1, 2 \ldots m$. The transverse variance is

$$\sigma_z^2 = \frac{1}{N} \sum_{i=1}^{m} n(z_i)(z_i - \bar{z})^2 \tag{5.68a}$$

where N = total number of drogues; m = total number of segments;

$n(z_i)$ = number of drogues in the segment centred on z_i and

$$\bar{z} = \frac{1}{N} \sum_{i=1}^{m} z_i n(z_i) \tag{5.68b}$$

Note that $n(z_i)$ is the discrete equivalent of $\bar{s}(z)$ obtained by integrating the concentration distribution in the x direction, see equation (5.61). The variance estimated from equations (5.68) is very sensitive to the location of drogues near the edges of the cloud, which may be affected by snags, wind and surface eddies. A more robust estimate of the variance can be made by calculating the cumulative number of drogues to the left of each segment

$$n_c(z_j) = \sum_{i=1}^{j} n(z_i) \tag{5.69}$$

where $n_c(z_j)$ = cumulative number of drogues to the left of the streamtube centred on $z = z_j$. When $n_c(z_j)/N$ is plotted on arithmetic probability paper, a straight line can be fitted (by eye or regression). If necessary lower weight can be given to outliers at each end of the plot. The transverse variance is

$$\sigma_z^2 = \tfrac{1}{4}(z_{0.84} - z_{0.16})^2 \tag{5.70}$$

where $z_{0.84}$ and $z_{0.16}$ = transverse locations at which the cumulative proportion of drogues equals 0.84 and 0.16, respectively. Then

$$k_z = \frac{1}{2}\frac{\mathrm{d}\sigma_z^2}{\mathrm{d}t} = \frac{1}{2} V_x \frac{\mathrm{d}\sigma_z^2}{\mathrm{d}x} \tag{5.71}$$

where t = time of travel; and V_x = mean velocity.

Example 5.4

As a fund-raising venture, several hundred numbered surface floats (plastic ducks!) were released instantaneously from a point at mid-channel in the Waikato River near Hamilton. The first ducks to reach the finishing line (about 5 km downstream) won their 'owners' a prize. Oblique photographs were taken from two bridges 20 minutes after release from which an estimate was made of the transverse dispersion coefficient.

Longitudinal grid lines were drawn on the photographs at approximately 5 m intervals across the channel. The exact number of floats in each streamtube could not be resolved, but a ranking of the relative number of floats in each streamtube (from 0–6) was assigned by eye. Table 5.6 shows the transverse distribution of drogues. The effects of transverse velocity shear were apparent with drogues travelling downstream more quickly in mid-stream. It was assumed, however, that there had been insufficient time for transverse velocity shear to affect the rate of transverse mixing. There were few drogues near either bank and it was assumed that the flow was unbounded.

Table 5.6. Transverse distribution of surface drogues measured in each of 15 segments across the Waikato River

	Transverse distance*	Number of drogues†	Cumulative total	Cumulative fraction
1	15	0	0	0
2	20	0	0	0
3	25	1	1	0.043
4	30	2	3	0.130
5	35	1	4	0.174
6	40	3	7	0.304
7	45	6	13	0.565
8	50	4	17	0.739
9	55	3	20	0.869
10	60	1	21	0.913
11	65	1	22	0.956
12	70	0	22	0.956
13	75	1	23	1
14	80	0	23	1
15	85	0	23	1

* Mid-point of the streamtube.
† Qualitative assessment of the number of drogues on a scale of 0–6 made from oblique photographs.

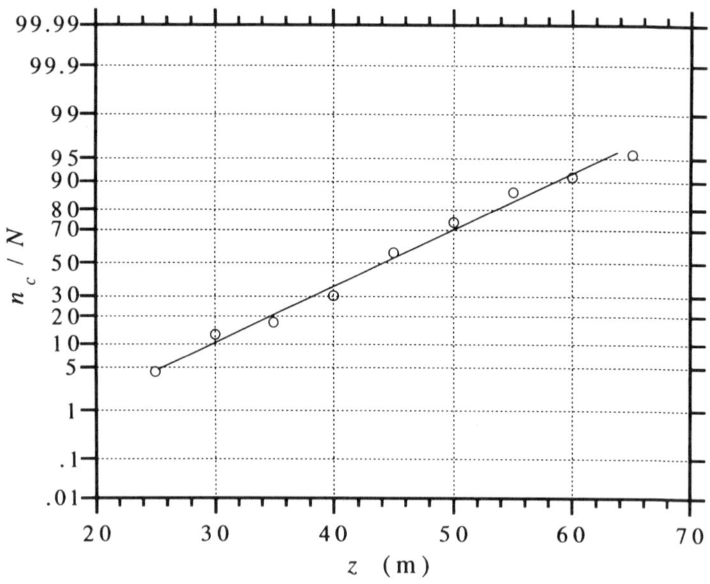

Figure 5.9. Cumulative frequency of surface floats plotted against transverse distance using probability paper. The line shown is fitted by regression. $z_{0.16} = 33$ m and $z_{0.84} = 57$ m

Figure 5.9 shows the cumulative proportion plotted against transverse location on cumulative probability paper. The transverse locations at which 16 and 84% of the surface floats occurred were estimated to be $z = 33$ and 57 m respectively. From equation (5.70)

$$\sigma_z^2 = \tfrac{1}{4}(z_{0.84} - z_{0.16})^2 = 144 \text{ m}^2$$

Then from equation (5.71)

$$k_z = \frac{1}{2}\frac{\mathrm{d}\sigma_z^2}{\mathrm{d}t} = 0.060 \text{ m}^2\,\text{s}^{-1}$$

This is comparable with estimates of $k_z = 0.058$–$0.076 \text{ m}^2\,\text{s}^{-1}$ made in the same reach using dye (see Example 5.3).

5.6 LONGITUDINAL DISPERSION COEFFICIENT

5.6.1 Method of Moments

The longitudinal dispersion coefficient is defined by the relationship

$$K_x = \frac{1}{2}\frac{\mathrm{d}\sigma_x^2}{\mathrm{d}t} \tag{5.72}$$

and given concentration versus distance profiles measured at times t_1 and t_2

$$K_x = \frac{1}{2}\frac{\sigma_x^2(t_2) - \sigma_x^2(t_1)}{t_2 - t_1} \tag{5.73}$$

where

$$\sigma_x^2(t_i) = \frac{\int_{x=-\infty}^{\infty} [x - \bar{x}(t_i)]^2 S(x, t_i)\,\mathrm{d}x}{\int_{x=-\infty}^{\infty} S(x, t_i)\,\mathrm{d}x} \tag{5.74a}$$

and

$$\bar{x}(t_i) = \frac{\int_{x=-\infty}^{\infty} xS(x, t)\,\mathrm{d}x}{\int_{x=-\infty}^{\infty} S(x, t)\,\mathrm{d}x} \tag{5.74b}$$

in which $S(x, t)$ = cross-sectional averaged tracer concentration; $\sigma_{\bar{x}}^2$ = longitudinal variance; and $\bar{x}$ = centroid of the spatial distribution. It is rare for experiments to yield concentration versus distance profiles as is required by equations (5.74). Fischer (1966) was able to show, however, that equation (5.73) can be rewritten

$$K_x = \frac{1}{2} V_x^2 \frac{\sigma_t^2(x_2) - \sigma_t^2(x_1)}{\bar{t}_2 - \bar{t}_1} \tag{5.75}$$

where

$$\sigma_t^2(x_i) = \frac{\int_{t=-\infty}^{\infty} (t - \bar{t}_i)^2 S(x_i, t) \, dt}{\int_{t=-\infty}^{\infty} S(x_i, t) \, dt} \tag{5.76a}$$

and

$$\bar{t}_i = \frac{\int_{t=-\infty}^{\infty} t S(x_i, t) \, dt}{\int_{t=-\infty}^{\infty} S(x_i, t) \, dt} \tag{5.76b}$$

The mean velocity is, of course

$$V_x = \frac{x_2 - x_1}{\bar{t}_2 - \bar{t}_1} \tag{5.77}$$

There are practical difficulties using equations (5.76). Firstly, to evaluate the integrals, complete concentration versus time curves must be collected at each site. This presents logistical problems in many natural channels where tracer profiles exhibit long tails and sampling must continue for a long time after the bulk of the tracer has passed. Secondly, small measurement errors in the low concentrations of the leading and trailing edges of the tracer profile are inevitable and these greatly influence the variance and hence the dispersion coefficient. Consequently values of K_x estimated using the method of moments can be unreliable and it is advisable to check any estimates using a more robust method (such as the routing procedure described in the following).

5.6.2 Routing Method

In Section 4.8.3 there is a discussion of how the concentration versus time profile measured at one site can be routed downstream using either the frozen cloud approximation [equation (4.41)] or the Hayami solution [equation (4.44)]. Observed and predicted concentrations can be compared at the downstream site and, if necessary, the mean velocity and longitudinal dispersion coefficient can be adjusted. The routing method is widely regarded as the most reliable way to estimate values of V_x and K_x (Fischer 1973). It is clear from equations (4.41) and (4.44) that the entire concentration versus time profile must be known at the upstream site. The routing procedure assumes that tracer is conservative so it is necessary to correct for tracer loss (caused by adsorption, decay or dilution by tributaries) before using equations (4.41) or (4.44). This usually means that it is necessary to measure the entire profile at the downstream site, calculate the mass flux and, if necessary, scale the downstream concentrations so that the mass fluxes are identical. An incomplete downstream profile (e.g. where the long tail is not sampled) can be used if tracer loss is known to be negligibly small. Values of V_x and K_x are varied by trial and error until a satisfactory match is achieved between observed and predicted profiles at the downstream site. The value of V_x is determined by matching the times of the observed and predicted peak concentration and the value of K_x by matching the magnitude of the observed and predicted peak concentrations as well as matching the width of the profile at lower concentrations. The two coefficients V_x and K_x are not independent because altering V_x changes the length of time during which dispersion can occur.

Example 5.5

Dye tests were conducted in the Manawatu River below Palmerston North, New Zealand, to measure the velocity and longitudinal dispersion coefficient. Figure 5.10 shows the cross-sectional averaged dye concentrations measured at two sites: Site C (5.0 km below the injection point) and Site D (6.4 km below the injection point). The length of the advective zone was estimated using equation (4.24) to be 5 km and Site C lies just within the equilibrium zone. The routing method was used to derive estimates of the mean velocity and longitudinal dispersion coefficient between Sites C and D.

A computer program (HAND) has been written which solves equation (4.41) numerically using trapezoidal integration. In this example $S(x_1, t)$ is the observed concentration at Site C. The program predicts $S(x_2, t)$, the concentration profile at Site D, and automatically plots the observed and predicted profiles. The user is then prompted for new values of V_x and/or K_x and iterates until a satisfactory match is achieved.

The broken curve in Figure 5.11 shows the predicted profile for $V_x = 0.41\ \mathrm{m\,s^{-1}}$ and $K_x = 22.6\ \mathrm{m^2\,s^{-1}}$, the initial guesses estimated by the program from the observed data using equation (4.4). The velocity is clearly an underestimate. In Figure 5.11 the solid curve is the profile predicted after six iterations with $V_x = 0.55\ \mathrm{m\,s^{-1}}$ and $K_x = 20\ \mathrm{m^2\,s^{-1}}$. This combination gives a fairly close match to concentrations above $20\ \mathrm{mg\,m^{-3}}$. A comparable

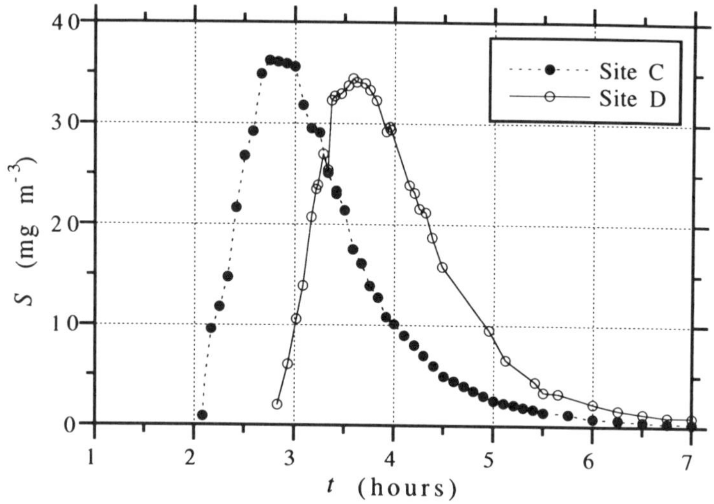

Figure 5.10. Observed dye concentrations at two sites in the Manawatu River, following an instantaneous dye release

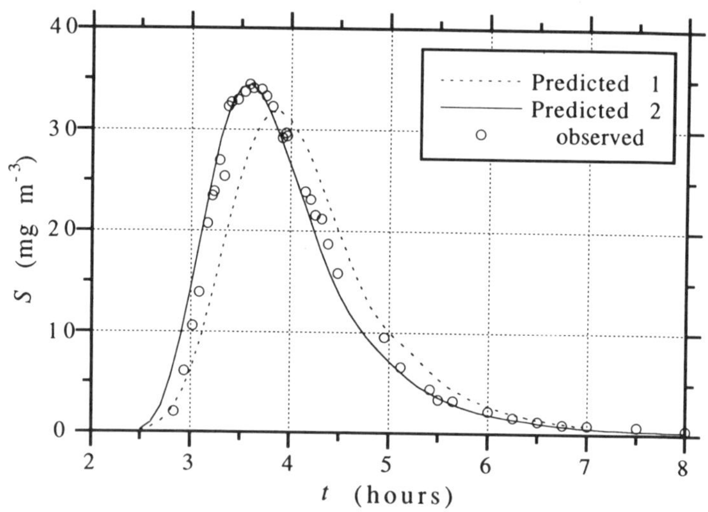

Figure 5.11. Comparison of observed dye concentrations at Site D with concentrations predicted from the measured profile at Site C using the frozen cloud model. Predicted 1 values used the initial guesses and Predicted 2 values the final calibrated values of V_x and K_x

fit was obtained with

$$0.5 < V_x < 0.6 \text{ m s}^{-1} \qquad \text{and} \qquad 17.5 < K_x < 25 \text{ m}^2 \text{s}^{-1}$$

It is worth noting that the observed dye profile arrived slightly later than the predicted profile. This behaviour has also been noted in the Waikato River (Rutherford *et al* 1980) and seems to be a common feature of experimental data. The match in the tail was fairly good in this example, but in many rivers the observed dye profile has a longer tail than the predicted profile.

Estimates of V_x and K_x were also made using a version of the routing program which solves Hayami's equation (4.44) rather than the frozen cloud equation (4.42). There was no significant difference in the concentration profiles predicted by the two models. This stems from the fact that at large values of $x_2 - x_1$ the two models are almost identical.

5.6.3 Chatwin's Method

In Chapter 4 we described the transformation which Chatwin (1971, 1980) suggested for analysing tracer results. Data which satisfy equation (4.4) plot as a straight line when transformed using equation (4.7). The slope and intercept of this line can be used to estimate V_x and K_x. There is a serious practical difficulty with this method. Equation (4.4) assumes that longitudinal dispersion satisfies the Fickian equation (4.3) from the moment tracer is released. However, we know that equation (4.3) is only valid in the equilibrium zone which starts some considerable distance from where tracer is released and consequently equation (4.4) only holds at asymptotically long times. Nevertheless, Chatwin's method can be used to make preliminary estimates of V_x and K_x and to test the applicability of equation (4.4).

Example 5.6

Dye concentration profiles measured at three sites in the Manawatu River are shown in Figure 5.12 plotted using Chatwin's transformation [see equation (4.7)]. For $t > t_{\max}$, the transformed variable

$$S^* = \sqrt{t \log_e \left(\frac{S_{\max} \sqrt{t_{\max}}}{S \sqrt{t}} \right)}$$

is taken to be negative. Figure 5.12 shows data from three sites. Site B is only 2.7 km below the source and is within the advective zone. Site C is 5 km below the source and is at the beginning of the equilibrium zone, whereas Site D is 6.4 km below the source and is well within the advective zone. There is marked curvature in each of the plots which means that none of these data comply with equation (4.4). This is not unexpected because the

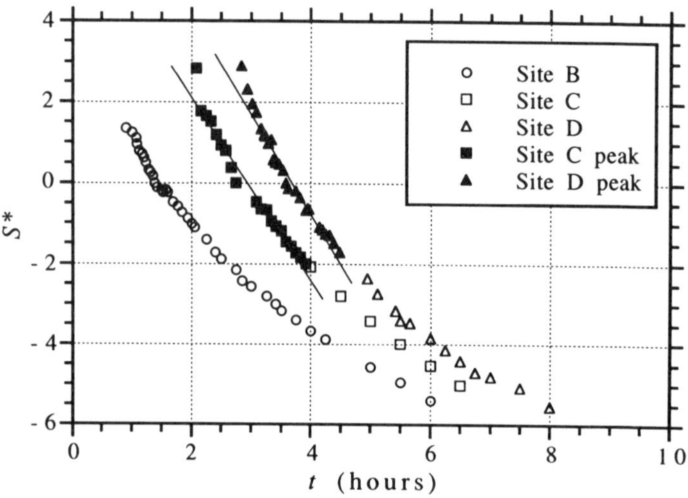

Figure 5.12. Measured dye concentrations at three sites in the Manawatu River plotted using Chatwin's transformation. The transformed variable S^* is defined in the text. Sites B, C and D are 2.7, 5 and 6.4 km below the injection point. The adjective zone is approximately 5 km long

sampling sites are fairly close to the source. As discussed earlier we might expect tracer profiles to become Gaussian at distances five to 10 times the length of the advective zone, which for the Manawatu River would be at distances 25–50 km below the source.

Approximate values of V_x and K_x were estimated by fitting a straight line to concentrations near the peak. In Figure 5.12 a straight line has been fitted to the solid points at Sites C and D. At Site C the intercept is 6.83 h^{-1}, which from equation (4.8) yields $K_x = 37\ m^2 s^{-1}$. The slope is $-2.31\ h^{-3/2}$ which for $K_x = 37\ m^2 s^{-1}$ gives $V_x = 0.47\ m s^{-1}$. At Site D the intercept and slope are $9.50\ h^{-1}$ and $-2.56\ h^{-3/2}$, respectively, which yield estimates of $K_x = 32\ m^2 s^{-1}$ and $V_x = 0.48\ m s^{-1}$.

The Chatwin transformation is ideally suited for making a visual examination of measured concentrations and a preliminary estimate of model coefficients. As few experimental data plot as a straight line, there is some subjectivity in estimating the slope and intercept. It is advisable, therefore, to check values of V_x and K_x estimated using the Chatwin transformation by the routing method. In Example 5.5 it is shown that the routing method yields the estimates:

$$0.5 < V_x < 0.65\ m s^{-1} \quad \text{and} \quad 17.5 < K_x < 25\ m^2 s^{-1}$$

Thackston *et al* (1967) developed an estimation technique similar to

Chatwin's by transforming equation (4.4) to give

$$\log_e(S\sqrt{t}) = \log_e A - \frac{(x - V_x t)^2}{4K_x t}$$

A plot of

$$\log_e(S\sqrt{t}) \text{ versus } \frac{(x - V_x t)^2}{t} \text{ has a slope } -\frac{1}{4K_x}$$

Thackston's method was also used to analyse the Manawatu River data and it yielded very similar estimates of V_x and K_x.

5.6.4 Other Methods

Equation (4.31) indicates that for a conservative tracer the peak concentration S_{max} decreases inversely as the square root of distance below the source.

$$S_{max} = \frac{M}{A\sqrt{4\pi K_x t_n}} = \frac{M}{A\sqrt{\dfrac{4\pi K_x x_{max}}{V_x}}} \qquad (4.31)$$

where M = tracer mass injected; t_n = time at which the concentration versus distance profile is measured; A = channel cross-sectional area; x_{max} = location of the maximum concentration [which for equation (4.4) is also the location of the centroid] and K_x = longitudinal dispersion coefficient. Equation (4.31) is derived from equation (4.4), which is only approximate (as is discussed in Section 4.7). Strictly equation (4.31) only holds for concentration versus distance profiles, but to a close approximation it also holds for concentration versus time profiles. It is clear from equation (4.31) that a plot of

$$S_{max} \quad \text{versus} \quad \frac{1}{\sqrt{x}} \qquad (5.78)$$

has a slope

$$\frac{M}{A\sqrt{4\pi K_x / V_x}} \qquad (5.79)$$

The slope and its standard errors can be estimated using linear methods (e.g. regression) and then used to estimate the value of K_x.

Equation (4.33) gives the length of the channel in which concentrations exceed a given concentration S_c

$$\Delta x_c = 4\sqrt{K_x t_n \log_e\left(\frac{S_{max}}{S_c}\right)} \qquad (4.33)$$

where $S_c < S_{max}$. We can make use of the frozen cloud assumption to recast equation (4.33) and develop a rapid method for estimating the longitudinal dispersion coefficient from a graph of concentration versus time data. The time during which concentrations exceed S_c is approximately

$$t_s \approx \frac{4}{V_x}\sqrt{K_x t_{\max} \log_e\left(\frac{S_{\max}}{S_c}\right)} \tag{5.80}$$

where we are approximating the centroid by the peak. Both the peak concentration $S_{\max}$ and the time of the peak $t_{\max}$ can be estimated quickly from the experimental data. Solving equation (5.80) for K_x gives

$$K_x \approx \frac{\left(\frac{t_s V_x}{4}\right)^2}{t_{\max} \log_e\left(\frac{S_{\max}}{S_c}\right)} \approx \frac{\left(\frac{t_s x}{4 t_{\max}}\right)^2}{t_{\max} \log_e\left(\frac{S_{\max}}{S_c}\right)} \tag{5.81}$$

For any given data set equation (5.81) can be evaluated at several different values of $S_{\max}/S_c$ (e.g. 1.5, 2, 4, etc.) and the average K_x calculated.

One advantage of these two estimation methods is that they do not require complete tracer versus time profiles although they do require that the tracer is conservative.

Example 5.7

Figure 4.1 shows measured dye concentrations in the Waikato River. It is possible to make preliminary estimates of the longitudinal dispersion coefficient using equations (5.78), (5.79) and (5.81). Information about the river channel is given in Example 4.1 and Table 4.3 (see pages 203–204).

Figure 5.13 shows data plotted using equation (5.78). Sites A and B lie within the advective zone and so are ignored. A straight line is fitted by linear regression to data from the remaining four sites which explains a high proportion of the variability ($r = 0.99$) and has a slope of 6260 mg m$^{-5/2}$. From Table 4.3 the mean cross-sectional area is $A = Q/V_x = 215$ m^2, the mass of dye injected is $M = 38$ kg, dye loss is negligibly small (Rutherford *et al* 1980) and the mean velocity is $V_x = 0.70$ m s^{-1}. From equation (5.79) $K_x = 44$ m^2 s^{-1}. Estimates based on the more exact routing procedure given in Table 4.2 were $K_x = 52$–67 m^2 s^{-1}. One advantage of this method of estimating K_x is that it requires only the peak concentration and not the entire tracer profile. It should be noted that the mass of dye injected, the channel area and the mean velocity must be known and dye loss must be negligible.

Estimates of K_x can also be made using equation (5.81). At Site D ($x = 10.9$ km below the source) the peak concentration is $S_{\max} =$ 48.4 mg m^{-3} which occurs 4.0 hours after injection. Concentrations of $S_c = 7.8$ mg m^{-3} occur 3.50 and 5.67 hours after injection and concentrations of $S_c = 38.8$ mg m^{-3} occur 3.75 and 4.42 hours after injection. For $S_c = 7.8$ mg m^{-3}, equation (5.81) gives $K_x = 83$ m^2 s^{-1} while for $S_c = 38.8$ mg m^{-3}, $K_x = 65$ m^2 s^{-1}.

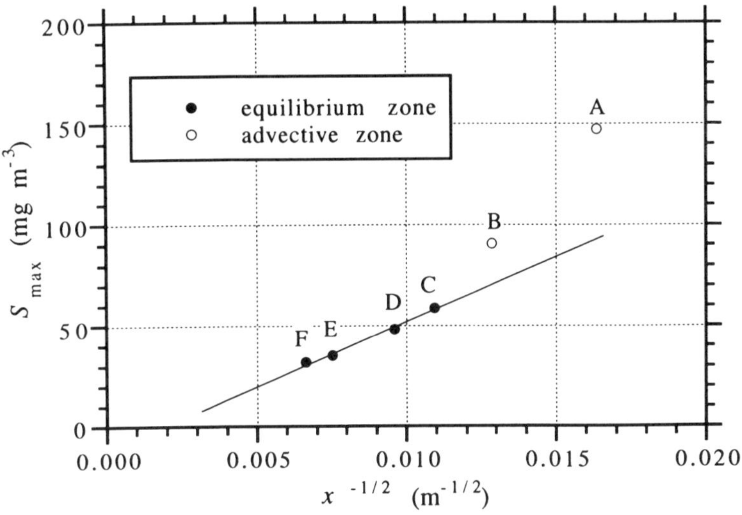

Figure 5.13. Plot of measured peak dye concentration versus time in the Waikato River, March 1975. Sites A and B lie in the advective zone. The regression line fitted to data at sites C–F (which are all within the equilibrium zone) has a slope of 6260 mg m$^{-5/2}$ with an $r = 0.99$

The first estimate is a little higher than the final value of $K_x = 52$–$67\ \mathrm{m^2\,s^{-1}}$ obtained using the routing method (see Table 4.2). I have found in other rivers, that equation (5.81) gives values of K_x which increase as S/S_{max} decreases. This suggests that equation (4.4) does not model low concentrations accurately. The second estimate of K_x is very close to the final value and indicates that equation (5.81) is a useful tool. It does not require *a priori* knowledge of the tracer mass, the complete tracer profile, is robust to tracer loss and is very quick to implement.

5.6.5 Use of Velocity Measurements

In Chapter 4 (Section 4.9) it is shown that the longitudinal dispersion coefficient can be calculated from the velocity and depth distribution within the channel. In large rivers tracer tests to measure K_x may not be attractive because the advective zone is very long and prohibitively large amounts of dye are required. It may be more cost effective to measure the flow distribution at several places along the channel and estimate K_x using equation (4.47).

Flow gaugings are commonly made at water level recording stations to derive rating curves between water level and discharge. Such data can be used to estimate K_x. Recorder sites are not always typical of the whole river and ideally gaugings should be made at a number of sites truly representative of the channel. Given estimates at several points the reach-averaged dispersion coefficient can be calculated using

$$\bar{K}_x = \frac{1}{L} \sum_{i=1}^{n} K_i \Delta x_i \qquad (4.49)$$

where K_i = dispersion coefficient at cross-section i, Δx_i = distance between cross-sections and L = reach length.

Example 5.8

There are five water level recorders on the Waikato River between Hamilton and Mercer at which gaugings are carried out regularly. As part of a study of pollutant flushing carried out during the 1970s estimates were made of the longitudinal dispersion coefficient using a selection of gauging data (Table 5.7). Gaugings were conducted from road bridges. At each site velocity measurements were made at three depths every 10 m across the channel and equation (4.47a) was evaluated numerically using trapezoidal integration. The transverse dispersion coefficient was assumed uniform across the channel. A high value of $k_z/hu^* = 1.0$ was assumed at Rangiriri where the flow was strongly convergent, whereas a low value of $k_z/hu^* = 0.25$ was assumed at Mercer where the channel was straight and uniform.

There is considerable variation in the estimated values of K_x at a given site (see Figure 5.14), which reflects measurement errors, changes in the river bed, and uncertainties in the estimates of k_z. Despite this variation, it is clear that the highest values of K_x occur at low flow. This is particularly marked at Hamilton, Huntly and Rangiriri and arises because at these sites bankside shallows result in strong transverse velocity shear at low flows. As the water level rises, the velocity increases markedly in these shallows, the transverse velocity shear diminishes and K_x decreases. The estimated values of K_x at Rangiriri are very high at low flows even though k_z/HU^* is high. There was an extensive area of shallows near the left bank at the time of these gaugings which resulted in very strong local velocity shear.

The reach-averaged value of K_x at mean flow can be estimated using equation (4.49). We assume that the tabulated values of K_x at mean flow are typical of the sub-reaches in which the gauging sites lie. If we take the sub-reach boundaries as the mid-point between gauging sites then equation (4.49) gives a reach-averaged value between Hamilton and Mercer of 530 $m^2 s^{-1}$. This assumes that K_x = 1500 $m^2 s^{-1}$ in a 20.5 km sub-reach near Rangiriri. From aerial photographs it appears that the islands and sandbanks which give rise to the high value of K_x at Rangiriri only cover about 5 km (and not 20.5 km). Adjusting the sub-reach boundaries to reflect this results in a reduction in the reach-averaged value to K_x = 285 $m^2 s^{-1}$.

Table 5.7. Longitudinal dispersion coefficients estimated from river gauging data at five sites on the Waikato River using equation (4.47)

Site	Flow rate (Q; $m^3 s^{-1}$)	Area (A; m^2)	Slope (s; $\times 10^{-4}$)	Velocity V_x ($m s^{-1}$)	Depth (H; m)	Shear velocity (u^*; $m s^{-1}$)	Transverse dispersion (k_z/Hu^*)	Longitudinal dispersion (K_x; $m^2 s^{-1}$)
1	148	138	1.0	1.07	1.84	0.043	0.30	388
1	195	162	1.2	1.20	2.08	0.050	0.25	114
1	255	207	1.7	1.23	2.59	0.066	0.25	98
1	274	227	1.8	1.20	2.68	0.069	0.25	48
1	330	265	2.0	1.24	3.00	0.077	0.25	72
1	372	296	2.1	1.25	3.30	0.083	0.25	154
1	392	328	2.2	1.19	3.66	0.089	0.25	74
2	121	247	1.0	0.48	1.93	0.044	0.50	197
2	187	320	1.1	0.58	2.19	0.049	0.50	119
2	360	458	1.2	0.78	2.97	0.059	0.50	113
2	490	548	1.22	0.89	3.45	0.064	0.50	136
2	701	710	1.3	0.98	4.17	0.073	0.50	172
3	195	344	0.7	0.57	1.51	0.030	0.50	424
3	265	382	0.8	0.69	1.69	0.036	0.50	233
3	384	520	0.95	0.74	2.21	0.045	0.50	266
3	484	644	1.0	0.75	2.67	0.051	0.50	286
3	794	946	1.45	0.84	3.86	0.074	0.50	214
3	886	1078	1.5	0.82	4.67	0.083	0.50	102
4	157	232	1.16	0.68	1.13	0.036	1.00	2518
4	302	386	1.2	0.78	1.77	0.046	1.00	1569
4	442	538	1.4	0.82	2.42	0.058	1.00	512
4	772	843	1.65	0.92	3.33	0.073	1.00	609
4	1050	1005	1.8	1.05	3.87	0.083	1.00	616
5	155	424	0.6	0.36	3.66	0.046	0.25	79
5	292	479	0.7	0.61	4.02	0.053	0.25	99
5	350	547	0.75	0.64	4.48	0.057	0.25	227
5	424	565	0.8	0.75	4.61	0.060	0.25	82
5	476	595	1.0	0.80	4.82	0.069	0.25	56
5	577	647	1.05	0.89	5.15	0.073	0.25	63
5	643	688	1.1	0.93	5.42	0.076	0.25	173

Table 5.8. Longitudinal dispersion coefficients in each sub-reach and averaged over the Hamilton–Mercer reach using equation (4.49)

Site	Distance from sea (km)	Reach average K_x ($m^2 s^{-1}$)	Sub-reach length* (km)	Sub-reach length† (km)
1. Hamilton	116.3	100	10.2	10.2
2. Ngaruawahia	95.9	150	17.4	17.4
3. Huntly	81.5	300	15.3	25.3
4. Rangiriri	65.2	1500	20.5	5.5
5. Mercer	40.5	100	12.4	22.4

* Reach-averaged $K_x = 530\ m^2 s^{-1}$.
† Reach-averaged $K_x = 285\ m^2 s^{-1}$.

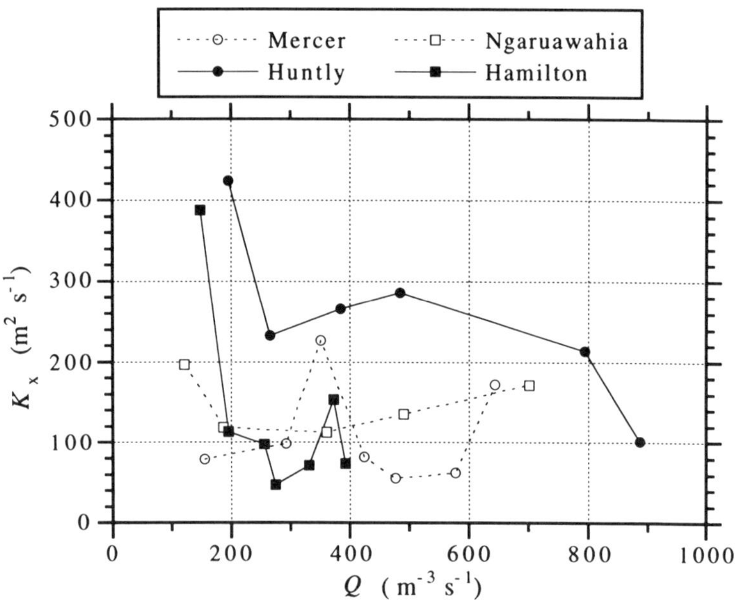

Figure 5.14. Longitudinal dispersion coefficients at four gauging sites on the Waikato River estimated from measured velocity distributions showing the variation with river flow

NOTATION

A	cross-sectional area
b	channel width
$\bar{b}$	reach-averaged channel width
c	tracer concentration
$c_{\max}$	maximum concentration over the depth
$\bar{c}$	width-averaged concentration
$\bar{c}_{\max}$	maximum value of c over the depth
D	factor of diffusion
e_y	vertical eddy diffusivity
$\bar{e}_y$	depth-averaged value of e_y
$f(x), g(x)$	integrals in the generalised method of moments
h	local depth
H	mean depth
$\bar{H}$	reach-averaged depth
k_x, k_z	longitudinal and transverse dispersion coefficients in depth-averaged mixing equations
K_x	longitudinal dispersion coefficient in cross-sectional averaged mixing equations
$\bar{K}_x$	reach-averaged value of K_x
K	width-averaged transverse dispersion coefficient
K_q	width-averaged factor of diffusion
L	half width of the tracer cloud
L	reach length
L_y	distance required for vertical mixing
L_z	distance required for transverse mixing
m	mass inflow rate of tracer
m	total number of stream tubes
M	mass of tracer injected
n	Manning's friction coefficient
$n_c\ (z_i)$	cumulative total number of drogues at z_i
N	total number of drogues
$n(z_i)$	number of drogues in the streamtube centred on z_i
P_{m}	degree of mixing $= c_{\min}/c_{\max}$ or $s_{\min}/s_{\max}$
q	cumulative flow
q_{o}	source location
$\bar{q}$	centroid with respect to cumulative flow
Q	river flow
r	concentration ratio $s_{\max}/s$
R	hydraulic radius
s	depth-averaged concentration
$s_{\max}$	maximum concentration over the width

$\bar{s}$	tracer mass per unit width
$\bar{s}_{max}$	maximum tracer mass per unit width
s	channel slope
S	cross-sectional averaged concentration
S^*	Chatwin's transformed concentration
S_c	critical concentration
S_{max}	maximum concentration
s_c	normalised cumulative concentration
t	time
$\bar{t}_1, \bar{t}_2$	time of passage of centroid at Sites 1 and 2
t_s	time during which $S > S_c$
u_x, u_y, u_z	velocities in the x, y, z directions
u^*	local shear velocity
U^*	cross-sectional averaged shear velocity
v_x, v_z	depth-averaged velocities in the x, z directions
V_s	velocity measured using surface floats
V_x	cross-sectional averaged velocity
$\bar{V}_s$	average velocity of a number of drogues
$\bar{V}_x$	reach-averaged velocity
x, y, z	longitudinal, vertical and transverse distances
Δx_c	length of channel in which $S > S_c$
x_1, x_2	locations of Sites 1 and 2
$\bar{z}$	centroid of a tracer cloud
z_o	source z location
$z_{0.84}, z_{0.16}$	transverse locations at which $S_c = 0.84$ and 0.16
β	non-dimensional mixing distance coefficient
$\varphi(z)$	transverse variation in the dispersion coefficient
σ_t^2	temporal variance at a fixed site
σ_q^2	transverse variance with respect to cumulative flow
σ_x^2, σ_z^2	spatial variances in the x and z directions
σ_3^2	transverse variance of the longitudinal tracer flux
τ	dummy time variable of integration
ψ	shape factor

Chapter 6

Mixing Between Sediments and River Water

6.1 INTRODUCTION

River sediments are porous and consist of a mixture of solid particles surrounded by pore water. We are often interested in predicting the concentration of various chemicals in the pore water (e.g. dissolved oxygen, ammonia, hydrogen sulphide, heavy metals and radionuclides). The sediment surfaces contain sites where a number of important chemical and biochemical reactions occur which remove or add compounds to the pore water. The rate of mixing between the pore water and the overlying river water is fairly slow and consequently pore water concentrations differ significantly from concentrations in the overlying water. There is, however, some exchange of water between the bed and the overlying flow and consequently there is a net flux of tracer across the sediment–water interface. To predict pore water concentrations we need to quantify the rate of mass transfer between the bed and the overlying flow (termed the benthic mass transfer rate). If the pore water concentration is significantly different from the concentration in the overlying flow then benthic mass transfer causes changes in river water quality. Thus to predict river water quality we need to predict both pore water concentrations and the rates of benthic mass transfer. For a given benthic flux (units $g\,m^{-2}\,d^{-1}$) the rate of change of concentration in the overlying water ($g\,m^{-3}\,d^{-1}$) varies inversely with the water depth and consequently benthic processes strongly influence water quality in shallow rivers (Wuhrmann 1972).

One problem which requires an assessment of benthic mass transfer is the estimation of residence time following an accidental tracer spill. If the bed is impermeable then the residence time is affected by the balance between turbulent mixing and velocity variations within the cross-section. If the bed is porous, some tracer diffuses into the sediments where downstream advection is orders of magnitude slower than in the overlying river water. Tracer remains in the bed for some time after tracer in the overlying flow has been carried downstream. Subsequently it diffuses back out of the bed and contributes to the long tails observed in longitudinal dispersion experiments, as discussed in Chapter 4. If the tracer is more dense than river water and the bed is highly porous then a substantial amount may enter the sediments, as is discussed by Thibodeaux (1979).

A second problem concerns chemical reactions within the bed. Sediment particles contain many adsorption sites for ions and the river bed can remove (either permanently or temporarily) pollutants such as heavy metals, pesticides and radionuclides. The fate of these pollutants depends on the rates of benthic mass transfer, diffusion within the sediment, adsorption, desorption and chemical transformation. There is widespread interest in this problem (see, for example, O'Connor *et al* 1983, Bencala 1984, Jackman *et al* 1984, O'Connor 1988, Cerling *et al* 1990).

A third problem relates to organic pollution and dissolved oxygen depletion. Wuhrmann (1972) showed that in shallow rivers benthic metabolism greatly exceeds aquatic metabolism. Benthic mass transfer supplies organic matter and dissolved oxygen to microbial communities in the sediment and carries away respiratory products. Microbes derive energy by oxidising carbon and a significant change occurs in sediment chemistry when the rate of respiration exceeds the rate of supply of oxygen. When oxygen is present the by-products of respiration are carbon dioxide and water, which are harmless and inoffensive. Under anaerobic conditions microbes reduce nitrate, sulphate and carbon dioxide and in so doing produce ammonia, hydrogen sulphide and methane, which are offensive smelling and can be toxic. Thus the balance between oxygen supply and consumption rates determines whether the sediments are aerobic or anaerobic and hence affects the types of organisms inhabiting the sediments. Porous gravels in upland streams which receive low organic loadings tend to be well aerated and support a diversity of macro-invertebrates, whereas fine sediments in lowland streams which receive high organic loadings tend to have low dissolved oxygen concentrations and only support pollution tolerant species.

In this chapter we concentrate on the physical aspects of benthic mass transfer. The chemical and biological processes which occur in river sediments are numerous, complex and beyond the scope of this book. The interested reader should consult texts such as Berner (1980) and Thibodeaux (1979). There are four physical processes which affect benthic mass transfer in rivers: molecular diffusion, turbulent dispersion, pumping and sediment transport. The use of Fickian models for molecular diffusion and turbulent dispersion in river sediments is discussed in Section 6.2. It is shown that the turbulent dispersion coefficient must be determined experimentally and varies with depth so that the model equations must be solved numerically. Section 6.3 discusses the pumping of water through sediments which results from uneven pressure distributions on the bed and a simplified model is given for calculating the benthic oxygen flux in a polluted stream. In Section 6.4 the effects of saltation, dune turnover and redeposition are discussed.

6.2 FICKIAN MODELS FOR BENTHIC MASS TRANSFER

6.2.1 Molecular Diffusion

In stagnant water Fick's law states that tracer diffuses from a region of high concentration to a region of low concentration at a rate proportional to the concentration gradient (see Chapter 1). If tracer concentration varies only in the vertical direction then the tracer flux and the rate of change of tracer concentration are given by

$$J_y = e_m \frac{\partial c}{\partial y} \tag{6.1a}$$

$$\frac{\partial c}{\partial t} = \frac{\partial}{\partial y}\left(e_m \frac{\partial c}{\partial y}\right) \tag{6.1b}$$

where y = vertical distance; J_y = vertical tracer flux resulting from molecular diffusion; c = tracer concentration; t = time; and e_m = molecular diffusion coefficient.

Consider now the situation where stagnant water overlies a layer of sediment and because of chemical or biological removal, concentrations are lower in the pore water than in the overlying water (Figure 6.1). There is a vertical concentration gradient and because of random movement of water molecules, tracer will diffuse from the overlying water through the laminar boundary layer (described in the following), across the sediment–water interface and then downwards through the sediment. Fick's law can be used to quantify the

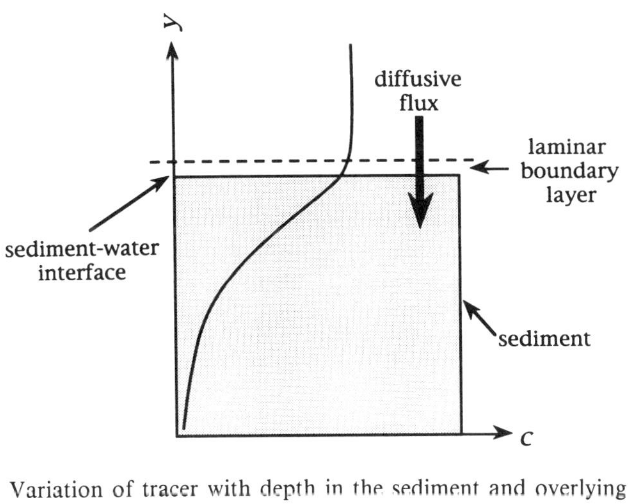

Figure 6.1. Variation of tracer with depth in the sediment and overlying water

diffusive flux within the sediment but it must be modified to take account of two factors: the porosity and tortuosity of the sediments.

Diffusion only occurs in the pore water which fills the void spaces between sediment particles. The sediment porosity θ is defined as the ratio of the void volume to total sediment volume. Equation (6.1) relates to the pore water and dealing with the sediment as a whole (voids plus sediment) we can write

$$J_y = \theta e_\mathrm{m} \frac{\partial c}{\partial y} \tag{6.2a}$$

$$\frac{\partial c}{\partial t} = \frac{1}{\theta} \frac{\partial}{\partial y} \left(\theta e_\mathrm{m} \frac{\partial c}{\partial y} \right) \tag{6.2b}$$

where c = tracer concentration in the pore water and θ = porosity.

Tracer must follow a tortuous path around the sediment grains as shown in Figure 6.2. The tortuosity ζ is defined

$$\zeta \approx \frac{\Delta y'}{\Delta y} \tag{6.3}$$

where Δy = net displacement in the y direction, $\Delta y'$ = total distance travelled by the particle (i.e. the path length). If we adopted a coordinate system in which distance y' was measured along the tortuous paths followed by the

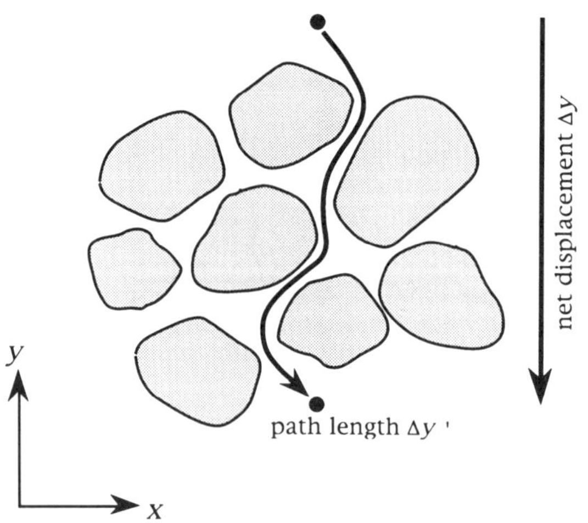

Figure 6.2. Tortuous path followed by a tracer particle in the bed. In this example the net x displacement is zero and tortuosity is defined as $\Delta y'/\Delta y$

tracer then it would be valid to rewrite equation (6.2b)

$$\frac{\partial c}{\partial t} = \frac{1}{\theta} \frac{\partial}{\partial y'} \left(\theta e_m \frac{\partial c}{\partial y'} \right) \tag{6.4}$$

Now when Δy and $\Delta y'$ become infinitely small equation (6.3) becomes

$$\frac{\partial}{\partial y'} = \frac{1}{\zeta} \frac{\partial}{\partial y} \tag{6.5}$$

Substituting equation (6.5) in (6.4) and adding a source–sink term yields

$$\frac{\partial c}{\partial t} = \frac{1}{\theta \zeta} \frac{\partial}{\partial y} \left(\frac{\theta e_m}{\zeta} \frac{\partial c}{\partial y} \right) + r \tag{6.6}$$

where r = rate of change of pore water concentration as a result of chemical or biological reactions. If we assume that ζ, θ and e_m are constant then equation (6.6) simplifies to

$$\frac{\partial c}{\partial t} = e_s \frac{\partial^2 c}{\partial y^2} + r \tag{6.7}$$

where e_s = effective diffusivity within the sediment

$$e_s = \frac{e_m}{\zeta^2} \tag{6.8}$$

As $\zeta > 1$ the effective diffusivity is lower in sediments than in say straight capillary tubes.

6.6.2 Steady-state Model with First-order Kinetics

Consider the situation where biological and chemical processes remove or transform tracer at a rate proportional to the concentration (termed first-order kinetics)

$$r = -\alpha c \tag{6.9}$$

where α = first-order removal rate coefficient. If a steady-state has been reached then equation (6.7) becomes

$$e_s \frac{\partial^2 c}{\partial y^2} - \alpha c = 0 \tag{6.10}$$

To solve equation (6.10) we need two boundary conditions. If the concentration at the sediment–water interface is known and is constant then

$$c(0) = c^* \tag{6.11}$$

There are two other possible boundary conditions depending on whether tracer penetrates to the bottom of the porous sediment layer or not. If the sediment layer is very thick and the only source of tracer is diffusion from the overlying water, then deep within the sediment the solute concentration is zero and the appropriate boundary condition is

$$c(\infty) = 0 \tag{6.12}$$

The solution to equation (6.10) for the boundary conditions (6.11) and (6.12) is

$$c(y) = c^* \exp\left(-\sqrt{\frac{\alpha}{e_s}}\, y\right) \tag{6.13}$$

The tracer flux into the bed at the sediment–water interface is then

$$J_y(0) = -\left[\theta e_s \frac{\partial c}{\partial y}\right]_{y=0} = \theta\sqrt{\alpha e_s}\; c^* \tag{6.14}$$

It can be noted that at steady-state the surface flux equals the total rate of removal of tracer within the sediments.

Consider next the situation where the porous sediment layer is thin and is underlain by an impermeable layer (e.g. bedrock). In this situation tracer penetrates to the bedrock where the flux must be zero. Thus in place of equation (6.12) we use

$$\left[\frac{\partial c}{\partial y}\right]_{y=L} = 0 \tag{6.15}$$

where L = thickness of the permeable layer (assumed known). The solution to equation (6.10) with boundary conditions (6.11) and (6.15) is

$$c(y) = c^* \frac{\sinh\left[\sqrt{\frac{\alpha}{e_s}}\,(y - L)\right]}{\sinh\left(\sqrt{\frac{\alpha}{e_s}}\, y\right)} \tag{6.16}$$

and the surface flux is

$$J_y(0) = \theta\sqrt{\alpha e_s}\, \tanh\left(\sqrt{\frac{\alpha}{e_s}}\, L\right) c^* \tag{6.17}$$

Note that in equations (6.14) and (6.17) the surface flux varies linearly with the tracer concentration in the overlying water. In these two cases benthic uptake results in solute removal from the overlying water, which also follows first-order kinetics. These solutions are well known from work on biofilms (Harremoes 1978, Characklis and Marshall 1990).

6.2.3 A Steady-state Model with Zero-order Kinetics

If solute removal follows zero-order kinetics then

$$e_s \frac{\partial^2 c}{\partial y^2} - \beta = 0 \tag{6.18}$$

where β = zero-order removal rate. Some care must be taken in solving equation (6.18). There is little difficulty in obtaining the solution when the sediment layer is thin and the boundary conditions are given by equations (6.11) and (6.15). Equation (6.18) can be integrated twice to yield a quadratic whose coefficients are deduced from the boundary conditions

$$c(y) = \frac{\beta}{2e_s} y^2 - \frac{\beta L}{e_s} y + c^* \tag{6.19}$$

for which the surface flux is

$$J_y(0) = \theta\beta L \tag{6.20}$$

Again the surface flux equals the total consumption within the bed. Equations (6.19) and (6.20) are only valid provided $c(L) > 0$ which requires that

$$\frac{2e_s c^*}{\beta L^2} > 1 \tag{6.21}$$

A slight difficulty arises when the solute concentration drops to zero within the bed. In an iterative numerical solution of this problem it would be easy to set the solute removal rate to zero once the solute concentration reached zero. It is not so easy to do this in the analytical solution: the zero-order removal term will continue to remove solute even when the solute concentration reaches zero and will happily drive the concentration negative. It is not sufficient merely to neglect any negative concentrations because the concentration gradients (and consequently the solute flux) may be seriously overestimated at the sediment–water interface. The solution is found as follows. At some (as yet unknown) critical depth L_c the solute concentration is zero. Below this point the solute concentration will, in reality, remain zero and it follows that the concentration gradient will be zero at $y = L_c$. Thus in addition to equation (6.11) there are two bottom boundary conditions

$$c(L_c) = 0 \tag{6.22}$$

and

$$\left[\frac{\partial c}{\partial y}\right]_{y=L_c} = 0 \tag{6.23}$$

There are now apparently three boundary conditions where only two were used previously, but note that the depth at which the solute concentration equals zero, L_c, is unknown and its estimation requires the third equation. The solution then follows

$$c(y) = \frac{\beta}{2e_s} y^2 - \sqrt{\frac{2\beta c^*}{e_s}}\, y + c^* \tag{6.24}$$

and the flux into the bed is

$$J_y(0) = \theta\sqrt{2\beta e_s c^*} \tag{6.25}$$

Note that equation (6.24) predicts positive concentrations and a positive concentration gradient in the region $y > L_c$ whereas in reality we would expect zero concentrations. It might be thought that this will result in a spurious diffusive flux upwards, but equation (6.25) ensures that at $y = L_c$ the diffusive flux is zero. Tracer concentrations in the region $y > L_c$ have no effect on concentrations in the region $0 < y < L_c$ and can be ignored. Again equations (6.24) and (6.25) are well known from biofilm work (Harremoes 1978).

6.2.4 Advection in the Overlying River Water

Consider now the situation in which the overlying river water is moving downstream but the flow remains laminar within the sediment. If the source is steady and tracer is being removed by the bed then tracer concentrations in the river water will decrease with distance below the source (i.e. a longitudinal concentration gradient will develop). In the sediment there will be a strong vertical concentration gradient as we have just discussed. If transverse variations are neglected then strictly we need a two-dimensional (x, y) model to describe sediment and water concentrations. Vertical concentration gradients within the sediments are, however, much larger than longitudinal gradients within either the overlying water or the sediment. In addition, vertical concentration gradients within the water column are negligibly small by comparison with vertical gradients within the sediments. This allows the equations governing tracer concentration in the sediment and the overlying water to be decoupled and simplified. For example, consider the problem of a steady source, first-order sediment kinetics and an infinitely thick river bed. We shall neglect longitudinal and transverse variations in sediment concentration but will allow the concentration in the overlying water to vary with x. Consequently the concentration at the sediment–water interface varies with x

$$c^* = c^*(x) \tag{6.26}$$

If the longitudinal variation in c^* is small compared with the vertical variation in c then we can adapt equation (6.13) and write

$$c(x, y) = c^*(x) \exp\left(-\sqrt{\frac{\alpha}{e_s}}\, y\right) \tag{6.27}$$

where $c(x, y)$ = concentration in the pore water and $c^*(x)$ = concentration in the river water.

Now we can also assume that tracer is fully mixed across the channel and can use the one-dimensional mass balance equation to predict $c^*(x)$

$$\frac{\partial c^*}{\partial t} + V_x \frac{\partial c^*}{\partial x} = K_x \frac{\partial^2 c^*}{\partial x^2} + r^* \tag{6.28}$$

where V_x = cross-sectional averaged longitudinal velocity; K_x = longitudinal dispersion coefficient (assumed constant); and r^* = rate of change of concentration in the river water resulting from benthic mass transfer. Equation (6.14) gives the flux into the bed and dividing by the mean water depth H

$$r^* = -\frac{\theta\sqrt{\alpha e_s}\, c^*(x)}{H} \tag{6.29}$$

For a steady tracer source we can drop the unsteady term, neglect longitudinal dispersion and rewrite equation (6.28) as

$$V_x \frac{\partial c^*}{\partial x} = -\frac{\theta\sqrt{\alpha e_s}}{H}\, c^* \tag{6.30}$$

If the river water concentration c_o^* is known just downstream from the source and the coefficients V_x, α, θ, H and e_s are constant then the solution to equation (6.30) is

$$c^*(x) = c_o^* \exp\left(-\frac{\theta\sqrt{\alpha e_s}}{V_x H}\, x\right) \tag{6.31}$$

Equation (6.31) indicates that the river water concentration decreases exponentially with distance (or equivalently with time of travel) downstream from a steady source. It is often observed that pollutant concentrations behave in this manner. The values of $c^*(x)$ from equation (6.31) can be substituted in (6.13) to give the tracer concentration profile in the sediments $c(x, y)$.

In the case of a thin permeable layer we can follow the same procedure and show that

$$c^*(x) = c_o^* \exp\left[-\frac{\theta\sqrt{\alpha e_s}}{V_x H} \tanh\left(\sqrt{\frac{\alpha}{e_s}}\, L\right) x\right] \tag{6.32}$$

If the bed kinetics are zero-order and the sediment layer is thin, then from equation (6.20) the surface flux does not vary with distance downstream (until such times as tracer no longer penetrates to the bottom of the sediment layer). Equation (6.30) must be rewritten

$$V_x \frac{\partial c^*}{\partial x} = -\frac{\theta \beta L}{H} \tag{6.33}$$

Table 6.1. Summary of the equations describing sediment concentration, benthic flux and water concentration downstream from a steady source in a uniform channel for zero-order and first-order removal kinetics in the bed

Kinetics		Thick	Thin
Zero order	Sediment concentration	$c(x,y)=\frac{\beta}{2e_s}y^2-\sqrt{\frac{2\beta c^*(x)}{e_s}}\,y+c^*(x)$	$c(x,y)=\frac{\beta}{2e_s}y^2-\frac{\beta L}{e_s}y+c^*(x)$
	Flux	$J_y(x,0)=\theta\sqrt{2\beta e_s c^*(x)}$	$J_y(x,0)=\theta\beta L$
	River concentration	$c^*(x)=\left(\sqrt{c_o^*}-\frac{\theta}{V_xH}\sqrt{\frac{\beta e_s}{2}}\,x\right)^2$	$c^*=c_o^*-\frac{\theta\beta L}{HV_x}x \quad \text{for} \quad x<\frac{c_o^*HV_x}{\theta\beta L}$
First order	Sediment concentration	$c(x,y)=c^*(x)\exp\left(-\sqrt{\frac{\alpha}{e_s}}\,y\right)$	$c(x,y)=c^*(x)\frac{\sinh\left[\sqrt{\frac{\alpha}{e_s}}(y-L)\right]}{\sinh\left(\sqrt{\frac{\alpha}{e_s}}\,y\right)}$
	Flux	$J_y(x,0)=\theta\sqrt{\alpha e_s}\,c^*(x)$	$J_y(x,0)=\theta\sqrt{\alpha e_s}\tanh\left(\sqrt{\frac{\alpha}{e_s}}L\right)c^*(x)$
	River concentration	$c^*(x)=c_o^*\exp\left(-\frac{\theta\sqrt{\alpha e_s}}{V_xH}x\right)$	$c^*(x)=c_o^*\exp\left[-\frac{\theta\sqrt{\alpha e_s}}{V_xH}\tanh\left(\sqrt{\frac{\alpha}{e_s}}L\right)x\right]$

and hence concentrations decrease linearly with distance

$$c^* = c_o^* - \frac{\theta \beta L}{H V_x} x \tag{6.34}$$

and drops to zero at

$$x = \frac{c_o^* H V_x}{\theta \beta L} \tag{6.35}$$

If bed kinetics are zero-order and the sediment layer is thick then equation (6.30) becomes

$$V_x \frac{\partial c^*}{\partial x} = -\frac{\theta \sqrt{2\beta e_s c^*}}{H} \tag{6.36}$$

which yields the solution

$$c^*(x) = \left(\sqrt{c_o^*} - \frac{\theta}{V_x H}\sqrt{\frac{\beta e_s}{2}}\, x\right)^2 \tag{6.37}$$

for the river water concentration. The value of c^* from equation (6.37) can be substituted in equation (6.24) to give the tracer profile in the sediments.

Table 6.1 summarises the equations which describe river concentration, sediment concentration and the mass flux across the sediment–water interface.

Example 6.1

The Tarawera River, New Zealand, receives treated pulp-mill effluent from a point source about 20 km from the sea. The river experiences a high de-oxygenation rate which has been shown to arise because of a high rate of microbial metabolism within the bed sediments (Rutherford *et al* 1991). We can calculate river biochemical oxygen demand (BOD) concentrations [BOD is a standard measure of biodegradable organic matter; see, for example, Velz (1980)] downstream from the source using the models outlined in Section 6.2.2. Channel and river bed characteristics are summarised in Table 6.2.

The oxygen uptake rate by river sediments has been measured in well stirred glass bottles (see Rutherford *et al* 1991 for details). These data give

Table 6.2. Channel and sediment characteristics of the Tarawera River below Kawerau

Mean depth (H)	1.30 m
Mean velocity (V_x)	0.76 m s^{-1}
Sediment properties	
Porosity (θ)	0.36
Relative density (ρ)	1.3
Oxygen uptake rate (β)	350 mg kg d^{-1}

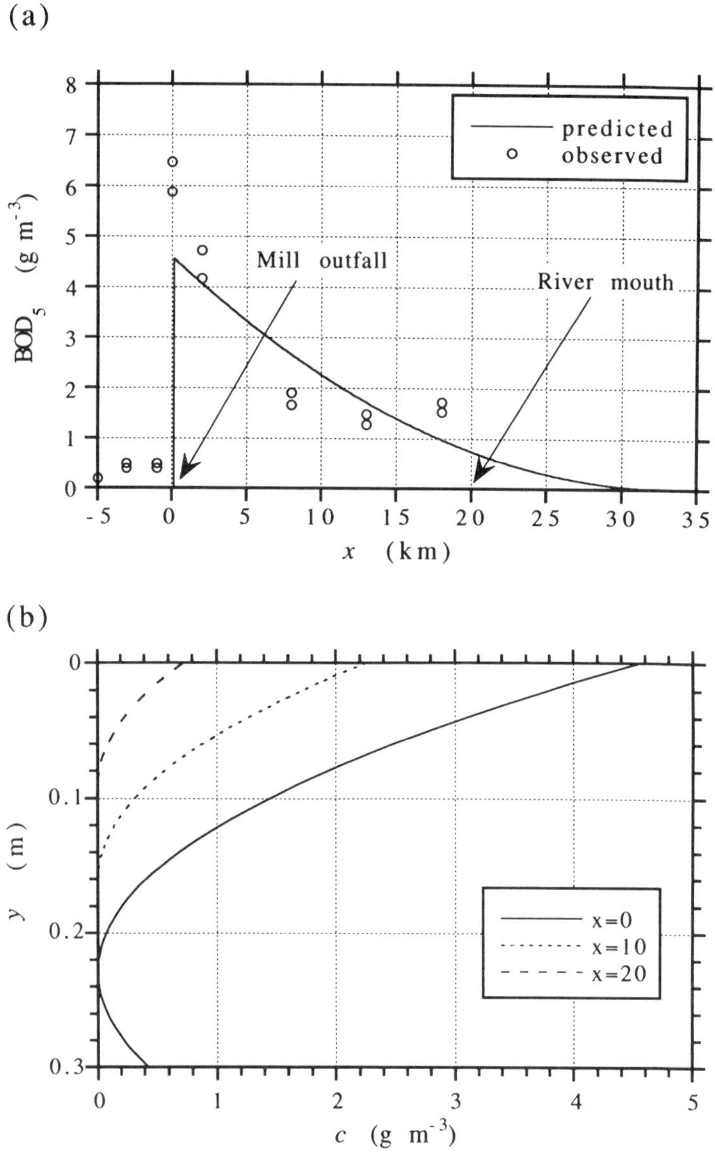

Figure 6.3. (a) Comparison of river water BOD concentrations predicted using equation (6.37) with concentrations measured on 7 March 1984. Note that close to the outfall the effluent may not be fully mixed. (b) BOD concentrations predicted using equation (6.24) within the sediment. Note that equation (6.24) predicts positive concentrations below $y = L_c$ (as shown for $x = 0$) but these concentrations have no effect on the benthic tracer flux and can be ignored (as shown for $x = 10$ and 20 km)

estimates of β (the zero-order removal rate). Accordingly we use equations (6.24) and (6.37) to predict sediment and river water concentrations. We shall assume that BOD transport into the bed can be modelled as turbulent dispersion and that, as a first approximation, the dispersion coefficient e_s is constant (this may not always be a good assumption as we will discuss shortly). Before we can solve equations (6.24) and (6.37) we must estimate the value of e_s.

A recirculating benthic chamber has been used to measure the benthic oxygen uptake rate in the Tarawera River (Hickey 1988). The oxygen flux into the sediments was found to be of the order of 30 $\mathrm{g\,m^{-2}\,d^{-1}}$ when the oxygen concentration was 7.5 $\mathrm{g\,m^{-3}}$. In equation (6.25) we set $J_y(0) = 30\ \mathrm{g\,m^{-2}\,d^{-1}}$, $\theta = 0.36$, $\beta = 350\ \mathrm{mg\,kg^{-1}\,d^{-1}}$ ($= 300\ \mathrm{g\,m^{-3}\,d^{-1}}$ given $\theta = 0.36$ and $\rho = 1.3$) and $c^* = 7.5\ \mathrm{g\,m^{-3}}$. This gives the dispersion coefficient $e_s = 2 \times 10^{-5}\ \mathrm{m^2\,s^{-1}}$. By comparison the molecular diffusivity in water is $2 \times 10^{-9}\ \mathrm{m^2\,s^{-1}}$ and in the sediment is likely to be even smaller, as shown in equation (6.8).

Substituting these coefficients into equation (6.37) we can deduce the river water BOD profile shown in Figure 6.3a. This analysis predicts that the BOD concentrations drop to zero about 30 km downstream from the outfall (i.e. there is still a significant BOD concentration at the river mouth), which is in general accord with measured BOD profiles.

Figure 6.3b shows the BOD concentrations within the sediment predicted using equation (6.24). Note that this equation predicts positive concentrations at depths greater than L_c but as discussed in the text these concentrations can safely be ignored. The predictions indicate that turbulent dispersion transports BOD about 20 cm into the sediment at $x = 0$ and about 10 cm at $x = 20$ km. This is in reasonable agreement with measurements which have been made of the depth of the oxic layer in the Tarawera River (Rutherford *et al* 1993). These calculations are rather simplistic because they ignore the fact that sediment properties vary along the length of the channel (which presumably affects the value of e_s) and they assume that e_s does not vary with depth (this point is discussed later). These calculations also assume that the sediments are BOD limited (i.e. that there is ample oxygen present in the sediments) which is not unreasonable given that there is evidence of oxygen penetration to depths of 10–20 cm (Rutherford *et al.* 1993).

6.3 ADVECTION WITHIN THE SEDIMENT

6.3.1 Effects of Channel Slope

In rivers the bed is sloping and gravity exerts a force on the pore water. This induces a laminas flow through the bed which has an average velocity of

$$w_d = sK \tag{6.38}$$

where w_d = Darcy velocity in the sediment; s = bed slope and K = hydraulic conductivity of the sediments. The Darcy velocity is defined as the flow per unit total area (voids plus particles). The average pore water velocity w_p is larger than the Darcy velocity

$$w_p = \frac{w_d}{\theta} \tag{6.39}$$

where θ = porosity. The hydraulic conductivity is a property of the sediment which depends on the particle size, sorting and degree of compaction. The value of K can be measured in the laboratory using apparatus such as that shown in Figure 6.4. The flow rate through a sediment-filled column is measured holding the water level in the reservoir constant. There is a pressure drop through the sediment column and it is found that the flow rate (and hence the Darcy velocity) varies linearly with the pressure gradient. Thus

$$w_d \propto \frac{\Delta h}{\Delta y} \tag{6.40}$$

where Δh = pressure head and Δy = length of the column of sediment. The

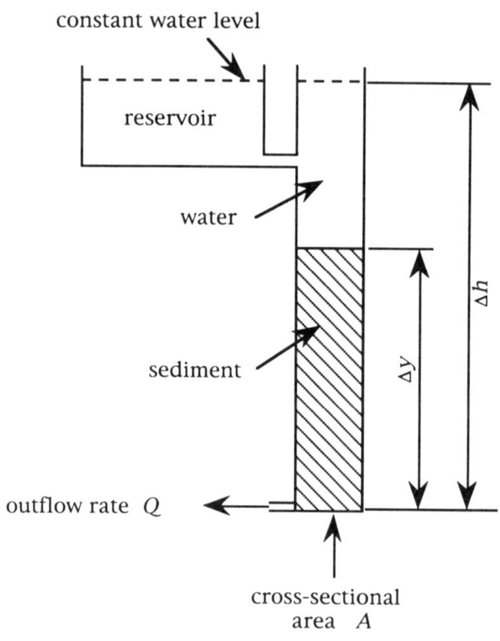

Figure 6.4. Experimental apparatus for measuring the hydraulic conductivity of sediments. The water level is held constant giving a pressure drop of Δh along a sediment column of length Δy. The Darcy velocity is Q/A

constant of proportionality is the hydraulic conductivity K. For the set-up shown in Figure 6.4

$$K = \frac{\Delta y}{\Delta h} \frac{Q}{A} \tag{6.41}$$

where Q = measured outflow rate and A = cross-sectional area of the sediment column.

When the river bed is flat the flow induced in the sediments by gravity [see equation (6.38)] is parallel with the bed and has little effect on benthic mass transfer. It is only when there is a significant flow vertically into and out of the bed that the rate of benthic mass transfer becomes high.

6.3.2 Turbulent Flow Over a Rough Bed

In most river channels the flow is turbulent (in Chapter 1 there is a discussion of laminar, transitional and turbulent flow) and the Reynolds number (the ratio of inertial to viscous forces) is large

$$Re = \frac{\rho u L}{\mu} \tag{6.42}$$

in which ρ = density of water; μ = dynamic viscosity; u = an appropriate velocity; and L = a length scale. When the river flow is viewed as a whole it is appropriate to set u to the cross-sectional averaged velocity V_x and L to the hydraulic radius R (which is approximately equal to the depth H). Experimental results indicate that open channel flow is laminar if this bulk Reynolds number is less than 500, is transitional if it is between 500 and 2000 and is fully turbulent if it is greater than 2000 (Simons and Senturk 1977).

A high bulk Reynolds number implies that the flow is turbulent over most of the channel. As we approach the bed, however, the velocity decreases because of the no-slip condition which applies at all solid boundaries. In addition, close to the bed the appropriate length scale becomes the distance from the bed rather than the total water depth. Equation (6.42) predicts that very close to the bed the Reynolds number is small, which implies that there is a region very close to the bed where viscous forces dominate inertial forces and the flow is laminar. This region is called either the viscous sub-layer or the laminar boundary layer. The thickness of the viscous sub-layer is denoted δ and is typically 10^{-6}–10^{-4} m.

Within the viscous sub-layer the flow is laminar and so

$$\tau = \mu \frac{du_x}{dy} \tag{6.43}$$

where $u_x(y)$ = longitudinal velocity; y = vertical distance; and τ = viscous shear stress. Because the viscous sub-layer is very thin it can be assumed that the shear stress is constant and equal to the bed shear stress. This allows equation (6.43) to be integrated to give

$$u_x(y) = \frac{\tau_o}{\mu}\, y \tag{6.44}$$

where τ_o = bed shear stress. Thus in the viscous sub-layer the velocity varies linearly with distance from the bed (Figure 6.6). The shear velocity is defined

$$u^* = \sqrt{\frac{\tau_o}{\rho}} \tag{6.45}$$

and hence equation (6.44) can be rewritten

$$u_x(y) = \frac{u^{*2}}{\nu}\, y \qquad 0 < y < \delta \tag{6.46}$$

where $\nu = \mu/\rho$ (kinematic viscosity); u^* = shear velocity; and δ = thickness of the viscous sub-layer.

In Chapter 2 (Section 2.3.2) it was shown that for two-dimensional turbulent flow the velocity varies logarithmically with depth. Thus

$$\frac{u_x(y)}{u^*} = \frac{1}{\varkappa} \log_e\left(\frac{y}{y_o}\right) \tag{2.13}$$

where y = distance from the bed; u^* = shear velocity; $\varkappa$ = von Karman's constant; and y_o = constant. Taken literally y_o is the distance from the bed at which $u_x = 0$. Equation (2.13) is derived using Prandtl's mixing length hypothesis for turbulent flow and is not valid in the viscous sub-layer. In many river channels $y_o < \delta$ and thus it is better to view y_o simply as a constant of integration.

The values of y_o and $\varkappa$ can be estimated by fitting equation (2.13) to measured vertical velocity profiles. It is found that $\varkappa$ is constant provided tracer concentrations are low (typically $\varkappa = 0.4$). The value of y_o, however, varies depending on the nature of the boundary roughness and the flow. Boundary roughness can be quantified by the roughness height y_r, which is the average height of the protrusions above the bed. In his classical experimental study Nikuradse roughened the walls of pipes with sand (see Schlichting 1968) and defined $y_r = d$ where d is the sand particle diameter. For a non-uniform bed material it is more usual to take $y_r = d_{65}$, the particle diameter exceeded by 65% of the sample by weight.

If the boundary has only small irregularities (e.g. a concrete lined channel) and the shear velocity is low, then the viscous sub-layer extends above the roughness elements (Figure 6.5) and the boundary is termed hydraulically smooth. This situation occurs when

$$0 < \frac{y_r u^*}{\nu} < 5 \tag{6.47}$$

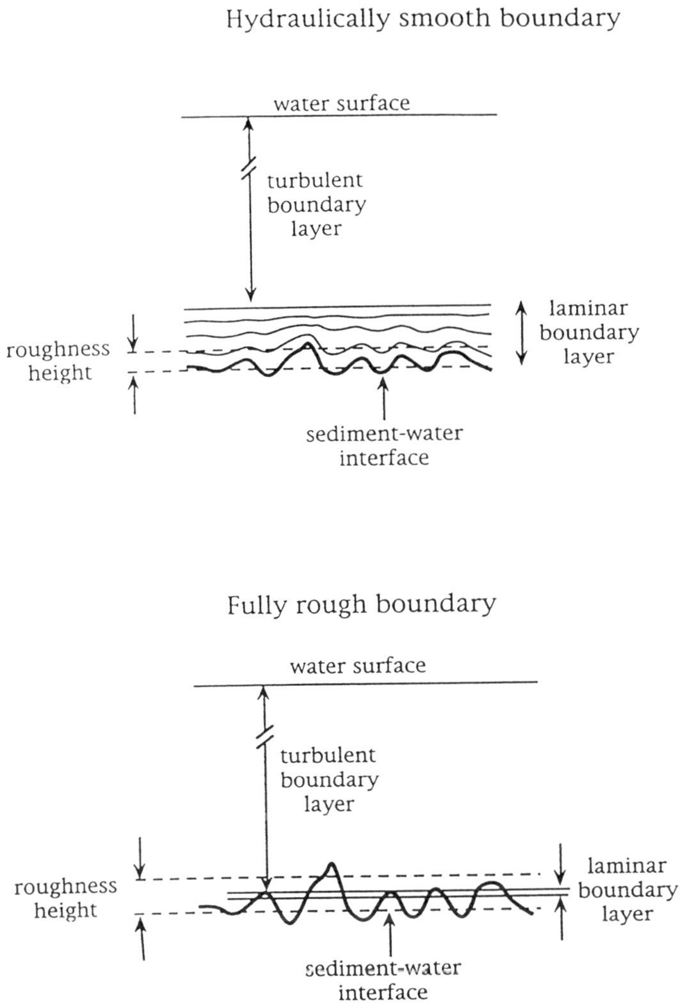

Figure 6.5. Hydraulically smooth and rough boundaries. In the case of a rough boundary sediment particles protrude through the viscous sub-layer (laminar boundary layer) into the turbulent boundary layer

For most alluvial channels

$$\frac{y_r u^*}{\nu} > 70 \tag{6.48}$$

and in this situation the roughness elements protrude through the viscous sub-layer into the turbulent boundary layer (Figure 6.5). The boundary is then termed fully rough. Nikuradse found that in fully rough pipe flow

$$y_o \approx \frac{y_r}{30} \tag{6.49}$$

and

$$\delta \approx 5y_o \approx \frac{y_r}{6} \tag{6.50}$$

These relationships give an indication of the thickness of the viscous sub-layer in natural river channels. Figure 6.6 shows the velocity distribution which is predicted by equations (2.13) and (6.46) for a fully rough boundary in a gravel bed stream. The roughness height is taken as $y_r = 5$ mm and the mean depth $H = 1$ m. From equation (6.49) $y_o = 0.17$ mm and the thickness of the viscous sub-layer (laminar boundary layer) $\delta = 0.83$ mm, which is substantially smaller than the roughness height.

There is a close analogy between momentum and mass transfer. In the main body of the flow both are transported vertically by turbulent eddies and for dilute tracer solutions the rates of momentum and mass transfer are similar (Reynolds analogy). As we approach the bed, however, turbulent eddies are affected by the presence of the solid boundary and the turbulent length scale decreases. This behaviour is clear in equation (2.18), which shows that the vertical diffusivity varies parabolically with depth and vanishes at the bed and

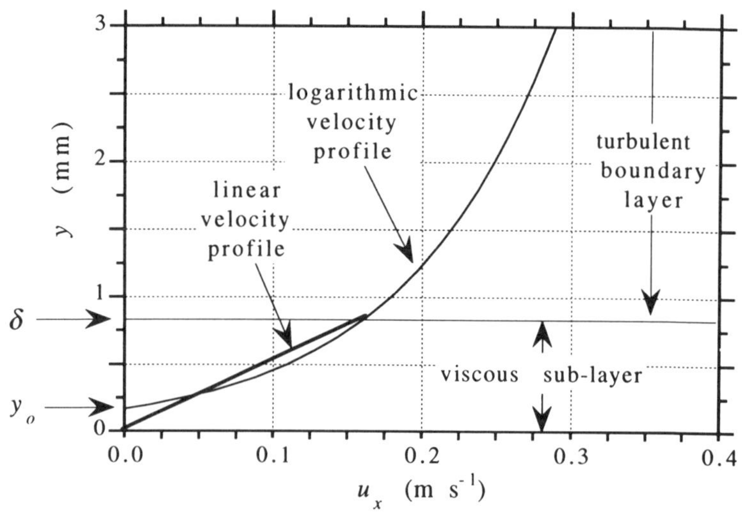

Figure 6.6. Vertical variation of mean velocity near the bed. The logarithmic velocity profile is calculated from equation (2.13). Parameters are $H = 1$ m, $u^* = 4$ cm s^{-1}, $y_r = 5$ mm, $y_o = 0.17$ mm and $\delta = 0.83$ mm

the water surface. Just as there is a thin viscous sub-layer close to the bed in which momentum transfer is effected primarily by viscosity, there is also a thin sub-layer in which mass transfer is effected predominantly by molecular diffusion. This layer is termed the diffusive sub-layer. For dilute tracer solutions the diffusive sub-layer is comparable in thickness with the viscous sub-layer.

Example 6.2

To illustrate the effect which the diffusive sub-layer has on benthic mass transfer, we need to estimate the time-scales of vertical diffusion in the overlying flow and the diffusive sub-layer.

Consider the hypothetical channel illustrated in Figure 6.6 in which the depth is $h = 1$ m and the shear velocity is $u^* = 0.04\ \text{m s}^{-1}$. From equation (2.19), the depth-averaged vertical diffusivity in the water column is $e_y = 0.067hu^* = 0.00267\ \text{m}^2\,\text{s}^{-1}$. Now the time-scale for vertical diffusion is $h^2/e_y = 6.3$ minutes.

From Figure 6.6 the thickness of the viscous sub-layer is $\delta = 0.83$ mm and we can assume that the diffusive sub-layer has much the same thickness. We can also assume that in the diffusive sub-layer the molecular diffusion coefficient is $e_m = 2 \times 10^{-9}\ \text{m}^2\,\text{s}^{-1}$. The time-scale for diffusion in the diffusive sub-layer is then $\delta^2/e_m = 5.7$ minutes.

In this example the time scales of diffusion are comparable in the water column and the diffusive sub-layer (even though the latter is very thin).

6.3.3 Turbulent Flow Over a Permeable Bed

In an alluvial river the bed is not a continuous solid boundary but consists of a series of discrete sediment particles with pore water filling the gaps. As water flows over the bed there is some transfer of momentum from the overlying flow to the pore water as a result of molecular viscosity and turbulent eddies. This induces a pore water flow in the downstream direction. As this flow arises as a result of shear stress at the bed it can be termed the shear-induced pore water flow (or pore water shear flow for short). Just above the sediment–water interface the eddy viscosity and turbulent shear stress are high, but within the sediment the turbulent length scale is small, viscous forces are high and turbulence is quickly damped. Consequently pore water shear flow is only significant in the surface sediment layer. The shear stress acts parallel to the bed but any water flowing downstream in the surface sediments must weave around sediment grains and consequently there is some mass transfer vertically across the sediment–water interface.

In most alluvial rivers sediment grains protrude through the viscous sub-layer into the overlying turbulent boundary layer (as shown in Figure 6.5). Water actually in contact with a sediment particle is at rest (because of the no-slip condition) and so there is a viscous sub-layer around each particle. The particles which protrude into the turbulent boundary layer shed eddies, some of which are advected downwards and impinge on the bed. In addition, bed features such as bends, sand dunes and obstacles generate large eddies which periodically impinge on the bed. These eddies cause low frequency turbulent pressure fluctuations on the bed which induce an intermittent flow into and out of the sediment (Heathershaw 1974). The resulting intermittent flows have a significant vertical component and cause mass transfer across the sediment–water interface.

The mass transfer which results from pore water shear flow and eddies impinging on the bed can collectively be termed turbulent dispersion as it results from small-scale advective flows into and out of the bed which arise as a result of turbulence in the overlying flow. Turbulence is rapidly damped within the sediment and consequently we would expect turbulent dispersion to be important only in the top few layers of sediment particles. This has been confirmed by Thibodeaux and Boyle (1987) who conducted laboratory experiments using gravel and found that there was significant turbulence in the sediment to a depth of about eight grain diameters. Deep within the bed the flow is laminar and the rate of mass transfer is small.

It is not yet possible to make useful theoretical predictions about the characteristics of turbulence in the vicinity of the sediment–water interface or its effect on the rate of benthic mass transfer. The rate of benthic mass transfer will almost certainly depend on the concentration gradient in the sediment and we can postulate a Fickian model for turbulent dispersion within the sediment. We assume that the effects of pore water shear flow and intermittent turbulent flows into and out of the bed can be lumped together and described by a turbulent dispersion coefficient k_s. If we neglect transverse and longitudinal changes then tracer concentration in the bed is governed by

$$\frac{\partial c}{\partial t} = \frac{1}{\theta} \frac{\partial}{\partial y} \left(\theta k_s \frac{\partial c}{\partial y} \right) + r \tag{6.51}$$

in which $k_s(y)$ = effective turbulent dispersion coefficient in the sediments and $r(y)$ = tracer production or uptake rate within the sediment. The value of k_s must be determined experimentally and is likely to vary considerably depending on the nature of the bed sediments and the characteristics of the overlying flow. In the top layers of the bed we can expect the turbulent dispersion coefficient k_s to be much larger than the effective molecular diffusion coefficient e_s. The value of k_s is, however, dependent on the local flow conditions and as turbulence is damped out fairly rapidly within the sediment k_s decreases with depth. [As both k_s and r vary with depth, numerical methods must be used to solve equation (6.51)]. At present we must infer the depth distribution from experimental results. McBride (1986) found that

$$k_s(y) = k_o \left(\frac{k_L}{k_o}\right)^{\frac{y}{L}} \tag{6.52}$$

where k_o and k_L = dispersion coefficients at $y = 0$ and L, respectively ($k_L \ll k_o$) and L = depth of the sediment layer. Given experimental measurements of $c(y)$, $r(y)$, θ and L, the values of k_o and k_L are adjusted until observed solute concentration profiles match those predicted by solving equations (6.51) and (6.52) numerically. If a satisfactory match is not achieved then an alternative to equation (6.52) must be postulated and the calibration process repeated. This approach yields good results provided $r(y)$, θ and L are known and the correct functional form is postulated for $k_s(y)$.

Example 6.3

A field and modelling study (Cooper 1984, McBride 1986) has been made of the ammonia flux into the bed of the Waikare Stream, New Zealand. This stream is small (base flow 0.15–0.40 $m^3 s^{-1}$), has a bed of porous gravel and receives a high input of ammonia from an anaerobic lagoon treatment system for piggery waste. The two sources of ammonia in the bed are dispersion from the overlying water and mineralisation of organic nitrogen deposited in the sediments. Nitrifying bacteria are active in the oxic layer (the top 5 cm of sediment) where they oxidise ammonia to nitrate.

At steady state the ammonia concentration is governed by the equation

$$\frac{1}{\theta}\frac{\partial}{\partial y}\left(\theta k_s \frac{\partial c}{\partial y}\right) + \beta - \frac{\alpha c}{\varepsilon + c}$$

where c = pore water ammonia concentration; θ = sediment porosity; k_s = dispersion coefficient in the sediment; α = maximum nitrification rate; β = zero-order mineralisation rate; and ε = half-saturation constant for nitrification. The values of α, β, ε and θ are known from laboratory and field studies (Cooper 1984). The last term in this equation is set to zero below the oxic depth (measured as 5 cm). The depth variation of k_s was described by equation (6.52) where $L = 6$ cm. The objective of the model calibration was to estimate values of k_o and k_L.

At the bottom boundary ($y = L$) the ammonia concentration was known from field studies, $c(L) = 6\ g\,m^{-3}$. At the sediment–water interface two alternative boundary conditions were investigated. Firstly, a concentration (or Dirichlet) boundary condition was used and the measured stream concentration $c(0) = 2.5\ g\,m^{-3}$ was specified. Secondly, a flux (or Neumann) boundary condition was used. The ammonia flux into the sediments was specified to match the flux measured in the field using a chamber

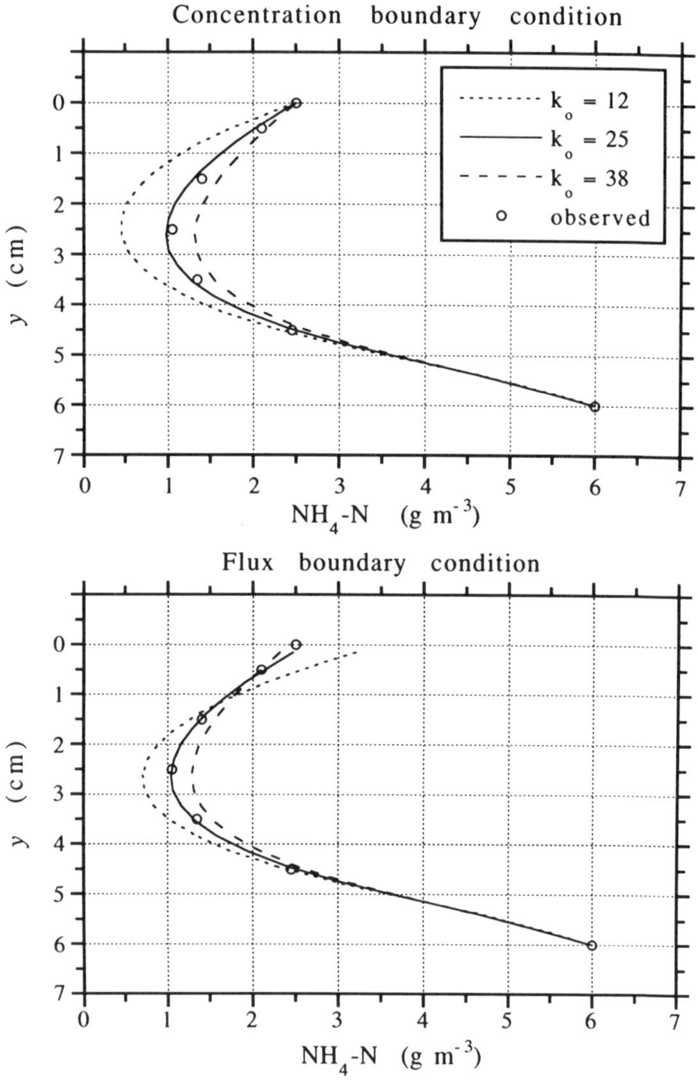

Figure 6.7. Comparison of observed and predicted ammonia concentrations in the top 6 cm of the Waikare Stream. Predictions were made using $L = 6$ cm, $\theta = 0.7$, $\alpha = 8\ \mathrm{g\,m^{-3}\,d^{-1}}$, $\beta = 18\ \mathrm{g\,m^{-3}\,d^{-1}}$, $\varepsilon = 0.14\ \mathrm{g\,m^{-3}}$ and $k_L = 7 \times 10^{-9}\ \mathrm{m^2\,s^{-1}}$

$J_y(0) = 0.155\ \mathrm{g\,m^{-2}\,d^{-1}}$. Numerical methods were used to solve the model equation, full details of which are given in (McBride 1986).

Figure 6.7 compares observed and predicted ammonia concentrations for three values of k_o with the two boundary conditions. These predictions are made with $k_L = 7 \times 10^{-9}\ \mathrm{m^2\,s^{-1}}$, which was found by trial and error to give the best fit (details omitted here). It can be noted that this value of k_L is three times the molecular diffusion coefficient of ammonia in water ($2 \times 10^{-9}\ \mathrm{m^2\,s^{-1}}$) and suggests that even at a depth of 6 cm there is sufficient random pore water flow to cause appreciable dispersion. Figure 6.7 shows that an excellent match between observed and predicted ammonia concentrations is achieved using a value of $k_o = 25 \times 10^{-9}\ \mathrm{m^2\,s^{-1}}$, which is about 12 times the molecular diffusion coefficient. It can be inferred that equation (6.52) provides an adequate description of the depth variation of k_s. It can be noted that an equally good fit is achieved using the concentration and flux boundary conditions. Predictions near the sediment–water interface are most sensitive to the flux boundary condition and deeper within the sediment are most sensitive to the concentration boundary condition.

6.4 PUMPING

6.4.1 Introduction

Dunes and other irregularities in the channel cause an uneven pressure distribution on the bed. If the bed sediments are porous then this induces a flow through the sediments, a process which can be termed pumping. Pumping velocities are smaller than the mean river velocity, typically by a factor of 10–1000 in permeable gravels (Thibodeaux and Boyle 1987) and by a higher factor in less permeable sands (Elliott 1991, Rutherford *et al* 1993). Pumping flows do, however, penetrate fairly deeply into the sediments, typically to depths several times the dune height. This is in marked contrast with the effects of turbulent dispersion (discussed in Section 6.3) which are confined to within a few grain diameters of the sediment–water interface (Thibodeaux and Boyle 1987). Thus pumping flows have the potential to enhance significantly benthic mass transfer, especially in porous gravels, and are of considerable interest in stream water quality studies.

6.4.2 Simplified Model for Pumping Flows

Elliott (1991) developed an approximate analytical model for pumping flows. The model assumes that the sediments are homogeneous, the bed is flat and transversely uniform, and longitudinal flows within the bed dominate transverse flows. Flow in the sediment is modelled in two dimensions using Darcy's law

$$w_x = -K \frac{\partial h}{\partial x}$$
$$w_y = -K \frac{\partial h}{\partial y} \tag{6.53}$$

where w_x and w_y = Darcy velocities in the sediments; h = pressure head; and K = hydraulic conductivity. Equation (6.53) is valid provided flow within the sediments is laminar, which requires the particle Reynolds number to be less than about 10 (Bear 1972). The particle Reynolds number is defined

$$Re_p = \frac{\rho w_d d}{\mu} \tag{6.54}$$

where d = particle diameter; ρ = fluid density; and μ = fluid viscosity.

The bed is assumed to be flat but a sinusoidal pressure head is specified at the sediment–water interface which mimics the pressure distribution observed on laboratory dunes

$$h(x, 0) = h_{max} \sin(\gamma x) \tag{6.55}$$

where h_{max} = half-amplitude of the measured pressure head distribution, $\gamma = 2\pi/\lambda$ and λ = dune wavelength. Felham (1985) measured the pressure head distribution over laboratory dunes and found that

$$h_{max} = 0.28 \frac{V_x^2}{2g}\left(\frac{Y_d}{0.34H}\right)^{3/8} \quad \text{if } \frac{Y_d}{H} < 0.34$$
$$h_{max} = 0.28 \frac{V_x^2}{2g}\left(\frac{Y_d}{0.34H}\right)^{3/2} \quad \text{if } \frac{Y_d}{H} > 0.34 \tag{6.56}$$

where V_x = mean velocity of the overlying flow; Y_d = dune height (crest elevation minus trough elevation); and H = water depth. The solution to equations (6.53) and (6.55) is

$$h(x, y) = h_{max} \sin(\gamma x) \exp(-\gamma y)$$
$$w_x(x, y) = -w_{max} \cos(\gamma x) \exp(-\gamma y)$$
$$w_y(x, y) = w_{max} \sin(\gamma x) \exp(-\gamma y) \tag{6.57}$$
$$w_{max} = \gamma K h_{max} = \frac{2\pi K h_{max}}{\lambda}$$

If a particle enters the bed at time $t = 0$ at a position $x = x_o$ (where $\lambda/4 < x_o < \lambda/2$) then it re-emerges from the bed after a time

$$t_e = \frac{\theta(2x_o - \lambda)}{w_{max} \cos(\gamma x_o)} \tag{6.58}$$

at a position

$$x_e = \lambda - x_o \tag{6.59}$$

At times when $t < t_e$ the particle is at

$$\begin{aligned} x &= x_o - \frac{w_{max} t}{\theta} \cos(\gamma x_o) \\ y &= \frac{1}{\gamma} \log_e \left[\frac{\cos(\gamma x)}{\cos(\gamma x_o)} \right] \end{aligned} \tag{6.60}$$

Figure 6.8 shows the particle paths predicted by equations (6.60) for a flat

(a)

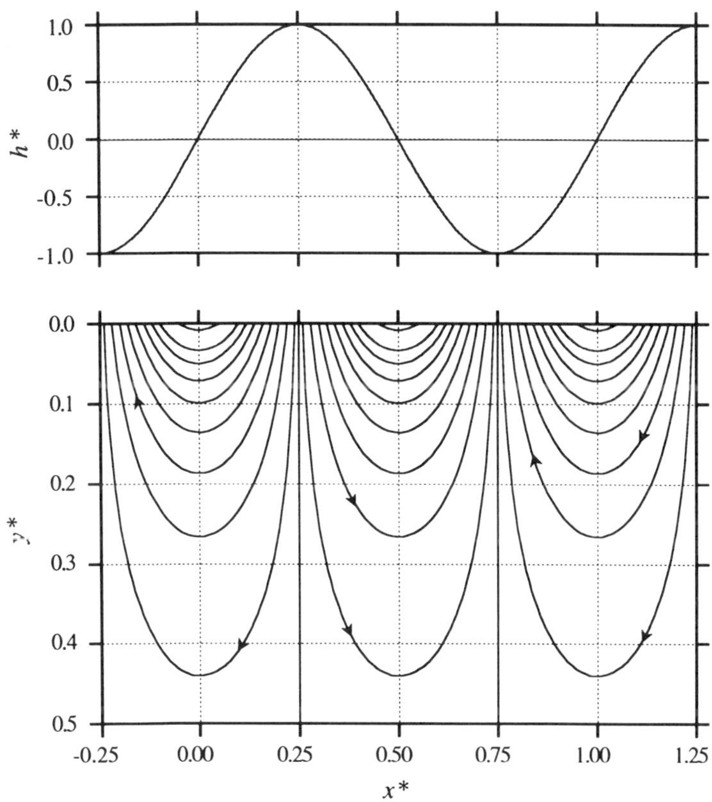

Figure 6.8. (a) Particle paths within the sediments predicted using Elliott's analytical model. The river bed is assumed flat and a sinusoidal pressure head distribution is specified at the sediment–water interface to mimic the effects of regular bedforms. (b) Particle paths within the sediments predicted using a numerical model. The bedforms are assumed triangular and the pressure head distribution is specified based on Felham's laboratory results. $x^* - x/\lambda$, $y^* - y/2\pi\lambda$, $h^* - h/h_{max}$

(b)

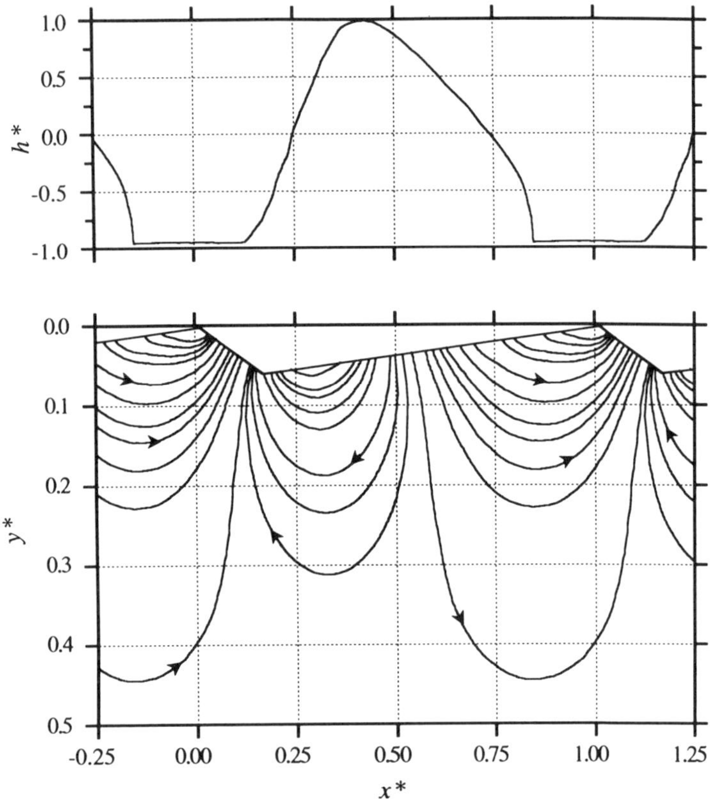

Figure 6.8. (*Continued*)

bed and a sinusoidal pressure distribution. Also shown are the streamlines predicted using a boundary element numerical model for the more realistic case of a triangular dune with a measured pressure distribution (Hatherell 1992). Both sets of results have been scaled so they can be compared directly. The patterns of streamlines predicted by the two models are similar and match experimental results fairly closely (Savant *et al* 1987, Thibodeaux and Boyle 1987). Thus the simple analytical model provides a fairly good approximation to the streamlines found in a homogeneous river sediment with regular dunes.

6.4.3 Simplified Model for Benthic Mass Transfer

A simplified model can be used to quantify the role of pumping on benthic mass transfer. The model has so far been applied only to benthic oxygen uptake, but has the potential to address other problems. The model makes four assumptions. Firstly, the rate of oxygen uptake within the bed is either constant (zero-order kinetics) or proportional to the biochemical oxygen demand (BOD) concentration (first-order kinetics). Secondly, the rate coefficients of BOD exertion within the bed (zero or first-order) are constant. Thirdly, the BOD and dissolved oxygen (DO) concentrations in the overlying water are assumed to be known *a priori*. Fourthly, the overlying water is assumed to be substantially free from suspended solids so that the hydraulic conductivity of the bed does not decrease over time as a result of clogging.

The concentrations of ultimate BOD and DO in the overlying water are denoted B_{in} and O_{in}, respectively. If $B_{\text{in}}/O_{\text{in}} < 1$ then the bed is BOD limited and the BOD concentration will drop to zero before all the oxygen is consumed. Conversely, if $O_{\text{in}}/B_{\text{in}} < 1$ then the bed is oxygen limited and DO will drop to zero before all the BOD is consumed. In the model BOD removal ceases once DO reaches zero.

6.4.3.1 *Zero-order bed kinetics*

If the kinetics of BOD removal in the bed are zero-order then the DO deficit and the BOD concentration in water leaving the bed are

DO limited ($B_{\text{in}} > O_{\text{in}}$)

$$D(t_e) = \beta t_e \quad B(t_e) = B_{\text{in}} - \beta t_e \qquad \text{if } t < \frac{O_{\text{in}}}{\beta}$$

$$D(t_e) = O_{\text{in}} \quad B(t_e) = B_{\text{in}} - O_{\text{in}} \qquad \text{if } t > \frac{O_{\text{in}}}{\beta}$$

BOD limited ($O_{\text{in}} > B_{\text{in}}$)

$$D(t_e) = \beta t_e \quad B(t_e) = B_{\text{in}} - \beta t_e \qquad \text{if } t < \frac{B_{\text{in}}}{\beta}$$

$$D(t_e) = B_{\text{in}} \quad B(t_e) = 0 \qquad \text{if } t > \frac{B_{\text{in}}}{\beta} \tag{6.61}$$

where $D(t_e)$ = DO deficit and $B(t_e)$ = BOD concentration in water as it leaves the bed, t_e = time of travel through the bed [see equation (6.58)] and β = zero-order decay rate ($\text{g m}^{-3}\,\text{s}^{-1}$). These concentrations can be expressed in terms of x_e rather then t_e by using equations (6.58 and 6.59).

The benthic oxygen uptake rate (BUR) is the total flux of DO deficit out of the bed divided by the length of the dune. Bearing in mind that the outflow occurs over only half of the dune

$$\mathrm{BUR} = \frac{2}{\lambda} \int_{x_c=\lambda/2}^{3\lambda/4} D(x_e) w_y(x_e, 0)\, \mathrm{d}x_e \tag{6.62}$$

Equation (6.62) cannot be evaluated analytically but it is easily evaluated numerically (e.g. using trapezoidal integration). If the dune is DO limited ($B_{in} > O_{in}$) and all the water leaving the dune is anoxic then the BUR can be calculated analytically

$$\mathrm{BUR} = \frac{2Kh_{max}O_{in}}{\lambda} = \frac{w_{max}O_{in}}{\pi} \tag{6.63}$$

and is equal to the flux of DO entering the bed.

Results can be summarised in non-dimensional form. The non-dimensional BUR is

$$B^* = \frac{\mathrm{BUR}}{C_{in}w_{max}} \tag{6.64}$$

where

$$\begin{aligned} C_{in} &= O_{in} \quad \text{if } B_{in} > O_{in} \text{ (DO limited)} \\ C_{in} &= B_{in} \quad \text{if } O_{in} > B_{in} \text{ (BOD limited)} \end{aligned} \tag{6.65}$$

For zero-order kinetics the non-dimensional length scale is

$$L^* = \frac{\lambda\beta}{C_{in}w_{max}} \tag{6.66}$$

Figure 6.9 summarises model predictions of BUR. For zero-order bed kinetics predictions for all combinations of coefficients fall on the same line when expressed in non-dimensionalised form. The benthic uptake rate attains a maximum value of $B^* = 1/\pi$, which is the value given by equation (6.63). For the DO limited dune ($B_{in}/O_{in} > 1$) the plateau on the right-hand side of Figure 6.9 corresponds to the outflow from the dune being completely anoxic, whereas for the BOD limited dune ($B_{in}/O_{in} < 1$) it corresponds to the outflow having zero BOD. The other limiting case is $L^* = 0$, $B^* = 0$, which corresponds to either a very small dune which has negligible advective flow, an unpolluted dune in which there is no DO consumption or a highly porous dune in which the residence time is too small for DO to be depleted. Moving from left to right up the sloping part of the curve in Figure 6.9 corresponds to an increasing proportion of the outflow becoming anoxic ($B_{in}/O_{in} > 1$) or devoid of BOD ($B_{in}/O_{in} < 1$).

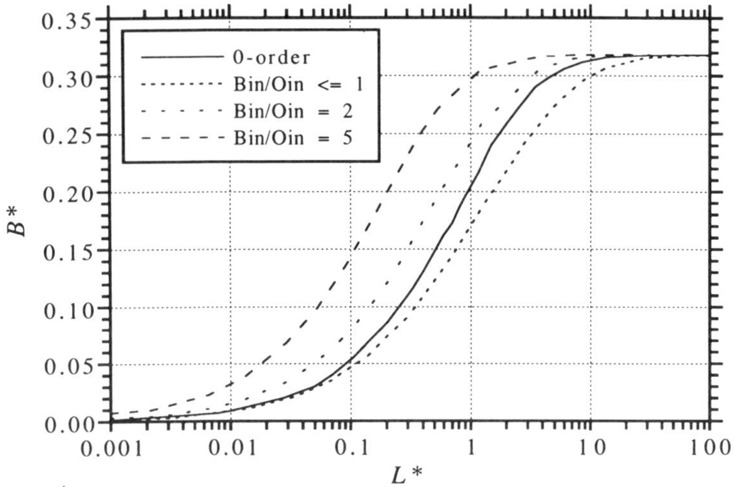

Figure 6.9. Benthic uptake rates resulting from pumping predicted assuming zero (solid line) and first-order (dashed lines) bed kinetics using a simplified model. See text for the definition of the non-dimensional variables B^* and L^*

6.4.3.2 First-order bed kinetics

If BOD decay within the bed follows first-order kinetics rather than zero-order kinetics then equations (6.61) are replaced by

DO limited ($B_{\text{in}} > O_{\text{in}}$)

$$D(t_e) = B_{\text{in}} e^{-\alpha t_e} \quad B(t_e) = B_{\text{in}} e^{-\alpha t_e} \qquad \text{if } t_e < \frac{1}{\alpha} \log_e\left(\frac{B_{\text{in}}}{O_{\text{in}}}\right)$$

$$D(t_e) = O_{\text{in}} \quad B(t_e) = B_{\text{in}} - O_{\text{in}} \qquad \text{if } t_e > \frac{1}{\alpha} \log_e\left(\frac{B_{\text{in}}}{O_{\text{in}}}\right)$$

BOD limited ($O_{\text{in}} > B_{\text{in}}$)

$$D(t_e) = B_{\text{in}} e^{-\alpha t_e} \quad B(t_e) = B_{\text{in}} e^{-\alpha t_e} \qquad \text{all } t_e \tag{6.67}$$

where α = first-order decay rate coefficient. For the first-order model the non-dimensional length scale becomes

$$L^* = \frac{\lambda \alpha}{w_{\text{max}}} \tag{6.68}$$

and the non-dimensional BUR is given by equation (6.64).

Figure 6.9 also shows model predictions of BUR for first-order bed kinetics. When the dune is BOD limited ($B_{\text{in}}/O_{\text{in}} < 1$) all the solutions fall on a single line which is similar to, but not identical with, the zero-order solution. When the dune is DO limited ($B_{\text{in}}/O_{\text{in}} > 1$) there is a separate curve for each value of B_{in}/O_{in}. As B_{in}/O_{in} increases (corresponding to increasing organic loading on the river) these curves move further to the left. This behaviour can be understood by considering the case where the bed parameters α, O_{in}, λ, w_{max} and L^* are constant. As B_{in} increases, the rate of oxygen uptake in the bed increases, a greater proportion of the dune becomes anoxic and so B^* increases.

Example 6.4

The Waiotapu River, New Zealand, has a low flow rate of 2.5 $m^3 s^{-1}$, receives treated dairy factory effluent and exhibits rapid deoxygenation (Hickey 1985). The irregular gravel bed does not have a well-defined dune structure but there are numerous irregularities such as holes, boulders, logs, pools and riffles. Measurements have been made of the oxygen consumption rate of the sediments (using well-stirred bottles), the overall deoxygenation rate of the river (estimated from the measured DO sag curve using the measured time of travel, reaeration rate and tributary inflows), the particle size distribution, and the porosity of the sediments (Hickey 1985). The hydraulic conductivity was estimated using (Bear 1972)

$$K = 6400 d_{90}^2$$

where d_{90} = particle size exceeded by 90% of the sediment by weight. Irregularities in the bed were assessed visually to average 1 m in length, 2.5 cm in height and to cover about 50% of the bed. Bed kinetics were assumed to be zero order.

Table 6.3. Summary of pumping benthic oxygen uptake rates in the gravel-bed Waiotapu River

Dune length (λ)	1.0 m
Dune height (Y_d)	0.025 m
Dune coverage	≈ 50%
Mean depth (H)	0.75 m
Mean velocity (V_x)	0.40 $m s^{-1}$
Pressure head (h_{max})	0.96 mm
Pumping velocity (w_{max})	1.50×10^{-4} $m s^{-1}$
Bulk sediment density (ρ)	1.05 $t m^{-3}$
Porosity (θ)	0.55
Hydraulic conductivity (K)	0.025 $m s^{-1}$
River DO (O_{in})	8.0 $g m^{-3}$
Volumetric DO uptake rate (β)	6.7 $mg m^{-3} s^{-1}$
Non-dimensional BUR (B^*)	0.28
Non-dimensional length (L^*)	3.06
Particle diameter (d_{90})	2.0 mm
Particle Reynolds number (Re_p)	2.9
Pumping BUR	14.6 $g m^{-2} d^{-1}$

The pumping BUR was estimated from equations (6.61) and (6.62) and found to be 15 $\mathrm{g\,m^{-2}\,d^{-1}}$ (Table 6.3). The maximum pumping velocity was low and irregularities covered only about 50% of the bed. However, a high proportion of water leaving the bed was anoxic ($L^* = 3.06$ and $B^* = 0.28$; see Figure 6.9) and the DO deficit flux out of the bed was high.

The BUR measured using a recirculating chamber was 28 $\mathrm{g\,m^{-2}\,d^{-1}}$ and the total BUR estimated from the overall deoxygenation rate of the river was 35 $\mathrm{g\,m^{-2}\,d^{-1}}$ (Hickey 1985). Thus pumping appears to make a substantial contribution (40–50%) to the BUR in the Waiotapu River. Other contributions come from turbulent dispersion into the bed and respiration by biofilm communities. Direct measurements of pumping flows and sediment DO concentrations are desirable to corroborate the estimate in Table 6.3.

6.5 SEDIMENT TRANSPORT

6.5.1 Introduction

When discussing diffusion, dispersion and pumping in Sections 6.2–6.4 we assumed that the bed was immobile. In many rivers the flow dislodges sediment particles from the bed and carries them downstream. Sediment begins to move when the bed shear stress exceeds a critical value which depends on the particle size, density, shape and cohesiveness of the bed material. Coarse gravels only move during floods whereas sands have a low critical shear stress and may be mobile even at low river flows. Although muds consist of very fine particles, they are cohesive and the critical shear stress required to initiate movement is fairly high. A detailed discussion of sediment movement is beyond the scope of this book but the interested reader may wish to consult texts such as Henderson (1966), Vanoni (1975), Simons and Senturk (1977), and Raudkivi (1990).

If the sediments are mobile then particles are periodically carried up into the turbulent boundary layer and this creates opportunities for benthic mass transfer. Three aspects of sediment transport are discussed here: saltation, dune turnover and redeposition.

6.5.2 Saltation

Consider a river with a bed of coarse sand which is mobile and forms dunes. Viewed on a macroscopic scale, the dunes migrate downstream at a celerity v_d which is typically of the order metres per day. Viewed on a smaller scale, however, the dunes do not migrate downstream as rigid bodies. Only the surface layer of particles is exposed to the bed shear stress and it is only these surface particles which move (see Figure 6.10). Particles are picked up by the

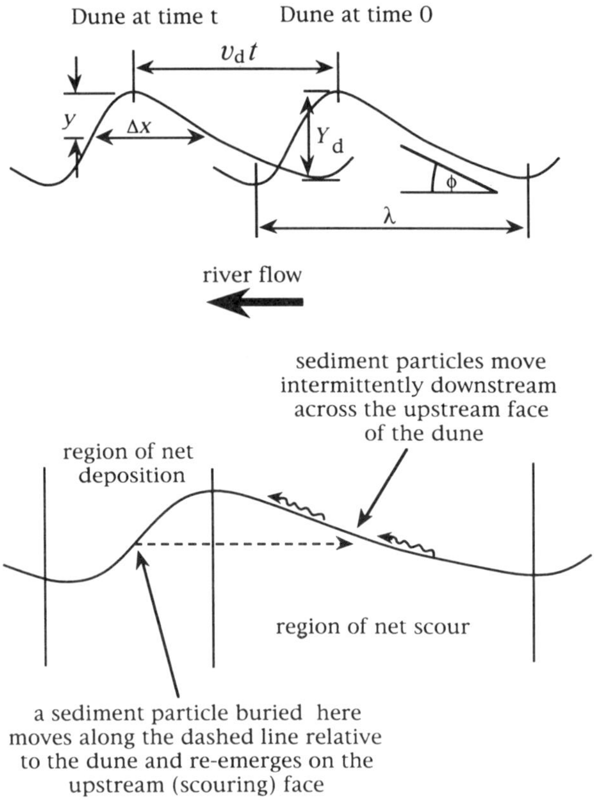

Figure 6.10. Dimensions of mobile bedforms (dunes) and the process by which they move downstream

flow on the upstream face of the dune and are eventually carried downstream over the dune crest. Here they deposit again and remain buried while the rest of the dune passes overhead. At some later time the particles are exposed again and may be mobilised.

If we label a particular sediment particle and watch it over time we find that its movement is not continuous. Laboratory and field observations indicate that from time to time a swarm of particles moves for a short distance along the top of the dune and then redeposits. In this way our labelled particle moves a short distance downstream and then stops. A short time later another swarm moves which may or may not include our labelled particle. This intermittent movement is termed saltation and continues until the particle is carried over the dune crest and is buried by the dune. There is considerable conjecture about how turbulent eddies act on sediment particles and initiate sediment movement. It is not yet possible to make reliable theoretical predictions about sediment movement but there is much published work on empirical methods for estimating sediment transport rates and dune sizes.

Consider now a sediment particle which is lifted off the bed and which moves along the top of the dune. The particle is surrounded by a thin diffusive sub-layer in which mass transfer is controlled by molecular diffusion. When the sediment particle is buried in the dune the concentration at the outer edge of this diffusive sub-layer is the pore water concentration. When the particle is lifted out of the bed into the turbulent boundary layer, the concentration at the outer edge of the diffusive sub-layer suddenly changes to that of the overlying water. This promotes a rapid exchange of solutes between the sediment surface (and any microbial populations there) and the overlying water. Thus saltation has an important role in helping to supply substrate and oxygen to microbial communities attached to sediment particles. Clearly only a very small number of sediment particles experience this high rate of mass transfer at any one time.

6.5.3 Dune Turnover

If sediment particles remain buried in the dune for a long time then the pore water composition changes. For example, in an organically enriched river, microbial activity often reduces the pore water dissolved oxygen concentration to zero. When sediment particles are exposed again by dune movement the surrounding pore water mixes with the overlying river water and reduces the river DO concentration. Thus dune turnover contributes to the BUR and can make a significant contribution to the deoxygenation rate of a polluted river.

A simple analytical model can be used to quantify the contribution to benthic oxygen uptake made by dune turnover. Dunes are assumed to be triangular, uniform and to move downstream at celerity v_d. At a distance y below the crest of the triangular dune, the length of the dune is (see Figure 6.10)

$$\Delta x = \frac{y\lambda}{Y_d} \tag{6.69}$$

where λ = dune wavelength and Y_d = dune height. Sediment particles which enter the dune at this height remain buried for a time

$$\Delta t = \frac{\Delta x}{v_d} = \frac{y\lambda}{v_d Y_d} \tag{6.70}$$

before re-emerging from the dune. The oxygen uptake rate within the sediment is assumed to be uniform and replenishment from the overlying water is neglected (i.e. turbulent dispersion is assumed to be negligibly small). Oxygen uptake ceases once the DO concentration reaches zero. The drop in DO concentration (i.e. the increase in DO deficit) between the time of burial and re-emergence is

$$\Delta D = \frac{\rho I y \lambda}{\theta Y_d v_d} \qquad \text{if } \frac{\rho I y \lambda}{\theta Y_d v_d} < O_{in}$$

$$= O_{in} \qquad \text{if } \frac{\rho I y \lambda}{\theta Y_d v_d} > O_{in} \tag{6.71}$$

where I = sediment uptake rate; ρ = sediment density and θ = porosity. If the DO concentration in the overlying waters is O_{in} then the sediment at a height below the dune crest

$$y_a = \frac{\theta Y_d O_{in} v_d}{\rho I \lambda} \tag{6.72}$$

will be anoxic along the upstream (scouring) face of the dune. The oxygen deficit flux normal to the upstream (scouring) face of the dune is

$$\Delta D v_d \sin \phi \tag{6.73}$$

where ϕ = dune angle (see Figure 6.10). To determine the total flux it is necessary to integrate over the length of the upstream face of the dune height. By resolving the flux in the horizontal direction this integration can be carried out over the height of the dune. Dividing the integrand by the length of the dune gives the benthic oxygen uptake

$$\text{BUR}_t = \frac{1}{\lambda} \int_{y=0}^{Y_d} \Delta D v_d \, dy$$

$$= \frac{1}{\lambda} \left(\int_{y=0}^{y_o} \frac{I \rho \lambda y}{\theta Y_d} \, dy + \int_{y=y_o}^{Y_d} O_{in} v_d \, dy \right)$$

$$= \frac{O_{in} v_d Y_d}{\lambda} \left(1 - \frac{O_{in} \theta v_d}{2 I \rho \lambda} \right) \tag{6.74}$$

where O_{in} = river water DO concentration; Y_d = dune height; v_d = dune celerity; λ = dune wavelength; θ = sediment porosity; ρ = sediment density; and I = sediment oxygen uptake rate within the dune.

Example 6.5

An assessment has been made of the contribution which dune turnover makes to benthic oxygen uptake in the Tarawera River, New Zealand (Rutherford *et al* 1993). The Tarawera is a mobile-bed river which receives treated pulp mill effluent and experiences a very high rate of deoxygenation. River bed sediments are of volcanic origin, geologically young and poorly sorted. The mean particle size decreases along the study reach (Table 6.4) but a high proportion of the bed is sand at all sites. At three study sites (numbered 7A, 3 and 1) the bed is highly mobile and dunes are well developed. Estimates of the gross rate of removal of oxygen from the river water by benthic microbial respiration average 34 g m^{-2} d^{-1} (Rutherford *et al* 1991).

Diver and remote video observations of sediment movement on the bed confirm that individual sediment particles are intermittently entrained into and drop out of the turbulent boundary layer. For much of the time the mobile layer contains only a single layer of particles. The contribution to BUR made by metabolism within the saltating layer is

$$\mathrm{BUR_s} = f_s \rho I_s y_s$$

where f_s = proportion of the bed covered by a saltating layer; ρ = sediment density; I_s = oxygen uptake rate by sediment particles in the saltating layer and y_s = thickness of the saltating layer. Oxygen uptake rates (per unit weight of sediment) in well-stirred bottles were found to average 250 mg kg^{-1} d^{-1}. Levels of turbulence within the stirred bottle are comparable with those in the saltating layer and it is reasonable to adopt this value for I_s. The thickness of the saltating layer is comparable with the sediment particle size and hence is of the order $y_s = 1\text{–}2 \times 10^{-3}$ m. Taking the upper bound estimate of $f_s = 1$ gives $\mathrm{BUR_s} = 1$ g m^{-2} d^{-1}, which is negligibly small compared to the total benthic deoxygenation rate which was estimated to be 34 g m^{-2} d^{-1}.

Table 6.5 summarises estimates of $\mathrm{BUR_t}$ caused by dune turnover made using equation (6.74). Dune height increases with distance downstream but this is offset by decreasing river DO concentration and consequently $\mathrm{BUR_t}$ is not significantly different at the three downstream sites. $\mathrm{BUR_t}$ was higher in December 1989 than June 1991 because the river DO was higher. The oxygen depletion rate deep within the bed has been measured by allowing a DO probe to become buried by a series of moving dunes. The average rate is 100 mg kg^{-1} d^{-1} (Rutherford *et al* 1993). When the oxygen consumption

Table 6.4. Dune dimensions and sediment characteristics in the Tarawera River

Site	Channel covered by dunes (%)	Number of dunes measured	Dune height (Y_d; m)	Dune wavelength (λ; m)	Particle size (d_{50}; mm)	Particle size (d_{90}; mm)
7A	75	12	0.20 ± 0.02	4.1 ± 0.2	1.5	5
3	>90	16	0.23 ± 0.02	4.5 ± 0.3	1	4
1	>90	14	0.28 ± 0.01	4.5 ± 0.5	1	3

Notes: mean ± 95% confidence interval; d_{50} = mean particle size; d_{90} = particle size exceeded by only 10% (by weight) of the sample.

Table 6.5. Estimated benthic oxygen uptake rate from dune turnover in the Tarawera River

Site	Date	O_{in} ($g\,m^{-3}$)	H (m)	λ (m)	BUR_t ($g\,m^{-2}\,d^{-1}$)	y_a (cm)
7A	December 89	7.6	0.20	4.1	9.2	9.5
	June 91	6.2			8.0	7.7
3	December 89	7.3	0.23	4.5	9.7	9.5
	June 91	5.8			8.1	7.6
1	December 89	7.0	0.28	4.5	11.4	11.1
	June 91	5.3			9.2	8.4

Notes: $I = 100\ mg\,kg^{-1}\,d^{-1}$; $\theta = 0.36$; $\rho = 1.28$.

rate is set to $I = 100\ mg\,kg^{-1}\,d^{-1}$ the depth of the anoxic layer is $y_a = 8$–11 cm. Setting $I = 250\ mg\,kg^{-1}\,d^{-1}$ gives $y_a = 3$–4 cm, indicating that the depth of the oxic layer is markedly smaller. Changing the value of I, however, only changes the average BUR_t from 9.3 to 10.6 $g\,m^{-2}\,d^{-1}$. The reason is that a substantial fraction of the dune is anoxic in both cases and decreasing the depth of the oxic layer has only a small effect on BUR_t. These calculations indicate that dune turnover is of the order 10 $g\,m^{-2}\,d^{-1}$ and makes a significant contribution (30%) to the observed deoxygenation rate of 34 $g\,m^{-2}\,d^{-1}$.

6.5.4 Redeposition

When the bed shear stress drops below the critical value for scour, sediment redeposits on the bed. There is a region of net redeposition on the downstream side of the dune crest, but as discussed earlier sediment particles are continually being lifted from the bed along the upstream face of the dune, carried a short distance downstream and then redeposited temporarily. Material which has been recently redeposited tends to be poorly compacted. As it is covered by other layers of sediment, however, the extra weight squeezes out some of the pore water and the material becomes more compacted. It remains compacted until the dune moves downstream and exposes the sediment to the flow on the upstream (scouring) face of the dune. Thus saltation gives rise to a poorly consolidated layer of sediment on the scouring face and the crest of the dune. A poorly consolidated layer is a feature of many mobile-bed sand rivers in New Zealand and presumably elsewhere. Anyone who has walked along an actively accreting ocean beach will have noticed poorly consolidated areas where sand has recently been deposited.

It is clear from a casual examination that recently redeposited sediment has a high water content (which implies a high porosity) and it probably has a low hydraulic conductivity. Unfortunately, the structure of recently redeposited sand is very fragile and it is not possible to sample it and conduct laboratory tests to measure the hydraulic conductivity without causing compaction. A high porosity and low hydraulic conductivity mean that water will flow more easily through recently redeposited sediment. Consequently the rate of benthic mass transfer with the overlying water is likely to be high for recently redeposited sediments.

Example 6.6

As part of the study of the Tarawera River described in Example 6.5, a preliminary investigation was made of recently redeposited sandy sediments. Divers found consistent differences in sediment texture between dune crests (where sediments were poorly consolidated) and dune troughs (where sediments were well compacted). The extent of the poorly consolidated layer was surveyed using a crude hand-held penetrometer. A 20 mm square graduated aluminium rod was held at arm's length by a diver lying on the river bed and the depth to which it penetrated the sediment under firm finger pressure was measured at 30 cm intervals along three 30 m transects.

It was found that the poorly consolidated layer covered about 75% of the upstream face and crest of the dune. In this region the aluminium rod penetrated 20–25 cm, whereas in the troughs between dunes it only penetrated about 10 cm. The difference suggests that the depth of the poorly consolidated layer is of the order of 10–15 cm.

The difference between the gross benthic deoxygenation rate (34 $\mathrm{g\,m^{-2}\,d^{-1}}$) and the sum of benthic oxygen uptake attributable to saltation (1 $\mathrm{g\,m^{-2}\,d^{-1}}$), pumping (1–2 $\mathrm{g\,m^{-2}\,d^{-1}}$) and dune turnover (10 $\mathrm{g\,m^{-2}\,d^{-1}}$) is of the order 20 $\mathrm{g\,m^{-2}\,d^{-1}}$ (Rutherford *et al* 1993). The poorly consolidated layer is very fragile and no *in situ* measurements of oxygen uptake have yet been made. Some inferences can be drawn, however, about the likely oxygen dynamics within this layer.

From the diver observations we can postulate the existence of a well-aerated and poorly consolidated layer over the crest of the dunes which is of the order $y_p = 10$–15 cm thick and covers about $f_p = 0.75$ of the bed surface. The rate of oxygen depletion measured in well-stirred bottles (see Example 6.5) averages $I_s = 250\ \mathrm{mg\,kg^{-1}\,d^{-1}}$. Sediment relative density and porosity are $\rho = 1.28$ and $\theta = 0.36$, respectively. The contribution which the poorly consolidated layer makes to benthic oxygen uptake is

$$\mathrm{BUR_p} = f_p \theta \rho I_s y_p$$

which with the figures given earlier is 9–13 $\mathrm{g\,m^{-2}\,d^{-1}}$. This estimate assumes a sediment relative density of 1.28, which may be a slight overestimate in the poorly consolidated layer. It is clear, nevertheless, that oxygen uptake in the poorly consolidated layer is capable of making a substantial contribution to benthic oxygen uptake. Clearly it is desirable to improve our understanding of this poorly consolidated layer, which will require developing experimental techniques (for use in either the field or the laboratory) to measure its dimensions, porosity, hydraulic conductivity, oxygen uptake rates and mass transfer characteristics.

NOTATION

A	cross-sectional area of channel
A	cross-sectional area of a sediment column
b	channel width
B	BOD_u concentration
B_{in}	BOD concentration in water entering the sediment
B^*	non-dimensional benthic oxygen uptake rate
BUR	benthic oxygen uptake rate
BUR_p	BUR of the poorly consolidated layer
BUR_s	BUR of the saltating layer
BUR_t	BUR resulting from dune turnover
c	tracer concentration
c	pore water tracer concentration
c^*	tracer concentration at the sediment-water interface
C_{in}	concentration of either BOD or DO entering the sediments
d	sediment particle diameter
d_{65}, d_{90}	diameter exceeded by 65% or 90% of the sediment by weight
D	oxygen deficit concentration
ΔD	increase in D_0 deficit
e_m	molecular diffusion coefficient
e_s	effective diffusivity within the sediment
f_p	proportion of the bed covered by a poorly consolidated layer
f_s	fraction of the bed covered by saltating sediment particles
g	gravitational acceleration
h	pressure head
Δh	difference in pressure head
h	local depth
h_{max}	half amplitude of pressure head distribution
H	mean water depth
I	sediment oxygen uptake rate
I_s	oxygen uptake by saltating sediment particles
J_y	vertical diffusive tracer flux
k_s	effective turbulent dispersion coefficient in the sediment
k_o, k_L	effective dispersion coefficients at depths 0 and L
K	hydraulic conductivity of the sediment
K_x	longitudinal dispersion coefficient in the river water
L	thickness of the sediment layer
L	length scale

L_c	depth at which tracer concentration reaches zero
L^*	non-dimensional length
m	maximum rate of nitrification of ammonia to nitrate
O_{in}	oxygen concentration in water entering the bed
Q	measured outflow rate from a sediment column
r	rate of change of pore water concentration resulting from biological or chemical reactions
r^*	rate of change of river water concentration resulting from biological or chemical reactions
R	hydraulic radius
Re	bulk Reynolds number of the flow
Re_p	particle Reynolds number
Q	river flow rate
s	channel slope
S	cross-sectional averaged concentration
t	time
t_e	time a fluid parcel takes to flow through the dune
Δt	time a sediment particle remains buried in the dune
u, u_x, u_y	water velocities
u^*	local shear velocity
v_d	dune celerity
V_x	cross-sectional averaged velocity
w_d	Darcy velocity within the sediment
w_p	pore water velocity within the sediment
w_x, w_y	Darcy velocities within the sediments
w_{max}	maximum Darcy velocity within the sediments
x, y	longitudinal and vertical distances
x_o, x_e	distances at which a particle enters and leaves the bed
$\Delta x(y)$	dune length at depth y
y_a	thickness of the oxic layer
y_o	constant of integration in logarithmic velocity formula
y_p	thickness of the poorly consolidated layer
y_r	roughness height
y_s	thickness of the saltating layer
y^T	distance along a tortuous path line
Δy	net displacement of particle
Δy	length of a sediment column
$\Delta y'$	total distance travelled by particle
Y_d	dune height
α	first-order removal rate coefficient
α	maximum rate of nitrification
β	zero-order removal rate coefficient

β	mineralisation rate of organic nitrogen to ammonia
γ	$2\pi/\lambda$
δ	thickness of the viscous sub-layer (laminar boundary layer)
ε	half-saturation coefficient for nitrification
ζ	tortuosity
θ	sediment porosity
λ	dune wavelength
μ	dynamic viscosity
ν	kinematic viscosity
$\varkappa$	von Karman's constant
ϕ	slope angle of the dune face
ρ	density
τ	viscous shear stress
τ_o	bed shear stress

Appendix

Summary of Analytical Solutions to the Mixing Equations

A.1 INTRODUCTION

A brief summary is presented of analytical solutions to a number of important mixing problems. It is assumed that the river channel is straight, the x axis is aligned with the main flow, the y axis is vertical and hence the z axis is transverse across the channel. The diffusion of tracer is governed by Fick's law with diffusion coefficients which are different in each direction (non-isotropic) but which do not vary spatially. The principal axes of the diffusion tensor are aligned with the coordinate axes so that the cross-product terms can be neglected. Making these assumptions, tracer concentration is determined by the equation (Holly 1985)

$$\frac{\partial c}{\partial t} + u_x \frac{\partial c}{\partial x} + u_y \frac{\partial c}{\partial y} + u_z \frac{\partial c}{\partial z} = e_x \frac{\partial^2 c}{\partial x^2} + e_y \frac{\partial^2 c}{\partial y^2} + e_z \frac{\partial^2 c}{\partial z^2} \tag{A.1}$$

where t = time; x = longitudinal distance; y = vertical distance; z = transverse distance; $c(x, y, z, t)$ = tracer concentration; u_x, u_y and u_z = velocities in the x, y and z directions; and e_x, e_y and e_z = turbulent diffusion coefficients in the x, y and z directions. We assume that u_x, u_y and u_z and e_x, e_y and e_z do not vary with x, y, z or t.

A.2 INSTANTANEOUS POINT SOURCE

If an instantaneous injection of mass M is made at point $x = 0$, $y = 0$, $z = 0$ and time $t = 0$ then subsequently tracer concentrations in an unbounded water body are given by (Fischer *et al* 1979)

$$c(x, y, z, t) = M \frac{\exp\left[-\frac{(x-u_x t)^2}{4e_x t}\right]}{\sqrt{4\pi e_x t}} \frac{\exp\left[-\frac{(y-u_y t)^2}{4e_y t}\right]}{\sqrt{4\pi e_y t}} \frac{\exp\left[-\frac{(z-u_z t)^2}{4e_z t}\right]}{\sqrt{4\pi e_z t}} \tag{A.2}$$

Note that if the injection is made at x_o, y_o, z_o and t_o then it is simply a matter of replacing x, y, z and t by $(x - x_o)$, $(y - y_o)$, $(z - z_o)$ and $(t - t_o)$ in equation (A.2). Equation (A.2) predicts that a time t after release, tracer occupies a cloud whose centroid is at $(x - u_x t)$, $(y - u_y t)$, $(z - u_z t)$. If a line is drawn through the centroid in the x, y or z directions then the tracer concentration along that line is distributed in the Gaussian bell-shaped curve shown in Figure 1.2. As written, equation (A.2) assumes that tracer does not impinge on any boundaries although the effects of boundaries can be accommodated using the method of images which is described in Section 2.6.1. Equation (A.2) is seldom used to solve practical problems in river mixing but it forms the basis for deriving other important mixing equations.

A.3 INSTANTANEOUS VERTICAL AND TRANSVERSE LINE SOURCES

Consider now a river channel which has a finite depth (i.e. the channel is vertically bounded) but where the banks are far apart (i.e. the channel is transversely unbounded). The vertical velocity $u_z = 0$ and for an instantaneous vertical line source, vertical concentration gradients are negligibly small. In this situation equation (A.2) becomes

$$c(x, z, t) = \frac{M}{h} \frac{\exp\left[-\frac{(x-u_x t)^2}{4e_x t}\right]}{\sqrt{4\pi e_x t}} \frac{\exp\left(-\frac{z^2}{4e_z t}\right)}{\sqrt{4\pi e_z t}} \tag{A.3}$$

Equation (A.3) describes transverse diffusion, longitudinal advection and longitudinal diffusion downstream from an instantaneous point source. Note that equation (A.3) assumes that u_x, e_x and e_z are all constant. As written equation (A.3) also assumes that tracer has not spread transversely far enough to reach either bank, although again the effects of the boundaries can be accommodated using the method of images as outlined in Section 3.4.1.

If the channel is very deep (i.e. vertically unbounded) and tracer originates from a transverse line source so that transverse concentration gradients are negligibly small, then we can exchange y for z and b for h in equation (A.3) and write

$$c(x, y, t) = \frac{M}{b} \frac{\exp\left[-\frac{(x-u_x t)^2}{4e_x t}\right]}{\sqrt{4\pi e_x t}} \frac{\exp\left(-\frac{y^2}{4e_y t}\right)}{\sqrt{4\pi e_y t}} \tag{A.4}$$

Equation (A.4) assumes that u_x, e_x and e_y are constant and that tracer does not impinge on either the water surface or the bed.

A.4 TIME-VARYING SOURCES AND SUPERPOSITION

If the tracer source is time-varying then the superposition principle can be used to calculate the concentration. A time-varying source is treated as if it were a (large) number of separate instantaneous sources. The concentration distribution from each instantaneous source is calculated separately using whichever of equations (A.2), (A.3) or (A.4) is appropriate. The separate solutions are then added together.

For example, consider the problem of a vertical line source at $x = 0$, $z = 0$ where the tracer mass inflow rate is $m(t)$ and the channel is unbounded. At time t over the time interval Δt the increment of mass injected is $m(t)\Delta t$. This gives rise to a concentration profile which from equation (A.3) is

$$\Delta c(x, z, t) = \frac{m(\tau)\Delta\tau}{h} \frac{\exp\left\{-\frac{[x-u_x(t-\tau)]^2}{4e_x(t-\tau)}\right\}}{\sqrt{4\pi e_x(t-\tau)}} \frac{\exp\left(-\frac{z^2}{4e_z(t-\tau)}\right)}{\sqrt{4\pi e_z(t-\tau)}} \tag{A.5}$$

We now integrate over all values of τ to get

$$c(x, z, t) = \int_{\tau=-\infty}^{\infty} \frac{m(\tau)}{h} \frac{\exp\left\{-\frac{[x-u_x(t-\tau)]^2}{4e_x(t-\tau)}\right\}}{\sqrt{4\pi e_x(t-\tau)}} \frac{\exp\left(-\frac{z^2}{4e_z(t-\tau)}\right)}{\sqrt{4\pi e_z(t-\tau)}} \, d\tau \tag{A.6}$$

For a steady vertical line source Sayre (1973) showed that

$$c(x, z) = \frac{m}{h} \frac{1}{2\pi\sqrt{e_x e_z}} \exp\left(\frac{x u_x}{2e_x}\right) K_o\left(\frac{u_x}{2e_x}\sqrt{x^2 + \frac{e_x}{e_z} z^2}\right) \tag{A.7}$$

where K_o = modified Bessel function of the second kind. For a steady transverse line source we can exchange y for z, b for h and write

$$c(x, y) = \frac{m}{b} \frac{1}{2\pi\sqrt{e_x e_y}} \exp\left(\frac{x u_x}{2e_x}\right) K_o\left(\frac{u_x}{2e_x}\sqrt{x^2 + \frac{e_x}{e_y} y^2}\right) \tag{A.8}$$

In river channels it is found that downstream from a steady source longitudinal dispersion is relatively unimportant (Sayre 1973, Thomann 1973). Consequently equations (A.7) and (A.8) can be rewritten

$$c(x, z) = \frac{m}{h\sqrt{4\pi e_z x u_x}} \exp\left(-\frac{z^2 u_x}{4e_z x}\right) \tag{A.9}$$

$$c(x, y) = \frac{m}{b\sqrt{4\pi e_z x u_x}} \exp\left(-\frac{y^2 u_x}{4e_y x}\right) \tag{A.10}$$

Again equations (A.7) and (A.9) are only valid provided tracer does not impinge on the banks, whereas equations (A8) and (A10) are valid provided that it does not impinge on the bed or the water surface.

Downstream from a steady point source the effects of longitudinal diffusion are also negligibly small and vertical and transverse spreading can be described by

$$c(x, y, z) = m \frac{\exp\left(-\frac{y^2 u_x}{4e_y x}\right)}{\sqrt{4\pi e_y x}} \frac{\exp\left(-\frac{z^2 u_x}{4e_z x}\right)}{\sqrt{4\pi e_z x}} \tag{A.11}$$

where x/u_x is, of course, the time of travel downstream from the source.

A.5 SPATIALLY DISTRIBUTED SOURCE AND SUPERPOSITION

The superposition principle can also be applied when the tracer source is distributed in space. Consider the problem where the tracer originates from a diffuser of finite length stretching across the channel from z_1 to z_2. We shall assume that tracer is vertically well mixed at the diffuser and that an instantaneous tracer release is made at time $t = 0$ such that the mass per unit width is $m(z)$. Now an increment of length along the diffuser Δz releases mass $m(z)\Delta z$. The concentration profile from this incremental release is given by equation (A.3) as

$$\Delta c(x, z, t) = \frac{m(\zeta)\Delta\zeta}{h} \frac{\exp\left[-\frac{(x - u_x t)^2}{4e_x t}\right]}{\sqrt{4\pi e_x t}} \frac{\exp\left[-\frac{(z - \zeta)^2}{4e_z t}\right]}{\sqrt{4\pi e_z t}} \tag{A.12}$$

Integrating along the length of the diffuser

$$c(x, z, t) = \int_{\zeta = z_1}^{z_2} \frac{m(\zeta)}{h} \frac{\exp\left[-\frac{(x - u_x t)^2}{4e_x t}\right]}{\sqrt{4\pi e_x t}} \frac{\exp\left[-\frac{(z - \zeta)^2}{4e_z t}\right]}{\sqrt{4\pi e_z t}} \, d\zeta \tag{A.13}$$

The superposition principle can be used to solve other mixing problems where the tracer mass inflow rate is known as a function of time and/or space. Thus if the tracer mass inflow rate from a vertical source of finite length is $m(z, t)$ then it follows by analogy with equations (A.6) and (A.13) that

$$c(x, z, t) = \int_{\tau=-\infty}^{\infty} \int_{\zeta=z_1}^{z_2} \frac{m(\zeta, \tau)}{h} \frac{\exp\left\{-\frac{[x - u_x(t-\tau)]^2}{4e_x(t-\tau)}\right\}}{\sqrt{4\pi e_x(t-\tau)}} \frac{\exp\left[-\frac{(z-\zeta)^2}{4e_z(t-\tau)}\right]}{\sqrt{4\pi e_z(t-\tau)}} \, d\zeta \, d\tau \tag{A.14}$$

A.6 LONGITUDINAL DISPERSION

Consider now the problem where tracer is fully mixed over the cross-section. If we assume that longitudinal dispersion can be modelled using Fick's law then

$$\frac{\partial S}{\partial t} + V_x \frac{\partial S}{\partial x} = K_x \frac{\partial^2 S}{\partial x^2} \tag{A.15}$$

If V_x and K_x are assumed constant then the solution for an instantaneous injection of mass M is

$$S(x, t) = \frac{M}{A\sqrt{4\pi K_x t}} \exp\left[-\frac{(x - V_x t)^2}{4K_x t}\right] \tag{A.16}$$

where A = cross-sectional area of the channel.

Now if at time $t = 0$ there is a sharp concentration front at $x = 0$ then

$$\begin{aligned} S(x, 0) &= 0 \qquad x > 0 \\ S(x, 0) &= S_o \qquad x < 0 \end{aligned} \tag{A.17}$$

By applying the superposition principle to equation (A.16) we find that

$$S(x, t) = \frac{S_o}{2}\left\{1 - \operatorname{erf}\left[\frac{(x - V_x t)}{\sqrt{4K_x t}}\right]\right\} \tag{A.18}$$

where erf = error function.

Equation (4.40) can be used to predict the concentration versus distance profile at time t_2 given the concentration versus distance profile at time t_1. Equations (4.41) or (4.44) can be used to predict the concentration versus time profile at distance x_2 given the concentration versus time profile at distance x_1 using the frozen-cloud (4.41) or Hayami (4.44) solutions.

References

Abbott, M.B. and Basco, D.R. (1989) *Computational Fluid Dynamics*, Longman, Harlow.

Abdelwahed, M.S.T. and Chu, V.H. (1981) Surface jets and surface plumes in cross-flows, *Tech. Rep. 81–1*, McGill University, Montreal.

Adams, E.E. (1982) Dilution analysis for unidirectional diffusers, *J. Hydr. Engin.*, **108** (HY3), 327–342.

Adams, E.E., Stolzenbach, K.D. and Harleman, D.R.F. (1975) Near and far field analysis of buoyant surface discharges into large bodies of water, *Ralph M. Parsons Lab. Rep. No. 205*, Massachusetts Institute of Technology.

Anonymous (1984) *System 21 HD (MK8) Users Guide*, Danish Hydraulic Institute, Horsholm, Denmark.

Anonymous (1989) *System 21 TD Users Guide*, Danish Hydraulic Institute, Horsholm, Denmark.

Banks, R.B. (1974) A mixing cell model for longitudinal dispersion in open channels, *Wat. Resour. Res.*, **10**, 357–358.

Barnett, A.G. (1983) Exact and approximate solutions of the advection-diffusion equation, in *XXth IAHR Congress, Moscow*, Hydro. Tech. Inst., Moscow, 5–9 September, 1983 Ba21, Vol. 3, 180–190.

Batchelor, G.K. (1953) *The Theory of Homogeneous Turbulence*, Cambridge University Press, London.

Batchelor, G.K. (1967) *An Introduction to Fluid Dynamics*, Cambridge University Press, London.

Bear, J. (1972) *Dynamics of Fluids in Porous Media*, Elsevier, New York.

Beer, T. and Young, P.C. (1984) Longitudinal dispersion in natural streams, *J. Environ. Engin.*, **109**, 1049–1067.

Beltaos, S. (1980a) Longitudinal dispersion in rivers, *Am. Soc. Civ. Engin.*, **106** (HY1), 151–172.

Beltaos, S. (1980b) Transverse mixing tests in natural streams, *Am. Soc. Civ. Engin.*, **106** (HY10), 1607–1625.

Beltaos, S. (1982) Dispersion in tumbling flow, *Am. Soc. Civ. Engin.*, **108** (HY4), 591–612.

Bencala, K.E. (1984) Interactions of solutes and streambed sediment. 2. A dynamic analysis of coupled hydrologic and chemical processes that determine solute transport, *Wat. Resour. Res.*, **20**, 1804–1814.

Bencala, K.E. and Walters, R.A. (1983) Simulation of solute transport in a mountain pool-riffle stream: a transient storage model, *Wat. Resour. Res.*, **19**, 718–724.

Berner, R.A. (1980) *Early Diagenesis: a Theoretical Approach*, Princeton University Press, Princeton, NJ.

Bowden, K.F. (1983) *Physical Oceanography of Coastal Waters*, Ellis Horwood, Chichester.
Carslaw, H.S. and Jaegar, J.C. (1959) *Conduction of Heat in Solids*, 2nd edn, Oxford University Press, London.
Cerling, T.E., Morrison, S.J. and Sobocinski, R. W. (1990) Sediment–water interaction in a small stream: adsorption of 137Cs by bed load sediments, *Wat. Resour. Res.*, **26**, 1165–1176.
Chang, Y.C. (1971) Lateral mixing in meandering channels, *PhD Thesis*, University of Iowa.
Characklis, W.G. and Marshall, K.C. (eds) (1990) *Biofilms*, Wiley-Interscience, New York.
Chatwin, P.C. (1970) The approach to normality of the concentration distribution of a solute in a solvent flowing along a straight pipe, *J. Fluid Mech.*, **43** (part 2), 321–352.
Chatwin, P.C. (1971) On the interpretation of some longitudinal dispersion experiments, *J. Fluid Mech.*, **48** (part 4), 689–702.
Chatwin, P.C. (1972) The cumulants of the distribution of concentration of a solute dispersing in solvent flowing through a tube, *J. Fluid Mech.*, **51** (part 1), 63–67.
Chatwin, P.C. (1980) Presentation of longitudinal dispersion data, *Am. Soc. Civ. Engin.*, **106** (HY1), 71–83.
Chatwin, P.C. and Allen, C.M. (1985) Mathematical models of dispersion in rivers and estuaries, *Ann. Rev. Fluid Mech.*, **17**, 119–149.
Chow, V.T. (1959) *Open-Channel Hydraulics*, McGraw-Hill, New York.
Chu, V.H. and Abdelwahed, M.S.T. (1990) Shore attachment of buoyant effluent in strong crossflow, *J. Hydr. Eng.*, **116**, 157–175.
Cleary, R.W. and Adrian, D.D. (1973) New analytical solutions for dye diffusion equations, *Am. Soc. Civ. Engin.*, **99** (EE3), 213–227.
Cotton, A.P. and West, J.R. (1980) Field measurement of transverse diffusion in unidirectional flow in a wide, straight channel, *Wat. Res.*, **14**, 1597–1604.
Cooper, A.B. (1984) Activities of benthic nitrifiers in streams and their role in oxygen consumption, *Microb. Ecol.*, **10**, 317–334.
Crank, J. (1975) *The Mathematics of Diffusion*, 2nd edn, Oxford University Press, London.
Csanady, G.T. (1967) Concentration fluctuations in turbulent diffusion, *J. Atmos. Sci.*, **24**, 21–28.
Csanady, G.T. (1973) *Turbulent Diffusion in the Environment. Geophys. Astrophys. Monog.*, D. Reidel, Boston.
Dagan, G. (1969) Dispersivity tensor for turbulent uniform channel flow, *Am. Soc. Civ. Engin.*, **95** (HY5), 1699–1712.
Demetracopoulos, A.C. and Stefan, H.G. (1983) Transverse mixing in a wide and shallow river: case study, *J. Environ. Engin.*, **109**, 685–699.
Day, T.J. and Wood, I.R. (1976) Similarity of the mean motion of fluid particles dispersing in a natural channel, *Wat. Resour. Res.*, **12**, 655–666.
Denton, R.A. (1990) Analytical asymptotic solutions for longitudinal dispersion with dead zones, *J. Hydr. Res.*, **28**, 309–329.
Dewey, R. and Sullivan, P.J. (1977) The asymptotic stage of longitudinal dispersion within a tube, *J. Fluid Mech.*, **80**, 293–303.
Dobran, H.B. (1982) Dispersion in mountainous natural streams, *Am. Soc. Civ. Engin.*, **108** (EE3), 502–514.
El-Hadi, N., Harrington, A., Hill, I., Lau, Y.L. and Krishnappan, B.G. (1984) River mixing—a state of the art report, *Can. J. Civ. Engin.*, **11**, 585–609.

Elder, J.W. (1959) The dispersion of marked fluid in turbulent shear flow, *J. Fluid Mech.*, **5**, 544–560.

Elliott, A.F. (1991) Transfer of solutes into and out of streambeds, *PhD Thesis*, California Institute of Technology, Pasadena.

Engelund, F. (1969) Dispersion of floating particles in uniform channel flow, *Am. Soc. Civ. Engin.*, **95**, 1149–1162.

Engmann, J.E.O. (1974) Transverse mixing characteristics of open and ice-covered channel flows, *PhD Thesis*, University of Alberta.

Engmann, J.E.O. and Kellerhals, R. (1974) Transverse mixing in an ice-covered river, *Wat. Resour. Res.*, **10**, 775–784.

Felham, H.M. (1985) Resistance components and velocity distributions of open channel flows over bedforms, *MSc Thesis*, Colorado State University, Fort Collins.

Fischer, H.B. (1966) Longitudinal dispersion in laboratory and natural streams, *Rep. No. KH-R-12*, California Institute of Technology, Pasadena.

Fischer, H.B. (1967) The mechanics of dispersion in natural streams, *Am. Soc. Civ. Engin.*, **93** (HY6), 187–215.

Fischer, H.B. (1968) Dispersion predictions in natural streams, *Am. Soc. Civ. Engin.*, (SA5), 927–943.

Fischer, H.B. (1969) The effects of bends on dispersion in streams, *Wat. Resour. Res.*, **5**, 496–506.

Fischer, H.B. (1973) Longitudinal dispersion and turbulent mixing in open channel flow, *Ann. Rev. Fluid Mech.*, **5**, 59–78.

Fischer, H.B. (1976) Mixing and dispersion in estuaries, *Ann. Rev. Fluid Mech.*, **8**, 107–133.

Fischer, H.B., List, E.J. Koh, R.C.Y. Imberger, J. and Brooks, N.H. (1979) *Mixing in Inland and Coastal Waters*, Academic Press, New York.

Fletcher, C.A.J. (1988) *Computational Techniques for Fluid Dynamics*, Springer Verlag, Berlin.

French, R.H. (1979) Transfer coefficients in stratified channel flow, *Am. Soc. Civ. Engin.*, **105** (HY9), 1087–1101.

Freund, J.E. (1974) *Modern Elementary Statistics*, 4th edn, Prentice-Hall, London.

Fukuoka, S. and Sayre, W.W. (1973) Longitudinal dispersion in sinuous channels, *Am. Soc. Civ. Engin.*, **99** (HY1) 195–217.

Glover, R.E. (1964) Dispersion of dissolved or suspended materials in flowing streams, *US Geol. Surv. Prof. Pap. No. 433-B.*

Godfrey, R.G. and Frederick, B.J. (1970) Dispersion in natural streams, *US Geol. Surv. Prof. Pap. No. 433-K.*

Gowda, T.P.H. (1983) Water quality prediction in mixing zones of rivers, *J. Environ. Engin.*, **110**, 751–769.

Harden, T.O. and Shen, H.T. (1979) Numerical simulation of mixing in natural rivers, *Am. Soc. Civ. Engin.*, **105** (HY4), 393–408.

Harremoes, P. (1978) Biofilm kinetics, in Mitchell, R. (ed.) *Water Pollution Microbiology*, Vol. 2, Wiley, Chichester.

Hatherell, T.V.J. (1992) Underground water seepage flow by complex variable boundary element method, *Project Rep.*, University of Exeter.

Heathershaw, A.D. (1974) Bursting phenomena in the sea, *Nature*, **242**, 394–395.

Henderson, F.M. (1966) *Open Channel Flow*, Macmillan, New York.

Henry, R.F. and Walters, R.A. (1991) *User Manual for PC TRIGRID*, Department of Fisheries and Oceans, Government of Canada.

Hickey, C.W. (1985) River oxygen uptake by benthic micro-organisms, *PhD Thesis*, University of Waikato, Hamilton.

Hickey, C.W. (1988) Benthic chamber for use in rivers: testing against oxygen mass balances, *J. Environ. Engin.*, **114**, 828–845.

Hinze, J.O. (1972) *Turbulence: an Introduction to its Mechanics and Theory*, 2nd edn McGraw-Hill, New York.

Holley, E.R. (1969) A unified view of diffusion and dispersion, *Am. Soc. Civ. Engin.*, **95** (HY2), 621–631.

Holley, E.R. and Abraham, G. (1973a) Field tests on transverse mixing in rivers, *Am. Soc. Civ. Engin*, **99** (HY12), 2313–2331.

Holley, E.R. and Abraham, G. (1973b) Laboratory studies on transverse mixing in rivers, *J. Hydr. Res.*, **11**, 219–253.

Holley, E.R., Siemons, J. and Abraham, G. (1972) Some aspects of analyzing transverse diffusion in rivers, *J. Hydr. Res.*, **10**, 27–57.

Holly, F.M. (1975) Two-dimensional mass dispersion in rivers, *Hydrol. Pap. No. 78*, Colorado State University, Fort Collins.

Holly, F.M. (1985) Dispersion in rivers and coastal waters. 1. Physical principles and dispersion equations, in Novak, P. (ed.) *Developments in Hydraulic Engineering*, Elsevier, London, 1–37.

Holly, F.M. and Nerat, G. (1984) Field calibration of stream-tube dispersion model, *J. Hydr. Engin.*, **109** (11), 1455–1470.

Jackman, A.P., Walters, R.A. and Kennedy, V.C. (1984) Transport and concentration controls for chloride, strontium, potassium and lead in Uvas Creek, a small cobble-bed stream in Santa Clara County, California, USA. 2. Mathematical modelling, *J. Hydrol.*, **75**, 111–141.

Jain, S.C. (1976) Longitudinal dispersion coefficients for streams, *Am. Soc. Civ. Engin.*, **102** (EE2), 465–474.

Jirka, G.H., Abraham, G. and Harleman, D.R.F. (1975) An assessment of techniques for hydrothermal prediction, *Ralph M. Parsons Lab. Rep. No. 203*, Massachusetts Institute of Technology.

Jobson, H.E. and Sayre, W.W. (1970a) Predicting concentration profiles in open channels, *Am. Soc. Civ. Engin*, **96** (HY10), 1983–1996.

Jobson, H.E. and Sayre, W.W. (1970b) Vertical transfer in open channel flow, *Am. Soc. Civ. Engin*, **96** (HY3), 703–724.

Kalinske, A.A. and Pien, C.L. (1944) Eddy diffusion, *Ind. Eng. Chem.*, **36**, 220–223.

Kitto, J. and Rutherford, J.C. (1982) Taranaki ring plain dye studies: preliminary results, in Rutherford, J.C. (ed.) *The River and Estuary Mixing Workshop, Hamilton, 7–18 November 1981*, Ministry of Works and Development, Wellington.

Lau, Y.L. and Krishnappan, B.G. (1977) Transverse dispersion in rectangular channels, *Am. Soc. Civ. Engin.*, **103** (HY10), 1173–1189.

Lau, Y.L. and Krishnappan, B.G. (1981) Modelling transverse mixing in natural streams, *Am. Soc. Civ. Engin.*, **107** (HY2), 209–226.

Lee, J.H.W. (1985) Comparison of two river diffuser models, *J. Hydr. Engin.*, **111**, 1069–1078.

Lee, J.W.H. and Jirka, G. H. (1980) Multiport diffuser as line source of momentum in shallow water, *Wat. Resour. Res.*, **16**, 695–708.

LeGrand-Marcq, C. and Laudelout, H. (1985) Longitudinal dispersion in a forest stream, *J. Hydrol.*, **78**, 317–324.

Liu, H. (1977) Predicting dispersion coefficients of streams, *Am. Soc. Civ. Engin.*, **103** (EEI), 59–69.

Liu, H. and Cheng, H. D. (1980) Modified Fickian model for predicting dispersion, *Am. Soc. Civ. Engin.*, **106** (HY6), 1021–1040.

Luk, G.K.Y., Lau, Y.L. and Watt, W.E. (1990) Two-dimensional mixing in rivers with unsteady pollutant source, *J. Environ. Engin.*, **116**, 125–143.

Lumley, J.L. and Panofsky, H.A. (1964) *The Structure of Atmospheric Turbulence*, Wiley-Interscience, New York.

Lung, W.-S. and O'Connor, D.J. (1982) Two-dimensional mass transport in estuaries, *J. Hydr. Engin.*, **110**, 1339–1357.

Marvioet, J.L. and van Craenenbroeck, W. (1986) Longitudinal dispersion in ship canals, *J. Hydr. Res.*, **24**, 123–132.

McBride, G.B. (1986) A procedure for predicting the flux of solutes across the sediment/water interface, in *Water Quality Modelling in the Inland Natural Environment*, BHRA, Cranfield, pp. 435–447.

McNulty, A.J. (1983) Dispersion of a continuous pollutant source in open channel flow, *PhD Thesis*, University of Canterbury, Christchurch.

McQuirk, J.J. and Rodi, W. (1978) A depth-averaged mathematical model for the near field of side discharges into open-channel flow, *J. Fluid Mech.*, **86**, 761–781.

McQuivey, R.S. and Keefer, T.N. (1974) Simple method for predicting dispersion in streams, *Am. Soc. Civ. Engin.*, **100** (EE4), 997–1011.

McQuivey, R.S. and Keefer, T.N. (1976) Dispersion—Mississippi River below Baton Rouge, La, *Am. Soc. Civ. Engin.*, **102** (HY10), 1425–1437.

Miller, A.C. and Richardson, E.V. (1974) Diffusion and dispersion in open channel flow, *Am. Soc. Civ. Engin.*, **100** (HY1), 159–171.

Millero, F.J., Gonzalez, A. and Ward, G.K. (1976) The density of seawater solutions at one atmosphere as a function of temperature and salinity, *J. Mar. Res.*, **34**, 61–93.

Munk, W. and Anderson, E.R. (1948) Notes on a theory of the thermocline, *J. Mar. Sci*, **7**, 276–295.

Naot, D. and Rodi, W. (1982) Calculation of secondary currents in channel flow, *J. Hydr. Engin*, **108** (HY8), 948–968.

Nokes, R.I. (1986) Problems in turbulent dispersion, *PhD Thesis*, University of Canterbury, Christchurch.

Nokes, R.I. and Wood, I.R. (1988) Vertical and lateral turbulent dispersion: some experimental results, *J. Fluid Mech.*, **187**, 373–394.

Nordin, C.F. and Troutman, B.M. (1980) Longitudinal dispersion in rivers: the persistence of skewness in observed data, *Wat. Resour. Res.*, **16**, 123–128.

Officer, C.B. (1976) *Physical Oceanography of Estuaries*, Wiley, London.

O'Connor, D.J. (1988) Models of sorptive toxic substances in freshwater systems. III streams and rivers, *J. Environ. Engin.*, **114**, 552–574.

O'Connor, D.J., Mueller, J.A. and Farley, K.J. (1983) Distribution of kepone in the James River Estuary, *J. Environ. Engin.*, **109**, 396–413.

Okoye, J.K. (1970) Characteristics of transverse mixing in open-channel flows, *Rep. No. KH-R-23*, California Institute of Technology, Pasadena.

Okubo, A. and Ozmidov, R.V. (1970) Empirical dependence of the coefficient of horizontal turbulent diffusion in the ocean on the scale of the phenomenon in question, *Izv. Atmos. Ocean. Phys.*, **6**, 308–309.

Orlob, G.T. (1959) Eddy diffusion in homogeneous turbulence, *Am. Soc. Civ. Engin.*, **85** (HY9), 75–101.

Owens, M., Edwards, R.W. and Gibbs, J.W. (1964) Some rearation studies in streams, *Air Wat. Pollut. Int. J.*, **8**, 469–486.

Parr, A.D. and Melville, J.G. (1981) Nearfield performance of river diffusers, *Am. Soc. Civ. Engin.*, **107** (EE5), 995–1008.

Pedersen, F.B. (1977) Prediction of longitudinal dispersion in natural streams, *Ser. Pap. No. 14*, Technical University of Denmark, Copenhagen, Denmark.
Prych, E.A. (1970) Effects of density differences on lateral mixing in open channel flows, *Rep. No. KH-R-21*, California Institute of Technology, Pasadena.
Raudkivi, A.J. (1990) *Loose Boundary Hydraulics*, 3rd edn, Pergamon Press, Oxford.
Rodi, W. (1980) *Turbulence models and their application in hydraulics*, International Association for Hydraulic Research, Delft.
Rutherford, J.C. (1980) *Handbook on Mixing in Rivers*, Ministry of Works and Development, Wellington.
Rutherford, J.C. and Williams, B.L. (1992) Transverse mixing and surface heat exchange in the Waikato River: a comparison of two models, *N.Z. J. Mar. Freshwater Res.*, **26**, 435–452.
Rutherford, J.C., Taylor, M.E.U. and Davies, J.D. (1980). Waikato River pollutant flushing rates, *Am. Soc. Civ. Engin.*, **106** (EE6), 1131–1150.
Rutherford, J.C., Wilcock, R.J. and Hickey, C.W. (1991) Deoxygenation in a mobile-bed river. I. Field studies, *Wat. Res.*, **25**, 1487–1498.
Rutherford, J.C., Latimer, G.J. and Smith, R.K. (1993) Bedform mobility and benthic oxygen uptake, *Wat. Res.*, **27**, 1545–58.
Sabol, G.V. and Nordin, C.F. (1978) Dispersion in rivers as related to storage zones, *Am. Soc. Civ. Engin.*, **104** (HY5), 695–708.
Savant, S.A., Reible, D.D. and Thibodeaux, L. (1987) Convective transport within stable river sediments, *Wat. Resour. Res.*, **23**, 1763–1768.
Sayre, W.W. (1968a) Dispersion of mass in open channel flow, *PhD Thesis*, Colorado State University, Fort Collins.
Sayre, W.W. (1986b) Discussion of: the mechanics of dispersion in natural streams by H.B. Fischer, *Am. Soc. Civ. Engin.*, **94** (HY6), 1549–1556.
Sayre, W.W. (1973) Natural mixing processes in rivers, in Shen, H.W. (ed.) *Environmental Impact on Rivers*, Shen, Fort Collins, pp. 6.1–6.37.
Sayre, W.W. (1979) Shore-attached thermal plumes in rivers, in Shen, H.W. (ed.) *Modelling in Rivers*, Wiley-Interscience, London, 15.1–15.44.
Sayre, W.W. and Chamberlin, A.R. (1964) Exploratory laboratory study of lateral turbulent diffusion at the surface of an alluvial channel, *US Geol. Surv. Circ. No. 484*.
Sayre, W.W. and Chang, F.M. (1968) A laboratory investigation of open channel dispersion processes for dissolved, suspended and floating dispersants, *US Geol. Surv. Prof. Pap. 433-E*.
Sayre, W.W. and Yeh, T. (1973) Transverse mixing characteristics of the Missouri River downstream from the Cooper Nuclear Station, *Rep. No. 145*, University of Iowa.
Schiller, E.J. and Sayre, W.W. (1973) Vertical mixing of heated effluents in open channel flow, *Rep. No. 148*, Iowa Institute of Hydraulic Research, University of Iowa.
Schiller, E.J. and Sayre, W.W. (1975) Vertical temperature profiles in open-channel flow, *Am. Soc. Civ. Engin.*, **101** (HY6), 749–761.
Schlichting, H. (1968) *Boundary-layer Theory*, 6th edn, McGraw-Hill, New York.
Schuster, J. (1965) Canal discharge measurements with radioisotopes, *Am. Soc. Civ. Engin.*, **91** (HY2), 101–124.
Simons, D.B. and Senturk, F. (1977) *Sediment Transport Technology*, Water Resources Publications, Fort Collins.
Smart, P.L. and Laidlaw, I.M.S. (1977) An evaluation of some fluorescent dyes for water tracing, *Wat. Resour. Res.*, **13**, 15–33.

Smith, R. (1979) Buoyancy effects upon lateral dispersion in open-channel flow, *J. Fluid Mech.*, **90** (part 4, 761–779.

Smith, R. (1982) Where to put a steady discharge in a river, *J. Fluid Mech.*, **115**, 1–11.

Smith, R. (1987) Shear dispersion looked at from a new angle, *J. Fluid Mech*, **182**, 447–466.

Somlyody, L. (1977) Dispersion measurement on the Danube, *Wat. Res.*, **11**, 411–417.

Somlyody, L. (1982) An approach to the study of transverse mixing in streams, *J. Hydr. Res.*, **20**, 203–220.

Spiegel, M.R. (1968) *Mathematical Handbook. Schaum's Outline Ser.*, McGraw-Hill, New York.

Stallard, R.F. (1987) Cross-channel mixing and its effect on sedimentation in the Orinoco River, *Wat. Resour. Res.*, **23**, 1977–1986.

Stefan, H. and Demetracopoulos, A.C. (1981) Cells-in-series simulation of riverine transport, *Am. Soc. Civ. Engin.*, **107** (HY6), 675–697.

Stelling, G.S. (1984) *On the Construction of Computational Methods for Shallow Water Flow Problems*, Rijkswaterstaat, Hague.

Stolzenbach, K.D. and Harleman, D.R.F. (1971) An analytical and experimental investigation of surface discharges of heated water, *Ralph M. Parsons Lab. Rep. No. 135.* Massachusetts Institute of Technology.

Sullivan, P.J. (1968) Dispersion in a turbulent shear flow, *PhD Thesis*, University of Cambridge.

Sumer, S.M. and Fischer, H.B. (1977) Transverse mixing in partially stratified flow, *Am. Soc. Civ. Engin.*, **103** (HY6), 587–600.

Taylor, G.I. (1921) Diffusion by continuous movements, *Proc. London Math. Soc., Ser. 2*, **20**, 196–212.

Taylor, G.I. (1953) Dispersion of soluble matter in solvent flowing slowly through a tube, *Proc. R. Soc. London, Ser. A*, **219**, 186–203.

Taylor, G.I. (1954) The dispersion of matter in turbulent flow through a pipe, *Proc. R. Soc. London Ser. A*, **223**, 446–468.

Tennekes, H. and Lumley, J.L. (1972) *A First Course in Turbulence*, MIT Press, Cambridge, MA.

Thackston, E.L., Hays, J.R. and Krenkel, P.A. (1967) Least squares estimation of mixing coefficients, *Am. Soc. Civ. Engin.*, 47–58.

Thackston, E.L. and Schnelle, K.B. (1970) Predicting effects of dead zones on stream mixing, *Am. Soc. Civ. Engin.*, **96** (SA2), 319–331.

Thibodeaux, L.J. (1979) *Chemodynamics*, Wiley, New York.

Thibodeaux, L.J. and Boyle, J.D. (1987) Bedform-generated convective transport in bottom sediment, *Nature*, **325**, 341–343.

Thomann, R.V. (1973) Effect of longitudinal dispersion on dynamic water quality response of streams and rivers, *Wat. Resour. Res.*, **9**, 355–366.

Thomas, I.E. (1958) Dispersion in open-channel flow, *PhD Thesis*, Northwestern University.

Thompson, J.F., Warsi, Z.U.A. and Mastin, C.W. (1985) *Numerical Grid Generator: Foundations and Applications*, North-Holland, New York.

Townsend, A.A. (1956) *The Structure of Turbulent Shear Flow*, Cambridge University Press, London.

Tsai, Y.H. and Holley, E.R. (1978) Temporal moments for longitudinal dispersion, *Am. Soc. Civ. Engin.*, **104** (HY12), 1617–1634.

Turner, J.S. (1973) *Buoyancy Effects in Fluids*, Cambridge University Press, London.

Valentine, E.M. (1978) The effects of channel boundary roughness on longitudinal dispersion, *PhD Thesis*, University of Canterbury, Christchurch.

Valentine, E.M. and Wood, I.R. (1977) Longitudinal dispersion with dead zones, *Am. Soc. Civ. Engin.*, **103** (HY9), 975–990.

Valentine, E.M. and Wood, I.R. (1979a) Dispersion in rough rectangular channels, *Am. Soc. Civ. Engin.*, **105** (HY12), 1537–1553.

Valentine, E.M. and Wood, I.R. (1979b) Experiments in longitudinal dispersion with dead zones, *Am. Soc. Civ. Engin.*, **105** (HY8), 999–1016.

Velz, C.J. (1980) *Applied Stream Sanitation*. Wiley, New York.

Vanoni, V.A. (ed.) (1975) *Sedimentation Engineering. Manuals & Reports in Engineering Practice No. 54*, American Society of Civil Engineers, New York.

Wallis, S.G., Guymer, I. and Bilgi, A. (1989a) A practical engineering approach to modelling longitudinal dispersion, in Falconer, R.A., Goodwin, P. and Matthew, R.G.S. (eds) *Hydraulic and Environmental Modelling of Coastal, Estuarine and River Waters*, Gower Technical, Aldershot, 1–22.

Wallis, S.G., Young, P.C. and Beven, K.J. (1989b) Experimental investigation of the aggregated dead zone model for longitudinal solute transport in stream channels, *Proc. Instn. Civ. Engin.*, **87** (part 2), 1–22.

Wang, S.T., McMillan, A.F. and Chen, B.H. (1978) Dispersion of pollutants in channels with non-uniform velocity distributions, *Wat. Res.*, **12**, 389–394.

Ward, P.B.R. (1974) Transverse dispersion in oscillatory channel flow, *Am. Soc. Civ. Engin.*, **100** (HY6), 755–772.

Webel, G. and Schatzmann, M. (1984) Transverse mixing in open channel flow, *J. Hydr. Engin.*, **110**, 423–435.

Wilkie, D.R. (1981) Design and construction of the outfall for the Hamilton city water pollution control plant, in Rutherford, J.C. (ed.) *Proceedings of the River and Estuary Mixing Workshop*, Ministry of Works and Development, Wellington.

Wuhrmann, K. (1972) Stream purification, in Mitchell, R. (ed.) *Water Pollution Microbiology*, Wiley-Interscience, New York.

Yih, C.-S. (1969) *Fluid Mechanics*, McGraw-Hill, New York.

Yotsukura, N. (1977) Derivation of solute-transport equations for a turbulent natural-channel flow, *J. Res. US Geol. Surv.*, **5**, 277–284.

Yotsukura, N. and Cobb, E.D. (1972) Transverse diffusion of solutes in natural streams, *US Geol. Surv. Prof. Pap. No. 582-C.*

Yotsukura, N. and Sayre, W.W. (1976) Transverse mixing in natural channels, *Wat. Resour. Res.*, **12**, 695–704.

Yotsukura, N., Fischer, H.B. and Sayre, W.W. (1970) Measurement of mixing characteristics of the Missouri River between Sioux City, Iowa and Plattsmouth, Nebraska, *US Geol. Surv. Wat. Supply Pap. No. 1899-G.*

Young, P.C. and Wallis, S.C. (1986) The aggregated dead zone model for dispersion, in *Water Quality Modelling in the Inland Natural Environment*, BHRA, Cranfield, 421–433.

Author Index

Subject Index